65 Springer Series in Solid-State Sciences

Edited by Peter Fulde

W0227317

Springer Series in Solid-State Sciences

Editors: M. Cardona   P. Fulde   K. von Klitzing   H.-J. Queisser

Volumes 1–39 are listed on the back inside cover

Peter Brüesch

Phonons: Theory and Experiments II

Experiments and Interpretation
of Experimental Results

With a Chapter by W. Bührer

With 123 Figures

Springer-Verlag Berlin Heidelberg New York
London Paris Tokyo

Dr. Peter Brüesch

Brown Boveri Research Center, CH-5045 Baden-Dättwil, and
Ecole Polytechnique Fédérale de Lausanne, CH-1015 Lausanne, Switzerland

Series Editors:
Professor Dr., Dr. h. c. Manuel Cardona
Professor Dr., Dr. h. c. Peter Fulde
Professor Dr. Klaus von Klitzing
Professor Dr. Hans-Joachim Queisser

Max-Planck-Institut für Festkörperforschung, Heisenbergstrasse 1
D-7000 Stuttgart 80, Fed. Rep. of Germany

ISBN 978-3-642-52265-9   ISBN 978-3-642-52263-5 (eBook)
DOI 10.1007/978-3-642-52263-5

Library of Congress Cataloging-in-Publication Data. Brüesch, Peter, 1934-. Phonons, theory and experiments. (Springer series in solid-state sciences ; 34, 65) Includes bibliographical references and index. Contents: Lattice dynamics and models of interatomic forces. 1. Phonons. I. Title. II. Series. III. Series: Springer series in solid-state sciences ; 34, etc. QC176.8.P5B78 1982 530.4'1 81-21424

2153/3150-543210

*Dedicated to my wife
and my family*

Preface

The first part of this three-volume treatment, *Phonons: Theory and Experiments I*, has been devoted to the basic concepts of the physics of phonons and to a study of models of interatomic forces. The present second volume, *Phonons: Theory and Experiments II*, contains a thorough study of experimental techniques and the interpretation of experimental results. In a third volume we shall treat a number of phenomena which are directly related to lattice dynamics.

The aim of this treatment is to bridge the gap between theory and experiment. Both experimental aspects and theoretical concepts necessary for an interpretation of experimental data are discussed. An attempt has been made to present the descriptive as well as the analytical aspects of the topics. Although emphasis is placed on the experimental and theoretical study of the dynamics of atoms in solids, most chapters also contain a general introduction to the specific subject. The text is addressed to experimentalists and theoreticians working in the vast field of dynamical properties of solids. It will also prove useful to graduate students starting research in this or related fields.

The choice of the topics treated was partly determined by the author's own activity in these areas. This is particularly the case for the chapters dealing with infrared, Raman and inelastic neutron spectroscopy, as well as for some newer developments such as the optical spectroscopy of thin films and adsorbates.

I am very grateful to my colleague W. Bührer who wrote Chapter 6 about "Inelastic Neutron Spectroscopy".

The author is indebted to BBC Brown, Boveri & Company Limited, Baden, Switzerland, for giving him the opportunity to carry out this work. The BBC Research Center provided the scientific atmosphere necessary to accomplish this goal. As in the first volume, several chapters of this second volume grew out of lectures the author gave at the "Ecole Polytechnique Fédérale de Lausanne" (EPFL) during the years 1975–1986 for graduate students in experimental physics in their last year of study. I am grateful to BBC and to the EPFL for giving me the chance to prepare parts of this book by lecturing.

The entire manuscript was strongly influenced by the detailed criticism and valuable suggestions of my colleagues Drs. Th. Baumann, J. Bernasconi, E. Cartier, T. Hibma, P. Pfluger, W.R. Schneider, H.J. Wiesmann and H.R. Zeller from the Brown Boveri Research Center, and by Dr. W. Bührer from the Institut für Reaktortechnik ETHZ in Würenlingen, Switzerland. My debt to them is indeed great. I am also grateful to Professor Ph. Choquard from EPFL for many interesting discussions in Lausanne.

I wish to express very grateful thanks to Mrs. M. Zamfirescu from BBC Research Center for the very skilful drawing of over 100 figures; without her immense work the goal of presenting a well-illustrated book could not have been accomplished. The author is also indebted to the late Mrs. E. Knotz, to Mrs. N. Bingham and to Mrs. E. Martens for their never-ending patience in typing the manuscript, and to W. Foditsch and P. Unternährer for photographing some of the figures. Finally, I am grateful to Professor P. Fulde for valuable suggestions and to Dr. H. Lotsch, Springer-Verlag, for good cooperation.

Baden, June 1986   *Peter Brüesch*

Contents

1. Introduction

The first part of this 3-volume treatment *Phonons: Theory and Experiments* [1.1], is devoted to the basic concepts of the physics of phonons. The present second volume contains a thorough study of experimental techniques and the interpretation of experimental results. In the third volume [1.2], we treat a number of phenomena which are directily related with phonons.

In the present volume the following topics are treated: Infrared absorption, Raman scattering, Brillouin scattering, diffuse X-ray scattering, inelastic neutron scattering, and some other techniques including ultrasonic methods, inelastic electron tunneling spectroscopy, point-contact spectroscopy and spectroscopy of surface phonons and adsorbates. These chapters contain a discussion of the experimental techniques as well as a presentation of the associated theoretical background necessary for an interpretation of the experimental data.

The author has tried to present both aspects of the subjects, descriptive and analytical. Simple models are often used to illustrate the basic concepts. More than 100 figures are included to illustrate both theoretical and experimental results. The interested reader will find lengthy derivations of important results in an appendix. Many chapters contain a number of problems with hints and results. They often contain additional information not contained in the text; it is therefore recommended that readers examine the problems, even if they do not intend to solve them. At the end of each chapter, references to the existing literature appear. Despite the inclusion of over 300 references, it is anavoidable that important papers have been omitted. Such omissions are unintentional and apologies are sincerely offered.

1.1 General Remarks

In [1.1] we were interested in the properties of the phonon system that is in equilibrium and in isolation from the rest of the universe. In order to see how a phonon system reacts with its surrounding we have to poke on it. In order to obtain meaningful information the "poke" must be controlled and the reaction must be carefully analyzed. From such an analysis much can be learned about the nature of phonons, such as dispersion curves, life times, dielectric properties due to phonons, etc. The poking can be done by means of suitable particles which interact with the system. Suitable particles are neutrons, photons, electrons, helium atoms or other phonons (i.e., externally produced phonons in the form of ultrasound or hypersonic waves).

The energy dispersion relations of these particles are given by the following expressions

Phonons: $E_{\text{Phon}} = \hbar\omega_j(\boldsymbol{q})$, (1.1)

Photons: $E_{\text{Phot}} = \hbar ck$ (1.2)

Neutrons, Electrons, Atoms: $E_{\text{part}} = \dfrac{\hbar^2}{2m} k^2$. (1.3)

Here $\boldsymbol{q}$ is the wave vector, and j the branch of the phonon; k is the magnitude of the wave vector $\boldsymbol{k}$, and m is the mass of the particle in question. The dispersion relation given by (1.3) refers to free particles, e.g. a beam of free electrons interacting with the system. For externally produced ultrasonic waves and for acoustic phonons described by the Debye model, the dispersion is given by $E = \hbar vq$, v being the velocity of sound. In Fig. 1.1 a log-log plot of the dispersion of these particles is presented. The phonon regime is illustrated by the two dotted lines marked by "optical phonons" and "acoustical phonons"; the two dotted lines indicate the dispersion of the optical branch of diamond [Ref. 1.1., Fig. 4.20], and the dispersion of a very low-energy acoustic branch such as the TA-branch of β-AgI along Δ [Ref. 1.1, Fig. 3.12], respectively. The figure also contains the regions in $E-k$ space which are relevant to the different experimental techniques, namely inelastic neutron scattering (INS), far-infrared spectroscopy (FIR), infrared spectroscopy (IR), Raman spectroscopy (R), Brillouin spectroscopy (B), diffuse X-ray scattering (X-RAYS), ultrasonic methods (ULTRASOUND), electron energy loss sepctroscopy (EELS) and inelastic molecular beam scattering (IMBS). Due to the small penetration depths of electrons and low-energy atoms or molecules, the two latter techniques probe surface phonons, while the other methods yield information about bulk phonons. Note that interactions of particles with phonons are also possible in regions of $E-k$ space where no crossing of the phonon and particle dispersion curves occur; examples are Raman, Brillouin and diffuse X-ray scattering as well as EELS.

The "99% ranges" in Fig. 1.1 present the energy and wave vector ranges required by the probe in order to study 99% of the phonons in a typical crystal. The essential feature of INS and IMBS is that these methods lie precisely in the region of overlap between these two 99% ranges. For this reason INS is the most important method to study bulk phonons. Although most of the other methods indicated in Fig. 1.1 yield mainly informations about phonons with small wave vectors $\boldsymbol{q}$, they are not unimportant. In contrary, they often yield very precious and complementary informations not available by INS. From FIR and IR experiments, for example, it is possible to obtain unique informations about dielectric properties of insulators and

2

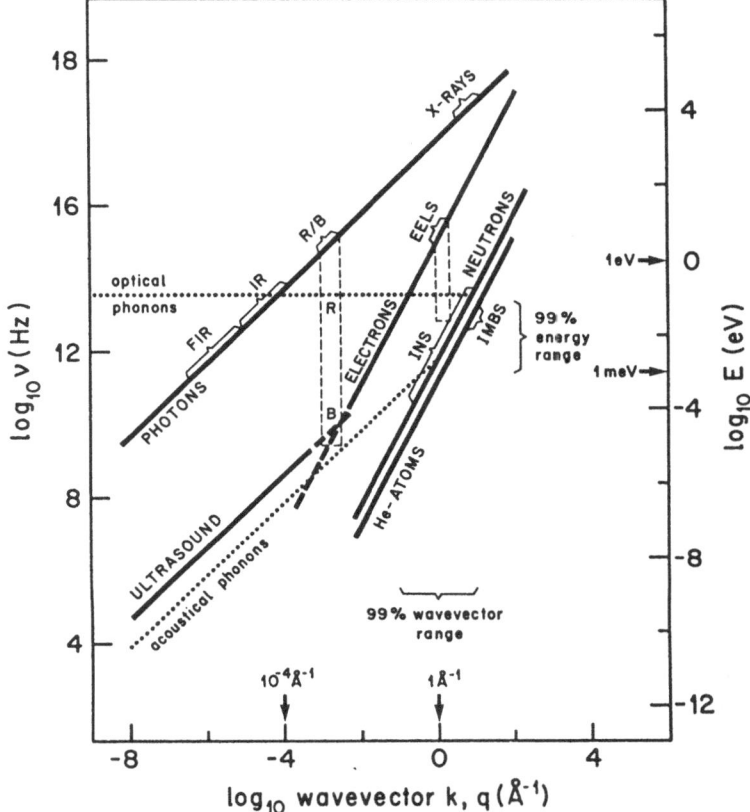

Fig. 1.1. Comparison of various radiation probes used in studies of the dynamical properties of solids. The figure shows the dispersion curves of photons, electrons, neutrons, He atoms and externally produced ultrasound with a sound velocity of $3 \cdot 10^3$ m/s. ($\cdots$) illustrate the dispersion of optical phonons (of diamond) and acoustical phonons (of β-AgI). The boundary of the first Brillouin zone is usually close to $q \cong 1\,\text{Å}^{-1}$. Also shown are regions on the $E - k$ map occupied by the different experimental methods *(see text)*

semiconductors. In the following we give a short discussion of the methods illustrated in Fig. 1.1. The experimental methods used to study phonons are summarized in Table 1.1.

1.2 Infrared Spectroscopy

Photons in the infrared and far-infrared region of the spectrum have energies comparable with phonon energies (regions marked IR and FIR in Fig. 1.1). Such photons can be strongly absorbed by certain TO phonons at $q \cong 0$. In one-phonon processes a photon can be absorbed only by those TO phonons

3

which induce a non-vanishing dipole moment; under special conditions it is also possible to observe LO phonons at $q \cong 0$. The strong interaction of photons with phonons leads to the concept of polaritons, which are excitations with mixed phonon-photon character. A photon can also couple with two or more phonons whose wavevectors are different from zero (multiphonon processes). The associated absorptions are generally weak compared with those of one-phonon processes. Infrared spectroscopy is therefore essentially limited to the study of TO phonons at $q \cong 0$. From a detailed infrared spectroscopic study it is possible to evaluate the complex frequency dependent dielectric constant $\varepsilon(\omega) = \varepsilon_1(\omega) + i\varepsilon_2(\omega)$, which describes the dispersive $(\varepsilon_1(\omega))$ and the absorptive $(\varepsilon_2(\omega))$ properties of the system.

Chapter 2 contains a thorough treatment of the experimental methods as well as the classical and quantum mechanical treatment of the dielectric properties of insulators and semiconductors in the infrared and far-infrared region.

1.3 Raman Spectroscopy

In Chap. 3 we consider the scattering of photons with energies of the order of 2–4 eV by optical phonons. The electric field E of the light induces an electronic dipole moment $M = \alpha E$, α being the electronic polarizability of the system. Due to the coupling between the electronic and nuclear motion, α depends on the configuration R of the nuclei: $\alpha = \alpha(R)$. If R changes due to thermal fluctuations, α changes, too. This is the origin of the Raman effect. Since according Fig. 1.1 (region marked R), the wave vector of visible light is still extremely small compared to the extension of the Brillouin zone, first-order Raman scattering yields information about TO and LO phonons at $q \cong 0$. If two or more phonons are involved in the scattering process, it is also possible to obtain informations about phonons with $q \neq 0$; the Raman lines associated with these higher order processes are, however, weak compared with those associated with first-order processes.

In Chap. 3 we are first considering the experimental techniques required for Raman scattering. In the classical theory we consider the basic model for Raman scattering, the polarizabilty tensor α and selection rules for the Raman effect. The quantum theory is introduced by a qualitative discussion of the scattering process: we then derive general expressions for the intensity of the scattered light which are explicitly evaluated on the basis of Placzek's approximation. Finally, we discuss Raman scattering by phonon-polaritons.

Since infrared absorption and Raman scattering processes are governed by different selection rules, the two methods yield complementary informations.

1.4 Brillouin Spectroscopy

From an empirical point of view Raman and Brillouin scattering differ only in that optical phonons are involved in Raman scattering and acoustical modes in Brillouin scattering. In both phenomena the intensity of the scattered light depends on the change in the electronic polarizability (or susceptibility) of the crystal which is induced by the phonons. Brillouin scattering applied to phonons yields information about the TA and LA modes at small wave vectors q (region marked B in Fig. 1.1). From the observed sound velocities it is possible to determine the elastic constants.

After a discussion of the experimental techniques, Chap. 4 contains a study of the kinematics and of the origin of Brillouin scattering. Consideration is given to the strain dependence of the dielectric constant as well as to an analysis of the intensities of the Brillouin components.

1.5 Interactions of X-Rays with Phonons

The scattering of X-rays by phonons is discussed in Chap. 5. X-rays have wave vectors comparable to those of phonons but their energies are extremely large compared with phonon energies (region marked X-rays in Fig. 1.1). Despite this huge discrepancy in energy it is possible to obtain limited informations about phonons by observing the intensities in diffuse X-ray scattering experiments.

The future of synchrotron X-radiaton together with multiple Bragg-reflection monochromators offer a very fascinating tool to study phonon dispersion curves. In this technique the scattered X-ray beam is directly energy analyzed as in the case of INS, Raman, Brillouin or EELS.

In Chap. 5 the following topics are discussed: Experimental techniques, interaction mechanism, scattering of X-rays by perfectly ordered crystals, thermal diffuse scattering of X-rays and the Debye-Waller factor. The chapter also serves as an introduction into inelastic neutron scattering presented in Chap. 6.

1.6 Inelastic Neutron Scattering

Inelastic neutron scattering (INS) is the most important technique for the study of bulk phonons. The advantage of INS compared with the other techniques stems from the fact that the de Broglie wavelength of a thermal neutron is of the same order of magnitude as the interatomic distances in crystals, and that the energy of a thermal neutron is comparable to phonon energies (region marked INS in Fig. 1.1).

Chapter 6 starts with the basic principles of neutron scattering (cross section, coherent and incoherent scattering). In Sect. 6.2 phonon dispersion relation measurements are discussed. This includes a discussion of the coherent cross-section, selection rules, spectrometer and examples of measured phonon dispersion curves. Section 6.3 is devoted to the measurements of phonon density of states.

A discussion of the similarities and differences between diffuse X-ray scattering and inelastic neutron scattering is especially instructive and helps to clarify the physical processes unterlying INS.

1.7 Other Techniques

In Chap. 7 we discuss some other experimental techniques, such as ultrasonic methods, inelastic electron tunneling spectroscopy, point contact spectroscopy and spectroscopic techniques for the investigation of surface phonons.

In the *ultrasonic method* considered in Sect. 7.1 a pulse of ultrasonic waves is injected into a block of the material, and the propagation behaviour of the stress waves is determined by the measurement of its velocity and attenuation. Stress waves can interact with various properties of the solid but we are confining our discussion to the interaction of the externally produced stress waves with thermal lattice vibrations. This kind of experiment gives information about sound velocities and phonon-phonon interactions.

Inelastic electron tunneling spectroscopy (IETS) is a technique that provides a versatile and sensitive method for measuring the vibrational properties of a thin insulating barrier sandwiched between two metals. A qualitative discussion if IETS is presented in Sect. 7.2.

During the last ten years a new spectroscopic method using point contacts between metals has been developed to study the interaction of electrons with elementrary excitations in metals, i.e. phonons, magnons, etc. Up to now most of the experiments peformed in this field deal with the investigation of electron-phonon interactions in pur metals or dilute alloys. *Point contact spectroscopy* is discussed in Sect. 7.3.

Although the emphasis of this textbook is on the study of bulk phonons, Sect. 7.4 contains a short discussion of the experimental study of surface phonons and adsorbates. Important experimental techniques for the study of surface phonons are *electron energy loss spectroscopy* (EELS) and *inelastic molecular beam spectroscopy* (IMBS). EELS allows the study of polar optical surface phonons (Fuchs-Kliewer modes), while acoustic surface phonons (i.e., Rayleigh modes) can be investigated by means of IMBS (Fig. 1.1). Infrared optical methods can be used to study surface phonons as well as the vibrational properties of thin films and adsorbates on metals. The dispersion of optical surface modes can be investigated by using a modification of the well-known *attenuated total reflection* method (ATR); a slight modification of this method is also known as *frustrated total reflection* (FTR). Thin films and adsorbates on metals can be studied by means of *infrared reflection absorption spectroscopy* (IRAS). Section 7.4 is devoted to a short discussion of EELS, IMBS, FTR and IRAS.

Table 1.1 contains the experimental method used to study phonons and also indicates the main informations obtained from these techniques.

Table 1.1. Experimental methods used to study phonons

Method	Abbreviations	Measurement of	Main Informations
Far-infrared and infrared spectroscopy	FIR IR	the intensity of transmitted or reflected light as a function of frequency	Infrared dielectric properties of insulators and semiconductors; optical phonons at $q \cong 0$
Raman spectroscopy	R	the intensity of scattered light as a function of frequency	Optical phonons at $q \cong 0$ Polaritons
Brillouin spectroscopy	B	as for Raman spectroscopy	Acoustic modes at small wave vectors q
Diffuse X-ray scattering		the intensity of scattered X-rays as a function of momentum transfer	Limited information about phonon dispersion
Inelastic neutron scattering	INS	the intensity of scattered neutrons as a function of energy and momentum transfer	Phonon dispersions, density of states
Ultrasonic methods	US	the velocity and attenuation of ultrasonic pulses	Sound velocities phonon-phonon interactions
Inelastic electron tunneling spectroscopy	IETS	current-voltage characteristics of metal-insulator-metal tunneling junctions	Vibrational properties of thin films and adsorbates
Point contact spectroscopy	PCS	current-voltage characteristics of point contacts between two metals	Electron-phonon interaction in metals and alloys
Electron energy loss spectroscopy	EELS	the intensity of backscattered electrons as a function of energy transfer	Optical surface phonons
Inelastic molecular beam spectroscopy	IMBS	the intensity of backscattered molecules as a function of energy and momentum transfer	Dispersion of acoustic surface phonons
Attenuated total reflection	ATR	the intensity of light totally reflected within an ATR crystal in direct contact with the sample as a function of frequency	Vibrational properties of insulators and semiconductors
Frustrated total reflection	FTR	the intensity of light totally reflected within a prism which is separated from the sample by a small air gap, as a function of frequency	Dispersion of optical surface phonons
Infrared reflection absorption spectroscopy	IRAS	the intensity of multiply reflected light between two metal plates covered with the thin film as a function of frequency	Vibrational properties of thin films and adsorbates

2. Infrared Spectroscopy

The chapter starts with a discussion of the principles of sepctrometers used in the infrared and far-infrared region. Reflectivity and transmission spectra of selected crystals serve to illustrate the experimental information. Methods are presented for the evaluation of optical constants, i.e. $\varepsilon_1(\omega)$ and $\varepsilon_2(\omega)$, from the observed reflectivity and transmission data. We then discuss the Kramers-Kronig relations and certain sum rules relevant for lattice vibrations. It follows a classical study of the interaction of lattice vibrations with light (polaritons) and the propagation of light in crystals. Finally, we present a discussion of one- and multi-phonon absorption processes on the basis of the quantum mechanical treatment of the dielectric constant. An extensive theoretical treatment of infrared absorption and Raman scattering by phonons of non-metals containing complementary aspects has been presented by *Bilz, Strauch* and *Wehner* [2.1].

2.1 Experimental Techniques

A variety of units are used in the infrared but the most useful are the micrometer (μm) for wavelength λ and the wavenumber $\tilde{\nu}$ in units of cm^{-1} for the frequency. The relation is

$$\tilde{\nu} = \frac{1}{\lambda} = \frac{\nu}{c} = \frac{\omega}{2\pi c}, \tag{2.1}$$

where ν and ω are the frequency and the angular frequency, respectively, and c is the velocity of light. $\tilde{\nu}$ is the number of wavelenghts per cm. The spectral range of the ordinary infrared extends from about 2.5–50 μm or from 4000–200 cm^{-1}. The far infrared region extends from about 50–3000 μm or from 200–3.3 cm^{-1}.

2.1.1 Grating Spectrometers

Figure 2.1 shows the diagram of the optical system of a Perkin-Elmer 680 Series double beam infrared spectrophotometer. The radiation emitted by the source ($T = 1100°$C) is divided into two beams. One beam is transmitted or reflected by the sample, whilst the other serves as a reference beam. In the photometer section the sample and reference beams are combined by the rotating sector mirror to form a single beam consisting of pulses of

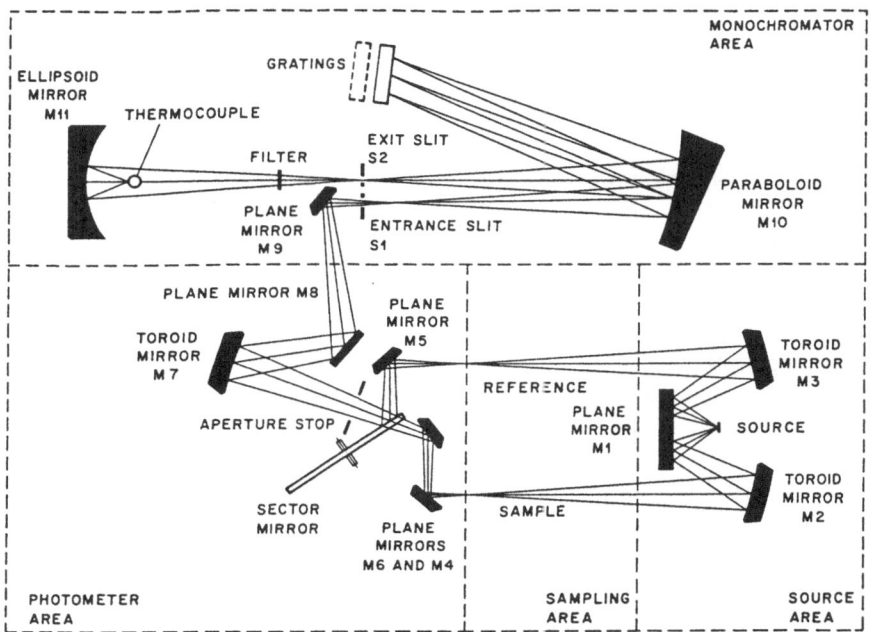

Fig. 2.1. Optical layout of the Perkin-Elmer 680 Series double beam spectrophometer. The 683-Model covers the range $200\,\text{cm}^{-1} \leq \tilde{\nu} \leq 4000\,\text{cm}^{-1}$

radiation from the sample and reference beams. This combined pulsed beam passes into the monochromator, where it is dispersed by the grating into the spectral components. As the grating is rotated the dispersed spectrum is scanned across the monochromator exit slit. The mechanical width of the monochromator slit determines the width of the wave-number band emerging from the monochromator. Decreasing the slit width decreases both the spectral bandwidth (i.e., improves the resolution) and the intensity of the emerging radiation (i.e., decreases the signal-to-noise ratio). After leaving the monochromator, the radiation passes through one of a set of optical filters. This filter rejects unwanted radiation diffracted from the grating at the same angle as the component of the desired wave-number. The transmitted radiation is finally focused onto a thermocouple detector.

For a detailed descritpion of infrared optical components and of spectrometers the reader is referred to [2.2, 3].

2.1.2 Fourier Interferometers

From Plank's law the intensity emitted from a black body at temperature T per unit frequency interval is given by

$$I(\omega) \sim \omega^3 [\exp(\hbar\omega/k_\text{B}T) - 1]^{-1}. \tag{2.2}$$

In the far infrared $\hbar\omega \ll k_BT$ for all thermal sources and therefore from (2.2) it follows that $I(\omega) \sim \omega^2 T$. Thus for constant resolution $\Delta\omega$ the intensity decreases with ω^2. Consider a black body at a temperature of $T = 5000\,K$ and a $1\,cm^2$ area. Its total emission is about $10^3\,W$; from this only about $10^{-2}\,W/cm^2$ is available in the whole region below $200\,cm^{-1}$. In such an energy-starved region, one of the disadvantages of grating spectrometer lies in the fact that the intensity is further decreased by the use of narrow slits to obtain high resolution. This is not the case if an interferometer is used. Another major advantage of interferometers over dispersion spectroscopy lies in the fact that with an interferometer the whole spectral range of interest is incident on the detector at one time, rather than a single resolution width. Assuming otherwise comparable efficiency, the interferometer can obtain the whole spectral range in the same time taken by the monochromator to obtain one resolution width. If there are N elements in the spectrum the interferometer will show an improvement of the order of $N^{1/2}$ in its signal/noise ratio, as compared to a single channel spectrometer of the same luminosity and resolving power. This signal/noise advantage is termed the *multiplex advantage* or the *Fellgett advantage*, which is particularly important when searching a wide spectral range at high resolution. For details of the spectroscopic technique the reader is referred to [2.4–6].

The principle of Fourier transform spectroscopy is simple. Figure 2.2 shows the optical layout of a Michelson interferometer; it can be used in the range from about 400 to about $20\,cm^{-1}$ ($\lambda = 25 - 500\,\mu m$), but interferometers are now commercially available which cover the infrared-visible- and ultraviolet regions as well. The radiation of the source (1) (high pressure

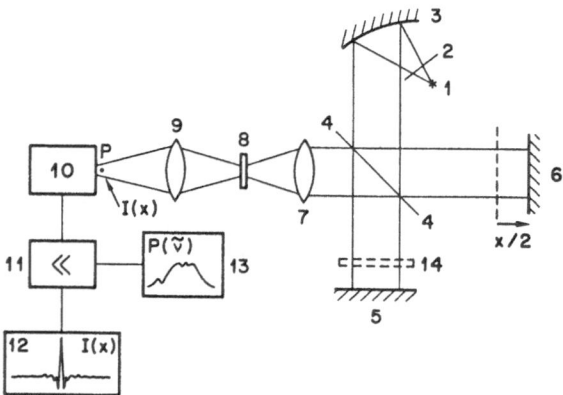

Fig. 2.2. Optical layout of a Michelson interferometer. The recorder (12) displays a typical double-sided interferogram and the scan on the computer (13) the corresponding spectrum $P(\tilde{\nu})$. In the asymmetric mode the sample either replaces the mirror (5) (for reflection measurements) or is inserted in one of the beams (sample 14) for transmission measurments) (Sect. 2.2.4)

mercury arc, $T \cong 5000\,\mathrm{K}$) is interrupted periodically by a rotating chopper (2) and reflected by a parabolic mirror (3). The parallel light beam is then split by a suitable beam splitter (4) (mylar film) into two beams of equal intensities; one of them is reflected by a fixed mirror (5), the other by a moving mirror (6), which is displaced by a variable distance $x/2$ with respect to the fixed mirror. After the beam splitter, the two beams are recombined to form a single beam which passes through the polyethylene lens (7) and is focused onto the sample (8). The light emerging from the sample is then focused by the lens (9) and a light pipe onto the detector (10) (Golay cell, pyroelectric detector or bolometer). The electrical signal from the detector is amplified by a Lock-In/ADC system (11). The interferogram is registered on a recorder (12) and simultaneously punched on tapes by a punch system or directly Fourier transformed by a dedicated computer (13). It is necessary to measure two interferograms, one for the sample and one for the background. Figure 2.3 shows an example of sample and background interferograms together with the corresponding spectra and the resulting ratioed reflectivity spectrum for CdCrSe$_4$ [2.6].

It is shown in Appendix A that for a symmetric interferogram $I(x) = I(-x)$ the desired spectrum $S(\tilde{\nu})$ can be approximated by the Fourier series

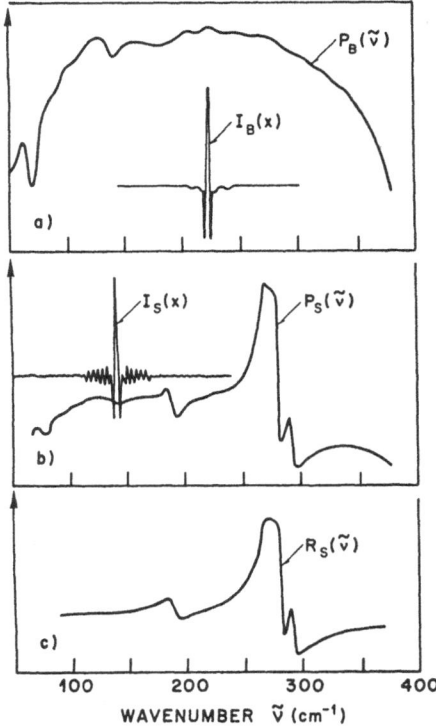

a)

b)

c)

WAVENUMBER $\tilde{\nu}$ (cm^{-1})

Fig. 2.3. Interferograms and corresponding spectra for **(a)** background, **(b)** sample, and **(c)** ratioed reflectivity spectrum of CdCrSe$_4$ [2.6]

$$S(\tilde{\nu}) = 4 \sum_{x=0}^{D} [I(x) - \frac{1}{2}I(0)](\cos 2\pi\tilde{\nu}x)\Delta x. \qquad (2.3)$$

Here $I(0)$ is the intensity at zero path difference, D is the maximum optical path difference and equals twice the total moving mirror displacements. $I(x)$ is sampled at discrete intervals Δx between $x = 0$ and $x = D$, thus $x = h\Delta x$ with $h = 0, 1, 2 \ldots D/\Delta x$.

In order to obtain all the information in a spectral range $0 < \tilde{\nu} < \tilde{\nu}_{max}$, it is necessary to sample points from the interferogram at intervals $\Delta\tilde{\nu} = 1/D$ [2.4].

Several computational refinements have been developed to improve the quality of the spectra. One of them corrects the disturbing effects which may arise by cutting off the interferogram at $x_{max} = D$ (apodization), another corrects the effects of unsymmetric interferograms (phase-error correction). The phase-error correction can be avoided by measuring the interferogram for an equal distance on each side of $x = 0$ (Fig. 2.3) and to compute the power transform of this double sided interferogram [2.4–6].

If $P(\tilde{\nu}) = GD(\tilde{\nu})S(\tilde{\nu})$ is the actual computed spectrum (G : amplifier gain, $D(\tilde{\nu})$: detector sensitivity), then the transmission of the sample is given by (Appendix A)

$$T_S(\tilde{\nu}) = G_B P_S(\tilde{\nu})/G_S P_B(\tilde{\nu}), \qquad (2.4)$$

where subscripts S and B refer to the sample and background spectra, respectively. An analogous relation holds for the reflectivity $R_S(\tilde{\nu})$; in this case a mirror is used at the place of the sample for the background spectrum (Fig. 2.3).

Finally, we mention the "Polarizing Interferometer" developed by *Martin* and *Puplett* [2.7]. This instrument is particularly useful for the investigation of low-frequency excitations because it is possible to extend the wave-number range down to less than $5\,\text{cm}^{-1}$.

2.2 Dielectric Properties: Classical Treatment

2.2.1 Reflectivity, Transmission, Absorptivity and Optical Constants

In optical experiments one generally measures the intensities of the incident (I_0), reflected (I_R) and transmitted (I_T) beams, as shown in Fig. 2.4.

From conservation of energy it follows

$$I_0 = I_R + I_T + I_A, \qquad (2.5a)$$

where I_A is the intensity absorbed by the sample. Dividing by I_0, we obtain

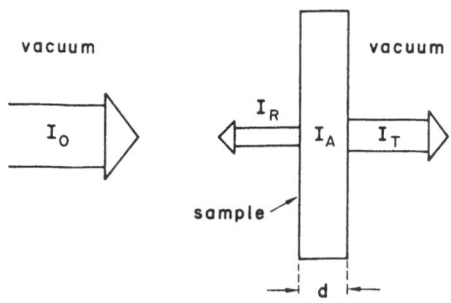

vacuum   vacuum

I_0

I_R

I_A I_T

sample

d

Fig. 2.4. The incident light beam of intensity I_0 is partly reflected (I_R), partly transmitted (I_T), and partly absorbed (I_A) by the sample. Possible scattering losses are neglected

the familiar relation

$$\tilde{R} + \tilde{T} + \tilde{A} = 1, \tag{2.5b}$$

where $\tilde{R} = I_R/I_0, \tilde{T} = I_T/I_0$, and $\tilde{A} = I_A/I_0$ are the "apparent" reflectivity, transmission and absorptivity respectively. (The "true" reflectivity R and absorptivity A refer to an infinitively or sufficiently thick sample such that the transmission can be neglected; the "true" transmission T is defined by (2.16)). From measurements or $\tilde{R}$ and $\tilde{T}$ it is possible to determine the optical constants. It is also possible to measure $\tilde{A}(\omega)$ by measuring the heating of the sample due to absorption of light [2.8,9]. In some cases it is more convenient to observe the thermal emission of the sample [2.10]. According to Kirchhoff's law the emissivity $\tilde{E}(\omega)$ (ratio of the spectral emissive power of the sample and a black-body at the same temperature) is equal to the absorptivity $\tilde{A}(\omega)$ [2.11].[1]

General expression for $\tilde{R}$ and $\tilde{T}$ for a parallel slab are given in Appendix B, see (B. 47, 48) [2.12]. For an opaque sample $\tilde{T} = 0$ and $\tilde{A} = 1 - \tilde{R}$. For a black body $\tilde{A} = 1$. For thick and highly reflecting metals $\tilde{R}$ is close to unity, $\tilde{T}$ is zero, and $\tilde{A}$ is correspondingly small.

In the following we summarize some important relations for the optical constants (Appendix B) [2.12, 13]. For the complex dielectric function we write

$$\varepsilon(\omega) = \varepsilon_1(\omega) + i\varepsilon_2(\omega). \tag{2.6}$$

We consider only optically isotropic materials (including crystals of cubic symmetry). In crystals of lower symmetry ε has to be identified with one of the principle values of the dielectric tensor. Let

$$n(\omega) = n_1(\omega) + in_2(\omega), \tag{2.7}$$

be the complex refractive index, where n_1 is the refractive index and n_2 the extinction coefficient. In terms of n_1 and n_2 we have (Appendix B)

$$\varepsilon_1 = n_1^2 - n_2^2, \tag{2.8}$$

---

[1] By definition, the absorptivity of a black-body is unity at all temperatures and frequencies, and its spectral emissive power is given by Planck's law

$$\varepsilon_2 = 2n_1 n_2. \tag{2.9}$$

The real part of the frequency dependent conductivity is given by

$$\sigma(\omega) = \frac{\omega}{4\pi} \varepsilon_2(\omega). \tag{2.10a}$$

If we introduce the wave-number $\tilde{\nu}$ ($\omega = 2\pi c \tilde{\nu}$), and use $1(\Omega\,\text{cm})^{-1}$ as the unit for σ [$1(\Omega\,\text{cm})^{-1} = 0.906 \cdot 10^{12}\text{s}^{-1}$], we can write (2.10a) in the practically useful form

$$\sigma(\tilde{\nu}) = 1.6638 \cdot 10^{-2} \tilde{\nu} \varepsilon_2(\tilde{\nu}). \tag{2.10b}$$

If light is reflected at near-normal incidence at the surface of a sufficiently thick crystal such that the transmission can be neglected, the reflectivity is given by (Appendix B)

$$R = \frac{(n_1 - 1)^2 + n_2^2}{(n_1 + 1)^2 + n_2^2}. \tag{2.11}$$

More general relations for the reflectivity of thick anisotropic crystals at non-normal incidence, as well as expressions for the reflectivity and transmission of plane parallel plates of finite thickness are given in Appendix B. For very thin films there is a strong reflection from the back surface of the crystal which is out of phase and cancels the strong reflection at the front surface at all frequencies except those close to the absorption frequencies, where the resonance absorption attenuates the waves in the crystal. The normal incidence reflection and transmission spectra will accordingly exhibit maxima and minima at the absorption frequencies, respectively.

Let $I(r)$ be the intensity of the transmitted beam a distance r apart from the surface of the parallel plate, and $I(0)$ the intensity at $r = 0$ penetrating into the medium. $I(r)$ is given by

$$I(r) = \frac{1}{2} \mid \boldsymbol{E} \mid^2, \tag{2.12}$$

where $\boldsymbol{E}$ is given by (B.7), namely,

$$\boldsymbol{E} = \boldsymbol{E}_0 \exp\{-i(\omega t - \boldsymbol{k} \cdot \boldsymbol{r})\}, \quad \text{and} \tag{2.13}$$

$$\boldsymbol{k} = \boldsymbol{k}_1 + i\boldsymbol{k}_2, \tag{2.14}$$

is the complex wave vector (B.30). Thus

$$I(r) = \frac{1}{2} E_0^2 \exp(-2k_2 r) = I(0)\text{e}^{-Kr}. \tag{2.15}$$

The "true" transmission is defined by

$$T = I(r)/I(0) = \text{e}^{-Kr}. \tag{2.16}$$

K is the absorption coefficient. From (2.15) and (B.30b) it follows that

$$K(\omega) = 2k_2(\omega) = 2\frac{\omega}{c}n_2. \tag{2.17}$$

The relation (2.16) is known as *Beer's law*.

The time average power absorption of the electromagnetic field per unit volume by the absorbing medium is given by

$$\overline{W} = \overline{J \cdot E}, \quad \text{where} \tag{2.18}$$

$$J = \sigma E, \tag{2.19}$$

is the current. Using (2.10a) and $E = E_0 \cos \omega t$ gives

$$\overline{W}(\omega) = \frac{1}{4\pi}\omega\varepsilon_2(\omega)E_0^2\overline{\cos^2\omega t}$$

and from $\overline{\cos^2\omega t} = 1/2$ we obtain

$$\overline{W}(\omega) = \frac{1}{8\pi}\omega\varepsilon_2(\omega)E_0^2(\omega) \tag{2.20}$$

2.2.2 Examples of Reflectivity and Transmission Spectra

In this subsection we present some illustrative reflectivity and transmission spectra which will be discussed below. Figure 2.5a displays the observed reflectivity R of a thick NaCl crystal at several temperatures [2.14].

The arrows at 61 μm and at 38 μm indicate the TO and LO frequencies at 300 K, respectively. Note that R is high between these two frequencies; the reason for this will be explained in Sect. 2.2. The two subsidiary structures at the short wavelength side of the main reflectivity band are due to anharmonic interactions (Sect. 2.3.7).

The transmission of thin NaCl layers of various thicknesses is shown in Fig. 2.5b [2.15]. The minimum occurs at 61 μm, which corresponds to the TO frequency at 164 cm^{-1}.

The reflectivity spectra of single crystals of LiF at various temperatures is shown in Fig. 2.6 [2.16a].

The main band which corresponds to the fundamental resonance is the low-frequency band, while the subsidiary band at higher wave numbers is again due to anharmonic interactions (Sect. 2.3.7). With increasing temperature, the reflectivity of the main band and side band decreases, and the structures shift to lower frequencies. The arrows indicate the TO and LO frequencies at room temperature at 306 and 675 cm^{-1}, respectively.

We have seen above that transmission experiments on thin films peformed at near-normal incidence yield the frequencies of the TO modes near $q \cong 0$. *Berreman* [2.16b] has shown that it is also possible to measure

15

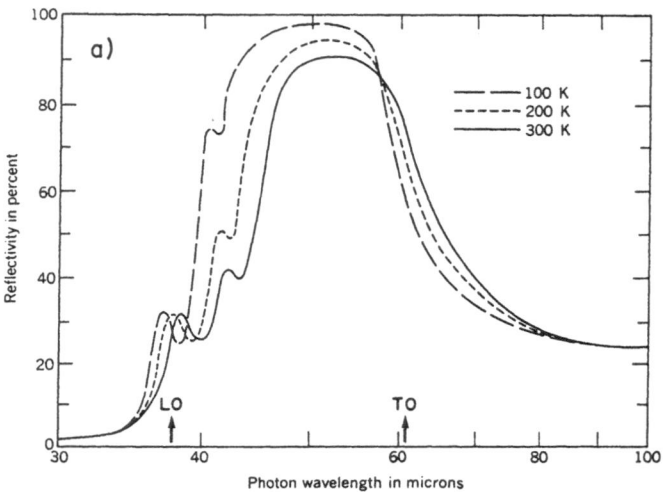

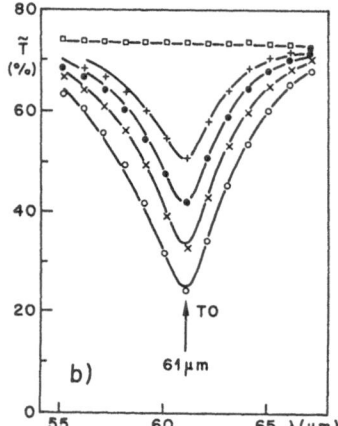

Fig. 2.5. (a) Reflectivity of a thick crystal of NaCl at several temperatures, versus wavelength [2.14]. The arrows indicate the positions of the TO and LO wavelengths. (b) Near-normal incidence transmission of NaCl layers of various thicknesses evaporated on Hostafan, versus wavelength. Thicknesses of the layers: $d = 0(\square)$, $10\,\mu m$ thick Hostafan; $d = 0.07\,\mu m(+)$; $d = 0.11\,\mu m(\bullet)$; $d = 0.17\,\mu m(\times)$; $d = 0.26\,\mu m(\circ)$ [2.15]

directly the LO frequencies when the incident radiation is not normal to the surface and if the light is polarized parallel to the plane of incidence. Consider first a very thin film of an alkali halide whose thickness d is much smaller than the wavelength λ of the infrared light. A film of such a crystal has two normal modes of polarized lattice vibrations with $\lambda \gg d$. In one mode the vibration is parallel to the film surface and corresponds to the TO mode near $q = 0$. In the other, the vibration is polarized normal to the film surface and corresponds to the LO mode; its frequency is higher than the ferquency of the TO mode due to the accumulation of charge at the surfaces of the film, which gives rise to the depolarization field $\boldsymbol{E}^{(M)} = -4\pi\boldsymbol{P}$ [Ref. 2.17, Fig. 4.5]. The most convenient experimental setup is shown in Fig. 2.7.

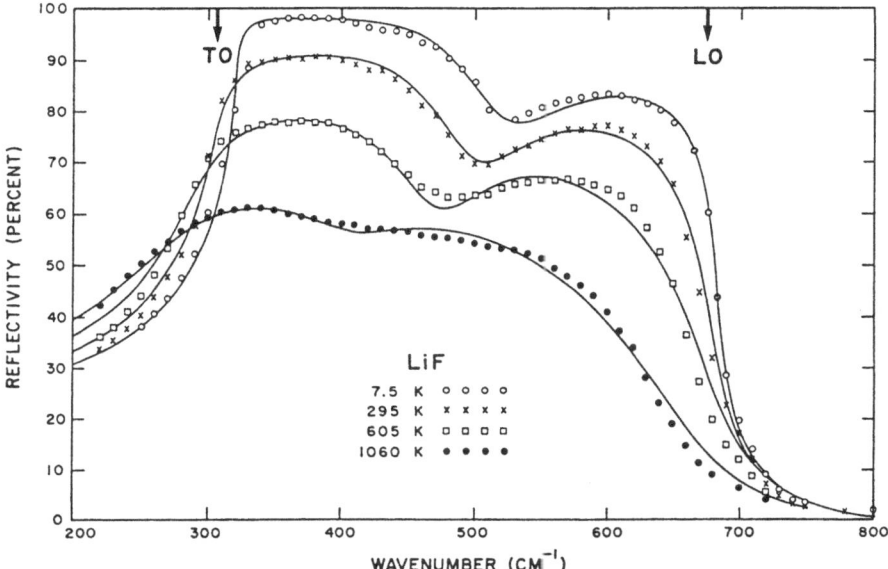

Fig. 2.6. Infrared reflectivity of LiF as a function of wavenumber and temperature. $T = 17.5\,\mathrm{K}(\circ)$; $T = 295\,\mathrm{K}(\times)$; $T = 605\,\mathrm{K}(\square)$; $T = 1060\,\mathrm{K}(\bullet)$. (——) are theoretical curves calculated on the basis of classical dispersion theory (Sect. 2.2.5) [2.16a]

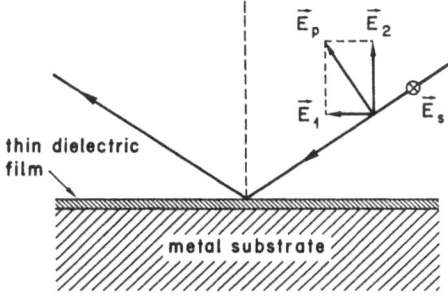

Fig. 2.7. Experimental configuration for the excitation of longitudinal lattice vibrations

    The thin film is evaporated on a metal substrate. E_s is the component of the electric field of the light polarized perpendicular (German: senkrecht), and E_p the component polarized parallel to the plane of incidence. The latter can be decomposed into the two components E_1 and E_2. The field E_s and E_1 can, in principle, excite TO vibrations; however, the underlying conductive medium practically eliminates the absorption by TO modes because almost no electric field can exist adjacent and parallel to a metallic surface. On the other hand, the field E_2 can excite the LO absorption and this absorption is not reduced by the conducting substrate. The experimental result for LiF is shown in Fig. 2.8.

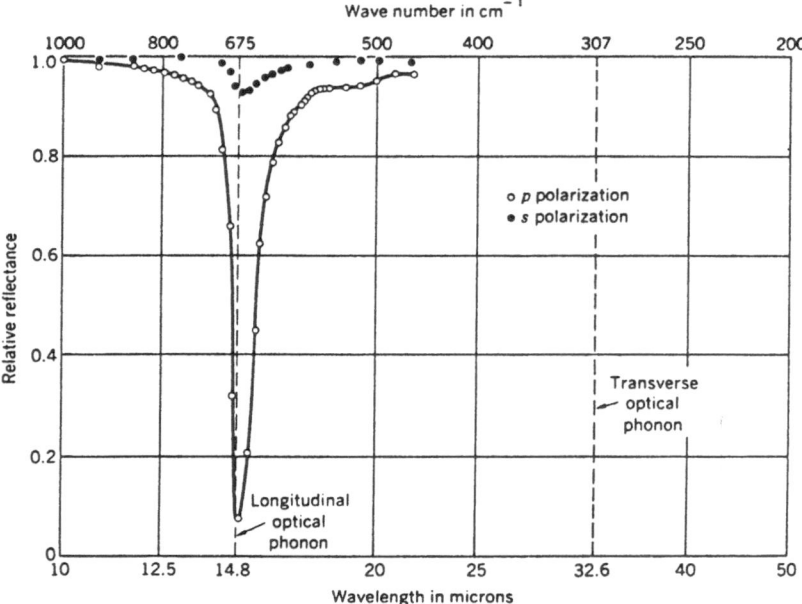

Fig. 2.8. Obseved reflectivity of $s-$ and $p-$polarized radiation by silvered glass slides with LiF films deposited on them, relative to that by silvered glass alone [2.16b]

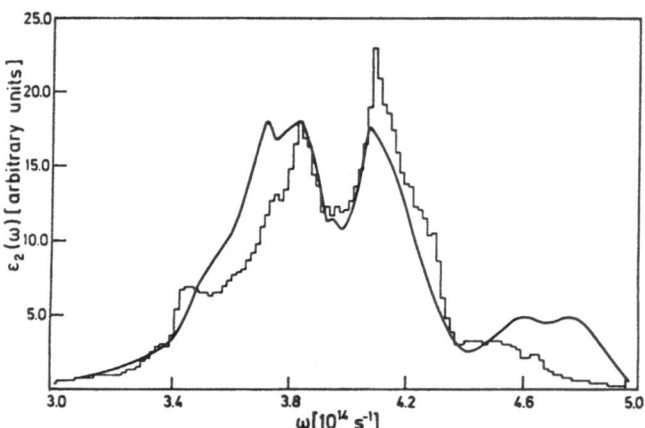

Fig. 2.9. Infrared absorption $\varepsilon_2(\omega)$ of diamond at 296 K. (——) is the experimental curve of [2.82]. The histogram is the absorption spectrum calculated with only one parameter [2.83]

There is high reflection for both s- and p-polarized light, except for p-polarized light near the wavelength of the LO phonon. At this wavelength the component $\boldsymbol{E}_2$ of the light couples strongly with the LO phonon, which gives rise to the sharp minimum at $\lambda = 14.95\,\mu$m $\left(\tilde{\nu} = 675\,\mathrm{cm}^{-1}\right)$.

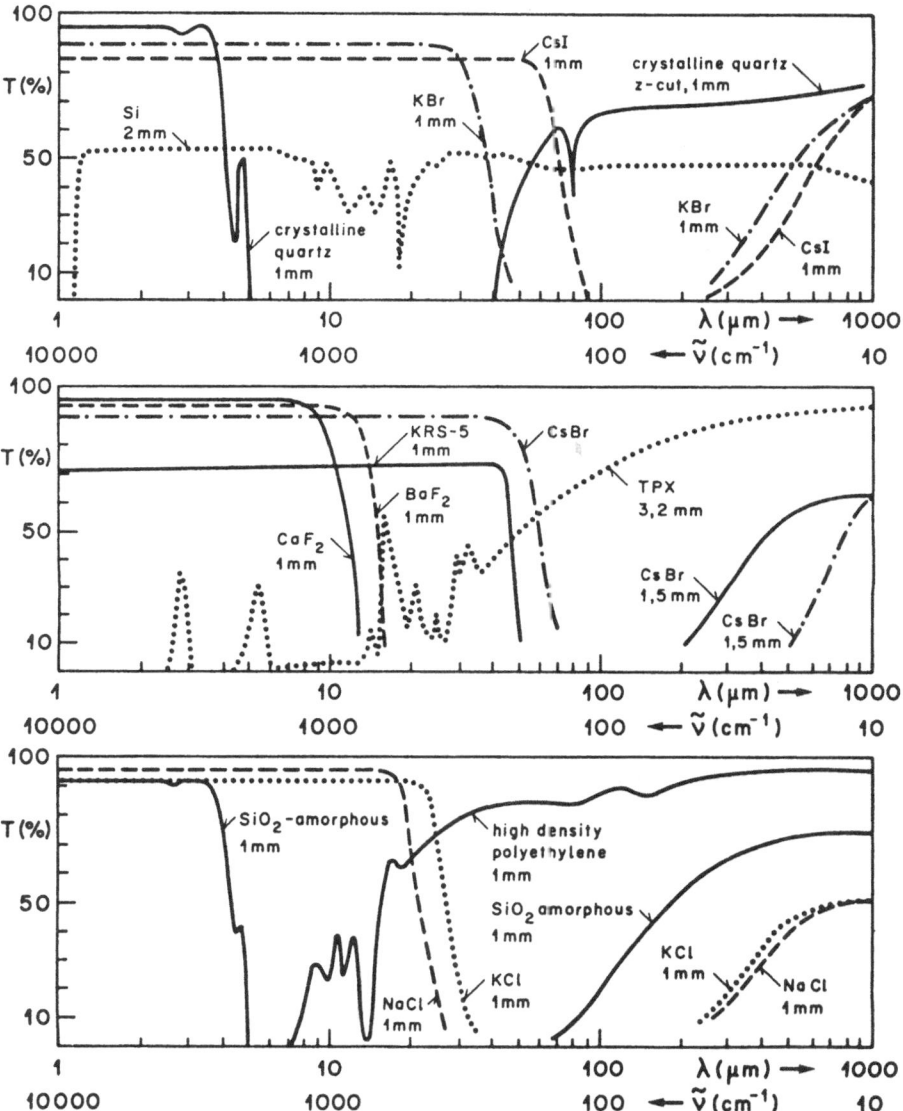

Fig. 2.10. Transmission spectra of a number of window materials used in the infrared and far infrared (with cortesy of ORIEL)

In this connection it should be mentioned that reflectivity experiments peformed with *p*-polarized light at non-normal incidence can also be used to detect LO modes in bulk materials if these are anisotropic. This has been demonstrated by *Decius* et al. in the case of calcite [2.18]. Furthermore, such experiments can also be used to excite electronic plasma oscillations which correspond to longitudinal electronic vibrations. An excellent example is the highly anisotropic electronic conductor KCP [2.19–21]. Finally, it is possible to excite the electron plasma frequencies of thin isotropic metal films [2.22] in experimental arrangements which are completely analogous to Berreman's technique.

At this point it should be recalled that multiple reflection of p-polarized light at a large angle of incidence can be used to study extremely thin oxide layers or adsorbate films on metals. Absorptions due to vibrational transitions in films with thicknesses down to 5 Å can be observed in favourable cases; for this reason the method has developed to an non-destructive surface analysis technique [2.23] (Sect. 7.4.5).

Until now we have discussed only ionic crystals in which the dipole moment induced by the TO phonons at $q = 0$ can directly couple with the electric field of the light leading to strong absorption at $\omega = \omega_{TO}$. In simple covalent crystals such as silicon, germanium and diamond, however, the TO phonons do not induce an electric dipole moment and therefore no direct coupling with the light is possible. Nevertheless, such crystals show a weak but distinctive absorption spectrum as is shown for the case of diamond in Fig. 2.9. This absorption is due to a coupling of the photons with two (or more) phonons having non-vanishing wave vectors q_1 and $q_2 = -q_1$ (Sects. 2.3.1, 6 and 7).

Figure 2.10 shows the transmission spectra of a number of window materials used in the infrared and far-infrared.

2.2.3 Evaluation of $\varepsilon(\omega)$ from a Kramers-Kronig Analysis of the Reflectivity

In terms of the amplitude reflection coefficient $\tilde{r}$, the reflectivity of a thick crystal is given by (Appendix B and [2.12])

$$R = \tilde{r}\tilde{r}^*, \quad \text{where} \tag{2.21}$$

$$\tilde{r} = |\tilde{r}| \, e^{i\phi} = \frac{n_1 + in_2 - 1}{n_1 + in_2 + 1}. \tag{2.22}$$

Here $|\tilde{r}| = R^{1/2}$ is the amplitude of $\tilde{r}$ and ϕ is the phase shift of the light wave caused by the reflection. From (2.21, 22) it follows that

$$n_1 = \frac{1 - R}{1 + R - 2R^{1/2}\cos\phi}, \tag{2.23}$$

$$n_2 = \frac{2R^{1/2}\sin\phi}{1 + R - 2R^{1/2}\cos\phi}. \tag{2.24}$$

The dispersion relation for $\phi(\omega)$ follows from the general Kramers-Kronig (KK) relations discussed in Sect. 2.2.6 and is given by [2.24, 25]

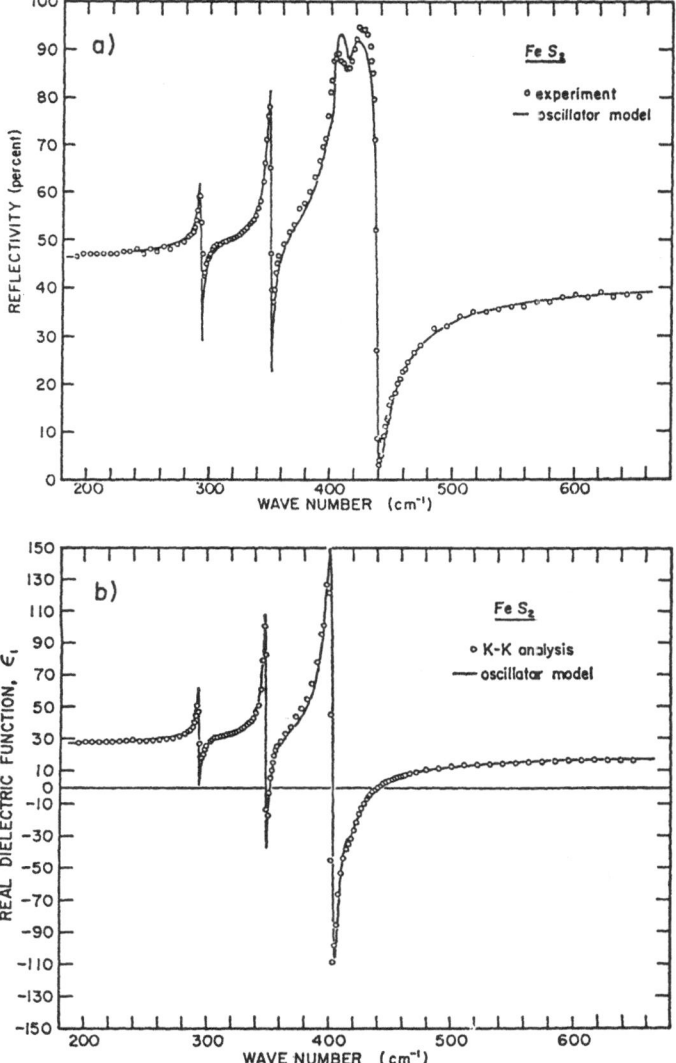

Fig. 2.11. (a) Reflectivity $R(\tilde{\nu})$ of FeS$_2$ as a function of wavenumber. (○) experimental values, (——) classical oscillator model. (b) $\varepsilon_1(\tilde{\nu})$; (○) from Kramers-Kronig analysis of $R(\tilde{\nu})$, (——) classical oscillator model

$$\phi(\omega) = \frac{\omega}{\pi} \int\limits_0^\infty \frac{\ln R(\omega') - \ln R(\omega)}{\omega^2 - \omega'^2} d\omega'. \qquad (2.25)$$

From (2.23–25) we can compute n_1 and n_2 and from (2.8, 9) we obtain ε_1 and ε_2; $\sigma(\omega)$ follows from (2.10a). In practice, $R(\omega)$ is known only in

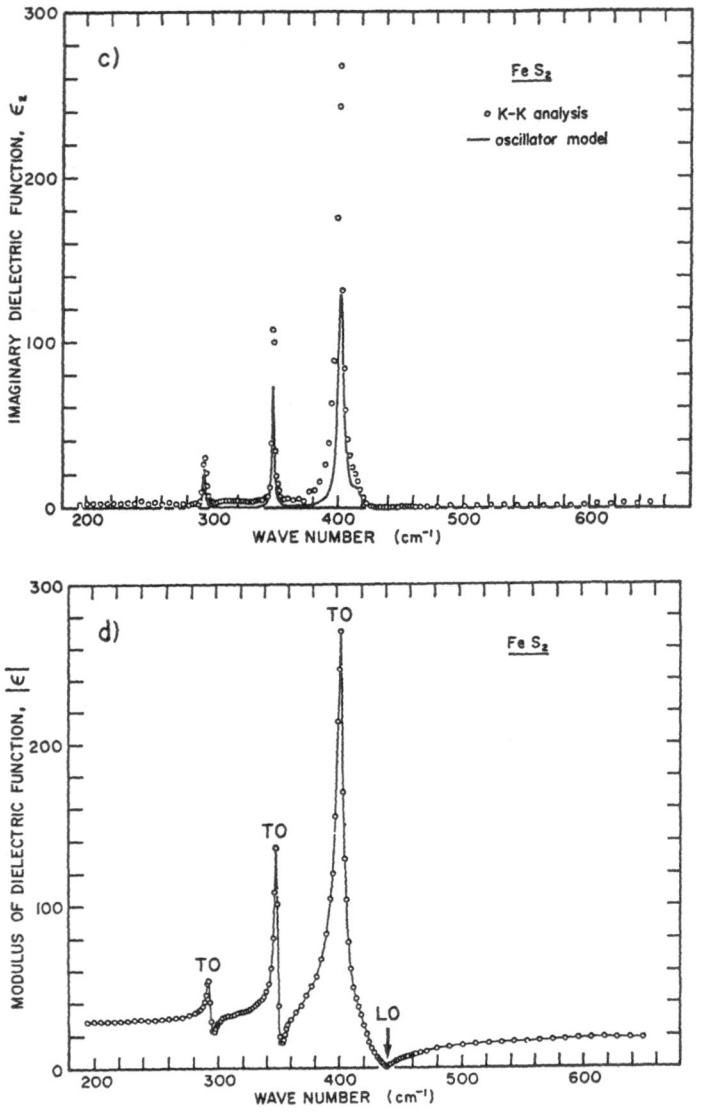

Fig. 2.11. (c) $\varepsilon_2(\tilde{\nu})$; ($\circ$) from Kramers-Kronig analysis of $R(\tilde{\nu})$, (——) classical oscillator model. (d) $|\varepsilon(\tilde{\nu})| = [\varepsilon_1^2(\tilde{\nu}) + \varepsilon_2^2(\tilde{\nu})]^{1/2}$; ($\circ$) from Kramers-Kronig analysis, the smooth curve connects the points but has no theoretical significance

a limited frequency range $\omega_1 \leq \omega \leq \omega_2$, and suitable extrapolations must be introduced outside this range [2.24].

As an example we consider the far-infrared optical properties of pyrite, FeS_2 [2.26]. Figure 2.11a shows the reflectivity spectrum, while Figs. 2.11b,c display ε_1 and ε_2 as obtained from the KK analysis based on (2.23–25). It has been shown [Ref. 2.17, Sect. 5.5.1] that for very small damping the TO frequencies are given in a good approximation by the peaks of $\varepsilon_2(\omega)$ [Ref. 2.17, Fig. 5.12] and that for vanishing damping the LO frequencies are obtained from the condition $\varepsilon(\omega) = 0$ [Ref. 2.17, Fig. 4.9]. *Chang* et al. [2.27] have shown that for damped phonons it is a better approximation to identify the maxima and minima of the modulus of the complex dielectric function, $| \varepsilon | = (\varepsilon_1^2 + \varepsilon_2^2)^{1/2}$, with the TO and LO frequencies, respectively. This is illustrated in Fig. 2.11d. It can also been shown that the TO frequencies can be obtained from the maxima of the frequency-dependent conductivity $\sigma(\omega)$, as defined by (2.10), while the LO frequencies are obtained from the maxima of the dielectric loss function, $-\mathrm{Im}\{1/\varepsilon\} = \varepsilon_2 / (\varepsilon_1^2 + \varepsilon_2^2)$ [2.28].

2.2.4 Direct Determination of $\varepsilon(\omega)$ from Experimental Data

The Kramers-Kronig analysis is not always applicable, for instance, if the necessary extrapolations can not be made [2.29]. In such cases it is possible to calculate ε_1 and ε_2 from combined reflectivity and transmission measurements or from transmission experiments at various thicknesses [2.29]. These calculations are based on general relations for the reflectivity and transmission of samples of finite thicknesses, as given by (B. 47, 48) [2.12] or on even more complicated expressions for the transmission and reflection of thin films on suitable supporting substrates with known optical properties [2.30, 31]. In theses calculations care must be exercised because of multiple solutions and critical points [2.32]. It is also possible to evaluate ε_1 and ε_2 from measurements of the reflectivity of thick crystals with polarized light as a function of the angle of incidence [2.33–37]. The relevant Fresnel formulas for anisotropic crystals belonging to the tetragonal, hexagonal, trigonal und orthorhombic systems are given in Appendix B (Fig. B.2).

It is also possible to measure both the real and imaginary part of the optical constants by the technique of dispersive or asymmetric Fourier spectroscopy [2.38–46]. In this technique the sample either replaces the mirror in one of the arms of a Michelson interferometer (for reflection measurements) or is merely inserted into the beam in one of the interferometer arms (for transmission measurements) (Fig. 2.2). The interferogram obtained with no sample in either beam is the Fourier transform of the spectral power density through the instrument. If a sample is placed in one beam each spectral component is reduced in amplitude and its phase is shifted. A Fourier transform of the sample and of the background interferogram gives the information for the determination of the alteration of the amplitude and the phase of each spectral component. In the reflection mode it is thus possible to measure both the amplitude $| \tilde{r}(\omega) |$ and the phase $\phi(\omega)$ of (2.22) which allows the determination of $n_1(\omega)$ and $n_2(\omega)$ from (2.23, 24).

2.2.5 $\varepsilon(\omega)$ from a Model Fit to the Observed Reflectivity

We have discussed $\varepsilon(\omega)$ for a system of damped harmonic oscillators [Ref. 2.17, Eqs. (5.136, 137)], as well as for the anharmonic crystal, for which we have introduced the response function [Ref. 2.17, Eqs. (5.141, 145)]. We may write

$$\varepsilon(\omega) = \varepsilon_\infty + \sum_j S_j R_j(0, \omega). \tag{2.26}$$

Here ε_∞ is the high frequency dielectric constant (at frequencies large compared with the lattice vibration frequencies but small compared with the electronic transition frequencies), S_j is the oscillator strength of mode j, and $R_j(0, \omega)$ the response function of that mode for wavenumber $q \cong 0$. In the pseudo-harmonic approximation $R_j(0, \omega)$ is given by [Ref. 2.17, Eq. (5.145)]

$$R_j(0, \omega) = [\tilde{\omega}_j^2(0) - \omega^2 - 2i\tilde{\Gamma}_j(0)\omega]^{-1}. \tag{2.27}$$

$\tilde{\omega}_j(0)$ and $\tilde{\Gamma}_j(0)$ are the eigenfrequencies and damping of the mode ($j, q \cong 0$), which depend on temperature [Ref. 2.17, Fig. 5.14]. In the following we use a simplified notation and write (2.26, 27) in the form

$$\varepsilon(\omega) = \varepsilon_\infty + \sum_j \frac{S_j}{\omega_j^2 - \omega^2 - i\gamma_j\omega}. \tag{2.28}$$

This expression is similar to the expression for $\varepsilon(\omega)$ of damped harmonic oscillators, but $\omega_j = \tilde{\omega}_j(0)$ and $\gamma_j = 2\tilde{\Gamma}_j(0)$ are regarded as the temperature dependent pseudo-harmonic frequency and damping, respectively. From (2.28) we obtain

$$\varepsilon_1(\omega) = \varepsilon_\infty + \sum_j \frac{S_j(\omega_j^2 - \omega^2)}{(\omega_j^2 - \omega^2)^2 + \gamma_j^2\omega^2}, \tag{2.29}$$

$$\varepsilon_2(\omega) = \sum_j \frac{S_j\gamma_j\omega}{(\omega_j^2 - \omega^2)^2 + \gamma_j^2\omega^2}. \tag{2.30}$$

From (2.8, 9, 11) we obtain for the reflectivity of a thick crystal

$$R = \frac{1 + |\varepsilon| - [2(\varepsilon_1 + |\varepsilon|)]^{1/2}}{1 + |\varepsilon| + [2(\varepsilon_1 + |\varepsilon|)]^{1/2}}, \tag{2.31}$$

where $|\varepsilon| = (\varepsilon_1^2 + \varepsilon_2^2)^{1/2}$. Based on (2.29–31) a fit of $R(\omega)$ given by (2.31) to the observed reflectivity can be obtained for a proper choice of the model parameters ω_j, S_j, γ_j, and ε_∞. Adjustment of the parameters is usually made either by trial and error or by least square fitting of (2.31) to the observed reflectivity. This method yields not only $\varepsilon(\omega)$ but also the model parameters

Table 2.1. Model parameters of the classical oscillator fit based on (2.29–31) to the observed reflectivity of FeS$_2$ [2.26]

mode j	$\tilde{\nu}_j$(TO)[cm^{-1}]	ϱ_j	γ_j[cm^{-1}]	$\tilde{\nu}_j$(LO)[cm^{-1}]
1	293	0.0258	1.5	294
2	348	0.065	1.9	352
3	402	0.27	5.2	441
4	415	0.022	8.3	439

which characterize the infrared active phonons. The results of the oscillator fit for FeS$_2$ are shown in Fig. 2.11. Table 2.1 contains the model parameters from the best fit to the measured reflectivity; $\varrho_j = S_j/4\pi\omega_j^2$ is the reduced oscillator strength. The longitudinal frequencies given in Table 2.1 have been determined on the basis of a method developed by *Chang* [2.27] which is derived from classical oscillator theory and takes into account damping.

From (2.29) the static dielectric constant is given by $\varepsilon_0 = \varepsilon_\infty + 4\pi\Sigma_j\varrho_j$. With $\varepsilon_\infty = 21.32$ [2.26], one finds $\varepsilon_0 = 26.31$, thus $\varepsilon_0/\varepsilon_\infty = 1.23$. On the other hand, the Lyddane-Sachs-Teller relation (LST) for a multimode crystal in the absence of damping is given by [Ref. 2.17, Eq. (4.196)], namely

$$\frac{\varepsilon_0}{\varepsilon_\infty} = \prod_j \frac{\tilde{\nu}_j^2(\text{LO})}{\tilde{\nu}_j^2(\text{TO})}. \tag{2.32}$$

For small damping a generalized LST relation has been derived [2.27, 47]. This relation has a similar form as (2.32) but the frequencies have to be evaluated on the basis of the method discussed in [2.27]. Using the values in Table 2.1, the right hand side of (2.32) gives a value of 1.22 in good agreement with $\varepsilon_0/\varepsilon_\infty = 1.23$.

The result of the oscillator fit to the reflectivity of LiF is shown in Fig. 2.6 [2.16a]. Two oscillators have been introduced to represent in a phenomenological way the fundamental reflectivity band and the subsidiary

Table 2.2. Model parameters of the classical oscillator fit based on (2.29–31) to the observed reflectivity of LiF at various temperatures [2.16a]

T [K]	$\tilde{\nu}_1$ [cm^{-1}]	$\tilde{\nu}_2$ [cm^{-1}]	γ_1 [cm^{-1}]	γ_2 [cm^{-1}]	ϱ_1	ϱ_2
7.5	320	520	~ 3.2	96	0.48	0.006
85	315	512	7.1	92	0.50	0.007
295	306	503	18.3	90	0.54	0.009
420	301	497	30.1	90	0.57	0.010
605	293	486	49	83	0.61	0.010
840	282	462	77	97	0.66	0.011
1060	271	430	104	105	0.69	0.013

Table 2.3. Infrared lattice vibration parameters of crystals with NaCl and CsCl stucture [2.48–50]. The data are derived from an oscillator fit based on (2.31 and 33) to the observed reflectivity

Crystal	ε_0	ε_∞	$\tilde{\nu}(TO)[cm^{-1}]$	$\tilde{\nu}(LO)[cm^{-1}]$	e_s^*/e	ϱ	γ
LiH	12.9	3.6	590	1120	0.52	0.74	
LiF	8.9	1.9	307	662	0.87	0.56	~18
LiCl	12.0	2.7	191	398	0.73	0.74	
LiBr	13.2	3.2	159	325	0.68	0.79	
NaF	5.1	1.7	239	414	0.93	0.27	~15
NaCl	5.9	2.25	164	264	0.74	0.29	~ 9
NaBr	6.4	2.6	134	209	0.70	0.30	~ 9
NaI	7.26	3.03	117	181	0.75	0.33	~21
KF	5.5	1.5	190	326	0.88	0.32	
KCl	4.85	2.1	142	214	0.81	0.22	~ 9
KBr	4.9	2.3	113	165	0.76	0.21	~ 4
KI	5.1	2.7	101	139	0.71	0.19	~ 5
RbF	6.5	1.9	156	286	0.95	0.37	
RbCl	4.9	2.2	116	173	0.84	0.21	~ 9
RbBr	4.9	2.3	88	127	0.84	0.21	~ 5
RbI	5.5	2.6	75	103	0.75	0.23	~ 4
CsCl	7.2	2.6	99	165	0.85	0.37	~10
CsBr	6.5	2.8	73	112	0.78	0.29	~ 3
CsI	5.65	3.0	62	85	0.67	0.21	~ 3
TlCl	31.9	5.1	63	158	0.80	2.13	~16
TlBr	29.8	5.4	43	101	0.82	1.94	~ 7
AgF	10.6	3.0	170	320	0.89	0.60	
AgCl	12.3	4.0	130	181	0.71	0.66	~52
AgBr	13.1	4.6	80	135	0.70	0.67	~24
MgO	9.8	2.95	400	730	0.88	0.54	
MnO	22.5	4.95	262	552	1.08	1.39	~24

band at higher wave numbers. The latter is due to anharmonicity, and in a theory based on anharmonic interactions it will be described by a response function of the type discussed in Sect. 2.3.7. Table 2.2 gives the parameters of the phenomenological two-oscillator fit for various temperatures [2.16a]. Note the pronounced temperature dependence of the frequencies and the damping parameters; ϱ_j also shows an increase with increasing temperature which is mainly due to the fact that $\varrho_j \sim \omega_j^{-2}$.

Finally, we consider the dielectric properties of simple diatomic crystals such as NaCl, CsCl and ZnS. In Problem 2.1 it is shown that if internal field effects and damping are included the dielectric constant is given by

Table 2.4. Infrared lattice vibration parameters of crystals with ZnS structure [2.48–50]

Crystal	ε_0	ε_∞	$\bar{\nu}_{TO}$ [cm^{-1}]	$\bar{\nu}_{LO}$ [cm^{-1}]	e_s^*/e	ϱ	γ
SiC	10	6.7	793	930	0.93	0.26	
AlSb	12.0	10.2	318	345	0.54	0.14	~ 6
GaP	10.2	8.5	366	401	0.57	0.14	
GaAs*	12.9	10.9	273	297	0.51	0.16	~ 2
GaSb*	15.7	14.4	231	241	0.36	0.10	~ 1.6
InP	12.6	9.6	307	351	0.69	0.24	~ 12
InAs*	15.1	12.3	218	243	0.56	0.22	~ 1.5
InSb*	17.9	15.7	185	197	0.42	0.17	~ 1.3
ZnS	8.3	5.0	274	350	0.48	0.26	~ 13.7
ZnSe	8.33	5.90	207	246	0.68	0.19	
ZnTe	9.86	7.28	177	206	0.64	0.20	
CuCl	7.0	3.61	141	196	0.59	0.27	
CuBr	6.6	3.71	125	165	0.61	0.23	
CuI	6.5	4.84	124	141	0.48	0.13	~ 15

* From measurements at liquid-helium temperature

$$\varepsilon(\omega) = \varepsilon_\infty + \frac{4\pi e_T^{*2}}{\mu v_a(\omega_{TO}^2 - \omega^2 - i\gamma\omega)}$$

$$= \varepsilon_\infty \frac{\omega_{LO}^2 - \omega^2 - i\gamma\omega}{\omega_{TO}^2 - \omega^2 - i\gamma\omega)} \qquad (2.33)$$

In (2.33) ω_{TO} is the frequency of the TO mode, μ the reduced mass, v_a the volume of the primitive unit cell, $e_T^* = (\varepsilon_\infty + 2)e_s^*/3$ the transverse effective charge, e_s^* the Szigeti charge, and γ the damping [Ref. 2.17, Sect. 4.3.2]. Comparing with (2.28) the oscillator strength of the model is given by

$$S = 4\pi\omega_{TO}^2\varrho = 4\pi e_T^{*2}/\mu v_a = (\varepsilon_0 - \varepsilon_\infty)\omega_{TO}^2. \qquad (2.34)$$

Table 2.3 gives infrared vibrational parameters of some crystals with NaCl and CsCl structure as obtained from an oscillator fit to the observed reflectivity on the basis of (2.31 and 33). Table 2.4 contains similar parameters

27

for ZnS type crystals. The origin and the physical meaning of the Szigeti charges contained in Tables 2.3 and 4 have been discussed before [Ref. 2.17, Sect. 4.3.1].

2.2.6 Kramers-Kronig Relations and Sum Rules

It is possible to express the real and imaginary part of a complex physical quantity in terms of dispersion relations, also called Kramers-Kronig (KK) relations. We have already made use of such a relation in Sect. 2.2.3. These relations are integral formula relating a dispersive process to an absorption process; they are based on linearity and therefore superposition, as well as on causality which means that there is no response by the system until a field is turned on. The KK relations for $\varepsilon_1(\omega)$ and $\varepsilon_2(\omega)$ are [2.24, 25]

$$\varepsilon_1(\omega) - 1 = \frac{2}{\pi} P \int_0^\infty \frac{\omega' \varepsilon_2(\omega')}{\omega'^2 - \omega^2} d\omega' \quad \text{and} \tag{2.35a}$$

$$\varepsilon_2(\omega) = -\frac{2\omega}{\pi} P \int_0^\infty \frac{\varepsilon_1(\omega') - 1}{\omega'^2 - \omega^2} d\omega', \tag{2.35b}$$

where P stands for the Cauchy principle value. We are interested in the KK relations modified for the analysis of the lattice vibrations of insulators or semiconductors. The resonance absorptions of light by phonons are then generally well separated in frequency from the optical processes associated with electronic transitions which take place at much higher frequencies (in the visible or ultraviolet part of the spectrum) (Fig. 2.12).

The integration in (2.35a) can therefore be divided into two parts, namely

$$\varepsilon_1(\omega) - 1 = \frac{2}{\pi} P \int_0^{\omega_c} \frac{\omega' \varepsilon_2(\omega')}{\omega'^2 - \omega^2} d\omega' + \frac{2}{\pi} P \int_{\omega_c}^\infty \frac{\omega' \varepsilon_2(\omega')}{\omega'^2 - \omega^2} d\omega'.$$

Since we are evaluating $\varepsilon_1(\omega)$ for $\omega \ll \omega_c$ where ω_c is well above the lattice frequencies, but well below the electronic frequencies, we can neglect ω in the second integral. The second term is therefore constant and can approximatively be replaced by $\varepsilon_\infty - 1$ (Fig. 2.12). This approximation is equivalent to the approximation $\varepsilon_2(\omega_c) = 0$. The KK relation (2.35a) modified for lattice vibrations is therefore

$$\varepsilon_1(\omega) - \varepsilon_\infty = \frac{2}{\pi} P \int_0^{\omega_c} \frac{\omega' \varepsilon_2(\omega')}{\omega'^2 - \omega^2} d\omega'. \tag{2.36a}$$

The inverse relation corresponding to (2.35b) is

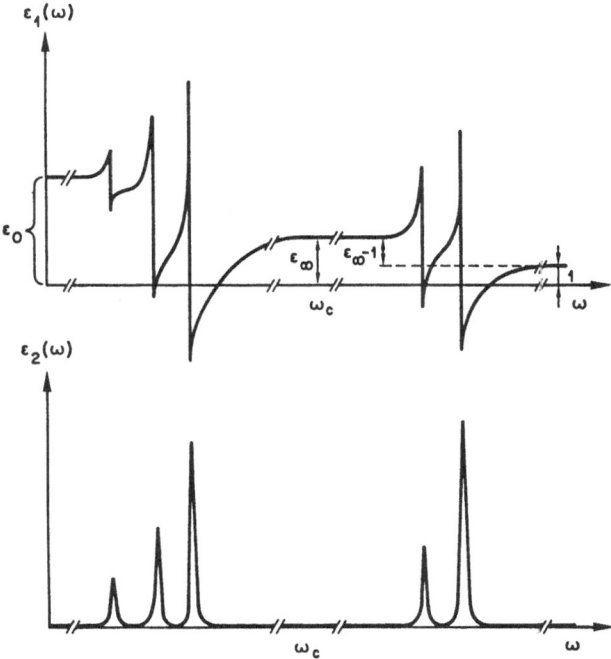

Fig. 2.12. Schematic representation of $\varepsilon_1(\omega)$ and $\varepsilon_2(\omega)$ of insulators and large gap semiconductors. $\omega < \omega_c$: absorption due to phonons; $\omega > \omega_c$: absorptions due to electronic transitions

$$\varepsilon_2(\omega) = -\frac{2\omega}{\pi} P \int\limits_0^{\omega_c} \frac{\varepsilon_1(\omega') - \varepsilon_\infty}{\omega'^2 - \omega^2} d\omega'. \tag{2.36b}$$

The relations (2.36) can be used to check the consistency of experimental data for $\varepsilon_1(\omega)$ and $\varepsilon_2(\omega)$ obtained by other techniques, such as from combined reflectivity and transmission measurments or from dispersive Fourier spectroscopy (Sect. 2.2.4). The static dielectric constant ε_0 can be obtained from (2.36a) by setting $\omega = 0$.

$$\int\limits_0^{\omega_c} \frac{\varepsilon_2(\omega)}{\omega} d\omega = \frac{\pi}{2}(\varepsilon_0 - \varepsilon_\infty). \tag{2.37}$$

This relation constitutes a *sum rule* which provides further information and checks for experimental data. Using the expression (2.30) for the classical oscillator model it is possible to derive a sum rule for the frequency dependent conductivity $\sigma(\omega) = \omega\varepsilon_2(\omega)/4\pi$ (Problem 2.4.2). The result is

$$\int\limits_{0}^{\omega_c} \sigma(\omega)d\omega \cong \frac{1}{8}\sum_{j} S_j = \frac{\pi}{2}\sum_{j} \varrho_j \omega_j^2. \tag{2.38a}$$

The cutoff of the integration at ω_c introduces an error of the order of $\gamma_j^2/\omega_j\omega_c$. For crystals with a single TO frequency, the relation reduces to

$$\int\limits_{0}^{\omega_c} \sigma(\omega)d\omega \cong \frac{1}{8}\varepsilon_\infty \omega_P^2, \tag{2.38b}$$

where $\omega_P^2 = \omega_{LO}^2 - \omega_{TO}^2$. ω_P is the ionic plasma frequency.

There exist a number of other useful sum rules for which the reader is referred to [2.51, 52]. An interesting one involves the real part of the refractive index $n(\omega)$,

$$\int\limits_{0}^{\infty} [n(\omega) - 1]d\omega = 0. \tag{2.39a}$$

The sum rule modified for lattice vibrations is

$$\int\limits_{0}^{\omega_c} [n(\omega) - n(\omega_c)]d\omega = 0, \quad \text{or}$$

$$\frac{1}{\omega_c}\int\limits_{0}^{\omega_c} n(\omega)d\omega = n(\omega_c) = \sqrt{\varepsilon_\infty} \tag{2.39b}$$

This sum rule asserts that the average value of the real refractive index over the lattice vibrational spectrum is equal to the high frequency refractive index $n(\omega_c) = \sqrt{\varepsilon_\infty}$.

2.2.7 Polaritons and Propagation of Light in Crystals

Up to this point we have described the electric interaction between ions by means of Coulomb forces, which have been assumed to act instantaneously between the ions. This is known as the electrostatic approximation which was the basis of the results obtained in [Ref. 2.17, Sects. 4.2–4]. It leads to well defined TO and LO frequencies as $q \to 0$ and to the Lyddane-Sachs-Teller (LST) relation. However, the Coulomb interaction does not act instantaneously, but propagates with the speed of light in the crystal. It is found that when the retardation of the Coulomb interaction is taken into account, the long-wave length TO modes can give rise to a transverse electromagnetic field. This field couples back with the TO modes which are its source; this coupling between phonons and photons changes the frequencies from what they would be in the absence of the coupling. Thus we expect the dispersion relations for the coupled phonon-photon system to be different from that of the uncoupled system as $q \to 0$. The coupled excitations phonon-photon are also called *polaritons*. The relevant theory has first been studied by *Huang* [2.53].

In Appendix B we have developed the necessary theoretical background for polaritons. The relation between the wave vector k of the light and the frequency ω for TO modes of long wave lengths is, see (B.23),

$$k^2 = \frac{\omega^2}{c^2} \varepsilon(\omega). \tag{2.40}$$

For a diatomic crystal $\varepsilon(\omega)$ is given by (2.33). Putting the damping $\gamma = 0$ we obtain

$$k^2 = \frac{\varepsilon_\infty}{c^2} \omega^2 \frac{\omega_{LO}^2 - \omega^2}{\omega_{TO}^2 - \omega^2}. \tag{2.41}$$

The dispersion relation for the longitudinal branch is given by (B.20), namely

$$\omega = \omega_{LO}, \tag{2.42}$$

independent on k. The dispersion curves for the two doubly-degenerate TO branches α and β predicted by (2.41) and of the LO branch given by (2.42) are shown in Fig. 2.13.

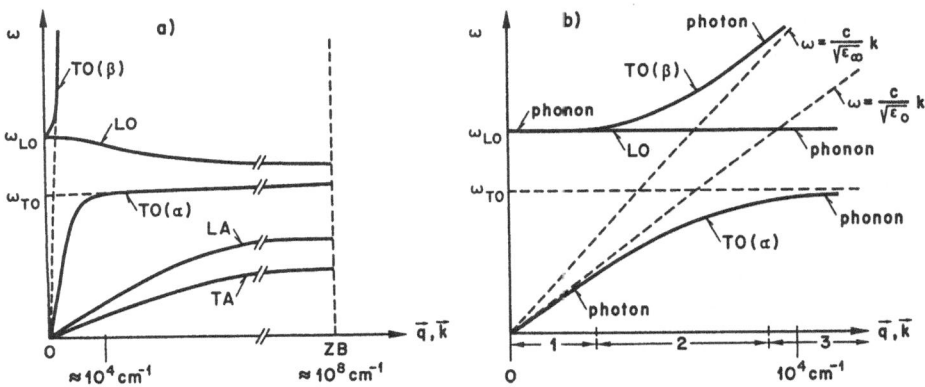

Fig. 2.13a,b. Dispersion curves for TO and LO polaritons in diatomic cubic crystals

From the figure it is seen that for $\omega < \omega_{TO}$ the polariton $TO(\alpha)$ is largely electromagnetic in nature (Region 1 in Fig. 2.13b). In such a mode the ion displacements are small, while the amplitude of the electromagentic field is large. The ion motions are low-frequency vibrations in this limit, and the electromagnetic wave sees a medium characterized by the static dielec-

31

tric constant ε_0 and propagates with the velocity $c/\sqrt{\varepsilon_0}$. In Region 2 the displacements of the ions gradually increase while the amplitude of the electromagnetic field decreases. In Region 3, ω approaches ω_{TO}; the polariton $TO(\alpha)$ is now essentially a mechanical vibration, i.e. the mode is characterized by large displacements of the ions and a small amplitude of the electromagnetic field. In the gap $\omega_{TO}<\omega<\omega_{LO}$, there are no polaritons, that is waves cannot propagate in this frequency region (see below). For $\omega \cong \omega_{LO}$, the polariton $TO(\beta)$ is essentially a mechanical vibration and as ω increases above ω_{LO} it changes gradually into an electromagnetic wave (Region 2). In Region 3 the polariton $TO(\beta)$ is basically an electromagnetic wave which propagates with the velocity $c/\sqrt{\varepsilon_\infty}$. In Region 3 the ions cannot vibrate because the frequencies are too high. These qualitative conclusions can be made quantitative by an analysis due originally to *Huang* [2.53].

From Fig. 2.13b we note that at $q=0$ the $TO(\beta)$ phonon is degenerate with the LO phonon. Thus the LST relation does not hold in Region 1 nor in Region 2 but only in Region 3. The limits of validity of the LST relation that follow from Huang's theory have been discussed by *Barron* [2.54]; he showed that it should hold for wavelengths $10^{-6}<\lambda<10^{-4}$ cm. The longest wavelength for which the LST relation is valid is at the boundary of Region 3 in Fig. 2.13b, while the short-wavelength limit arises because for shorter wavelengths the Huang theory is no longer valid.

Huang's theory implicitly assumes an infinite crystal. *Hardy* [2.55] has shown that for a finite crystal with long range forces Huang's theory is only applicable to optical modes whose wavelength is short compared with the size of the crystal, but large compared with the lattice constant.

Experimentally, polaritons can be observed by means of Raman scattering; this will be discussed in Chapter 3. Inelastic neutron scattering and infrared spectroscopy are not suitable techniques to observe polaritons [2.56]. The resolution of neutrons corresponds to q values of the order of 10^6 cm^{-1}. Since the LST relation holds in the Region 3 of Figure 2.13b, which extends down to about 10^4 cm^{-1}, we would not expect neutron scattering to give a deviation from the LST relation for the longest-wavelength phonons that are involved. In infrared experiments the absorption of infrared radiation is a maximum close to ω_{TO} and the corresponding wavenumber $k = \omega_{TO} n(\omega_{TO})/c$ is too large to allow the obsevation of polaritons and deviations from the LST relation.

If we allow for damping the dispersion relation for the longitudinal branch is given by the real part of (B.20), namely

$$\omega = \left(\omega_{LO}^2 - \frac{1}{4}\gamma^2\right)^{1/2}, \tag{2.43}$$

which is again independent on k. The dispersion relation for the transverse branches is given by, see (B.29),

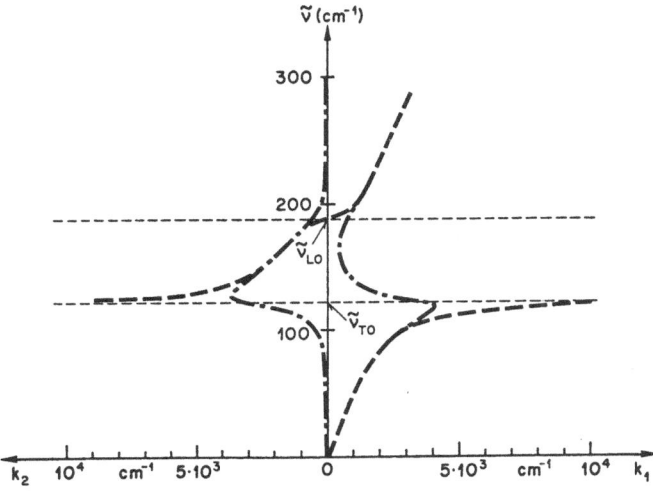

Fig. 2.14. Dispersion curves for the real- and imaginary part of the propagation vector of light, $k = k_1 + ik_2$, for AgCl at 7.5 K. (- - - - -) $\gamma = 0$; ($-\cdot-\cdot-\cdot-$) $\gamma = 17\,\mathrm{cm}^{-1}$

$$k^2 = \frac{\varepsilon_\infty}{c^2}\omega^2 \frac{\omega_{LO}^2 - \omega^2 - i\gamma\omega}{\omega_{TO}^2 - \omega^2 - i\gamma\omega}. \tag{2.44}$$

Figure 2.14 shows the real and imaginary parts of $k = k_1 + ik_2$ for AgCl. For $\gamma = 0$ we obtain from (2.44)

$$k_1^2 - k_2^2 = \frac{\varepsilon_\infty}{c^2}\omega^2 \frac{\omega_{LO}^2 - \omega^2}{\omega_{TO}^2 - \omega^2}, \quad k_1 k_2 = 0, \tag{2.45}$$

whence it follows that for $\omega < \omega_{TO}$ or $\omega > \omega_{LO}$

$$k_2 = 0, \quad k_1 = \frac{\omega}{c}\left(\varepsilon_\infty \frac{\omega_{LO}^2 - \omega^2}{\omega_{TO}^2 - \omega^2}\right)^{1/2}, \tag{2.46a}$$

and for $\omega_{TO} < \omega < \omega_{LO}$

$$k_1 = 0, \quad k_2 = \frac{\omega}{c}\left(\varepsilon_\infty \frac{\omega_{LO}^2 - \omega^2}{\omega^2 - \omega_{TO}^2}\right)^{1/2}. \tag{2.46b}$$

Since according to (B.30) $k_1 = \omega n_1/c$ and $k_2 = \omega n_2/c$ it follows that in the frequency gap between ω_{TO} and ω_{LO} the refractive index $n_1 = 0$ but the extinction coefficient $n_2 \neq 0$. From (2.11) we thus obtain $R = 1$ in the gap. This also follows from (B.31), which shows that wavelike solutions do not exist in the gap, thus R is expected to be high in this frequency region.

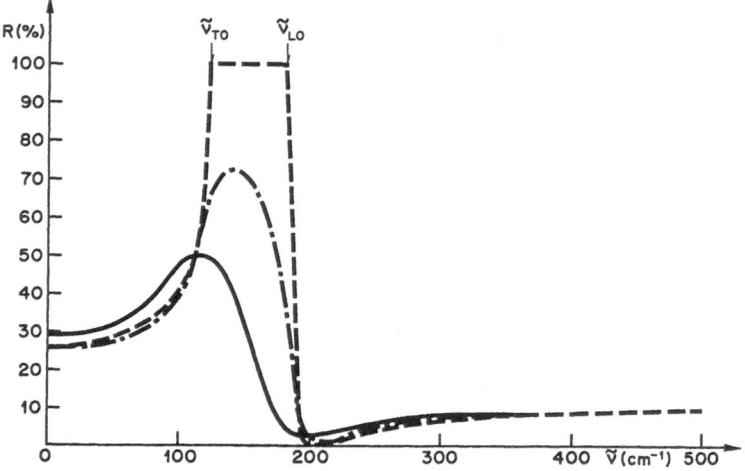

Fig. 2.15. Reflectivity of AgCl. (- - - - -): $\tilde{\nu}_{TO} = 122\,\mathrm{cm}^{-1}$, $\tilde{\nu}_{LO} = 188\,\mathrm{cm}^{-1}$, $\varepsilon_\infty = 4$, $\gamma = 0$. (—·—·—·—): $T = 7.5\,\mathrm{K}$, $\tilde{\nu}_{TO} = 122\,\mathrm{cm}^{-1}$, $\tilde{\nu}_{LO} = 188\,\mathrm{cm}^{-1}$, $\varepsilon_\infty = 4$. $\gamma = 17.1\,\mathrm{cm}^{-1}$. (———): $T = 295\,\mathrm{K}$, $\tilde{\nu}_{TO} = 103\,\mathrm{cm}^{-1}$, $\tilde{\nu}_{LO} = 171.6\,\mathrm{cm}^{-1}$, $\varepsilon_\infty = 4$, $\gamma = 41.2\,\mathrm{cm}^{-1}$

Alternatively we can say that for $\gamma = 0$ the absorptivity $A = 0$ for $\omega \neq \omega_{TO}$ and according to (2.5b) the reflectivity is $R = 1 - T$. Outside the gap, T is large and R is small; within the gap $T = 0$ and $R = 1$. For $\gamma \neq 0$ the absorptivity is different from zero in the neighbourhood of ω_{TO}, thus $R = 1 - A - T$. Outside the gap A is small, T is large, thus R is relatively small. Within the gap A is larger but T is small which leads to a comparatively large value of R (Fig. 2.15).

2.3 Quantum-Mechanical Treatment of the Dielectric Constant

2.3.1 Qualitative Discussion

It is the aim of this section to present a qualitative discussion of the interaction of infrared light with lattice vibrations. The Hamiltonian of the system is

$$H = H_0 + H', \quad \text{where} \tag{2.47}$$

$$H_0 = H_{\mathrm{h}} + H_{\mathrm{anh}} \tag{2.48}$$

is the Hamiltonian of the unperturbed system, containing the harmonic part H_{h} as well as the anharmonic contribution H_{anh}, and H' is the external perturbation caused by the interaction of light with the lattice vibrations.

Our treatment is semiclassical in the sens that the lattice vibrations are quantized but that the electromagnetic field is treated classically. For a full quantum mechanical treatment the reader is referred to [2.57].

The perturbation is given by

$$H' = -E \cdot M \tag{2.49}$$

where E is the electric field of the light and M is the total dipole moment of the crystal. For a light wave of the form

$$E = E_0 \exp[i(k \cdot r - \omega t)],$$

the wavelength λ in the infrared (and visible) is much larger than the lattice parameter, and hence $k = 2\pi/\lambda$ is negligibly small compared with the extension of the Brillouin zone. For this reason $k = 0$ and in first-order processes the interaction of light with lattice vibrations involves only phonons with wave vector $q = 0$ (Fig. 1.1). Furthermore, only TO phonons can be excited in first-order processes and under normal experimental conditions (transmission and reflectivity experiments at near-normal incidence). This follows from the facts that the interaction (2.49) involves the scalar product $E \cdot M$, that the light is transversal ($E \perp k$), and that for a TO phonon the ionic displacements and hence M are perpendicular to q which is parallel to k because of momentum conservation (Fig. 2.16).

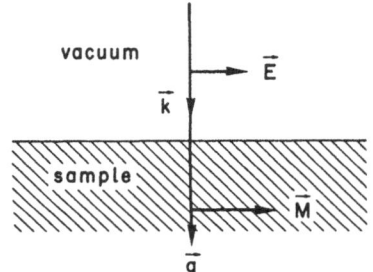

Fig. 2.16. In usual transmission or reflectivity experiments a light wave with wave vector k and electric field E falls at normal (or near-normal incidence) on the crystal and excites a TO phonon $q \cong 0$ which induces a dipole moment M

Note, however, that for suitable experiments at non-normal incidence, such as those discussed in Sect. 2.2.2, it is also possible to excite LO phonons.

The dipole moment appearing in (2.49) can be expanded in a Taylor series with respect to the ionic displacements. We write

$$M = M^{(0)} + M^{(1)} + M^{(2)} + \dots . \tag{2.50}$$

$M^{(0)}$ is the static dipole moment of the crystal, which is usually zero except in pyroelectric or ferroelectric crystals, but in any case is of no relevance for absorption processes. The first-order dipole moment $M^{(1)}$ is linear in the ionic displacements or normal coordinates and gives rise to *one-phonon processes*. The second-order dipole moment $M^{(2)}$ is quadratic in the displacements or normal coordinates and is one mechanism leading to *two-phonon absorption* ($M^{(2)}$ mechanism). The physical origin of $M^{(2)}$ will be discussed in Sect. 2.3.4. At this point we only remark that $M^{(2)}$ depends on the deformability of the electron shells and on anharmonic coupling parameters. As will be discussed below, there is another mechanism leading to two-phonon absorption, namely the cubic term $\Phi^{(3)}$ and higher-order terms of the potential energy contained in H_{anh} of (2.48) ($\Phi^{(3)}$ mechanism). We shall show in Sect. 2.3.2 that the absorpiton and hence the imaginary part of the dielectric constant, $\varepsilon_2(\omega)$, depends on the matrix elements $\langle n'|M|n\rangle$ for transitions from an initial vibrational state $n\rangle$ to a final state $n'\rangle$. In a systematic theory of multi-phonon absorption these matrix elements have to be calculated with the eigenfunctions of the Hamiltonian $H_0 = H_{\mathrm{h}} + H_{\mathrm{anh}}$. This has been done by *Szigeti* [2.58] using conventional peturbation methods but the treatment is by no means straightforward, and more recent developments have used many-body techniques [2.1, 59–63]. *Szigeti*'s results provide a satisfactory account of two-phonon absorption well away from the main one-phonon line at ω_{TO}, but breaks down in the case of the $\Phi^{(3)}$ mechanism if ω approaches Ω_{TO}. His study also shows that in principle the $\phi^{(3)}$ mechanism and the $M^{(2)}$ mechanism cannot be separated in ionic crystals, because there are cross terms arising from a coupling of the two mechanisms which also contribute to $\varepsilon(\omega)$. For our purpose, however, it is more instructive to discuss the $M^{(2)}$ and $\Phi^{(3)}$ mechanisms separately by neglecting either $M^{(2)}$ in (2.49) or H_{anh} in (2.48). If both, $M^{(2)}$ and H_{anh} are neglected we are dealing with one-phonon processes.

a) *One-Phonon Processes*

In this approximation the Hamiltonian is

$$H = H_{\mathrm{h}} - E \cdot M^{(1)}. \tag{2.51}$$

The matrix elements $\langle n'|M^{(1)}|n\rangle$ are calculated with the eigenfunctions of H_{h}. It will be shown in Sect. 2.3.5 that $\varepsilon(\omega)$ is identical with the classical result for a system of harmonic oscillators [Ref. 2.17, Eq. (5.132)]. The interaction processes are shown in Fig. 2.17.

In the photon-phonon conversion (Fig. 2.17a), a $k \cong 0$ photon is annihilated and a TO phonon with $q \cong 0$ is created; this corresponds to the absorption of a photon. In the phonon-photon conversion (Fig. 2.17b), a TO phonon with $q \cong 0$ is annihilated and a $k \cong 0$ photon is created; this corresponds to the induced emission of a photon. In calculating $\varepsilon(\omega)$ both these processes must be considered with the result that $\varepsilon(\omega)$ is independent on temperature. The conservation of energy requires

$$\omega(k) = \omega_{\mathrm{TO}}, \tag{2.52}$$

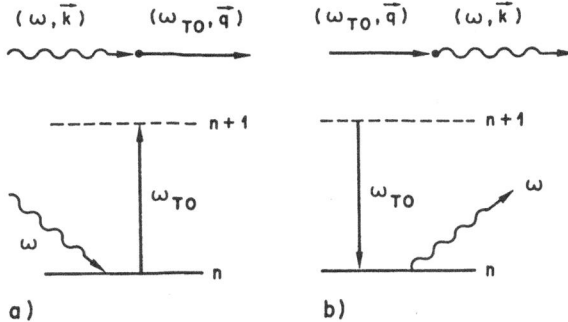

Fig. 2.17.a,b. One-phonon optical processes based on the first-order dipole moment $M^{(1)}$. (a) absorption of a photon $(\omega, \mathbf{k})$ and creation of a phonon $(\omega_{TO}, \mathbf{q})$; (b) annihilation of a phonon $(\omega_{TO}, \mathbf{q})$ and emission of a photon $(\omega, \mathbf{k})$

and the conservation of momentum is

$$\mathbf{k} = \mathbf{q} \cong 0. \tag{2.53}$$

For a diatomic cubic crystal such as an alkali-halide, the present approximation predicts a single and infinitely sharp absorption line at $\omega = \omega_{TO}$. In contrast, the experiments show rather broad absorption bands with weak subsidiary maxima (Figs. 2.5a, 6).

b) Two-Phonon Processes Due to the $\Phi^{(3)}$ Mechanism

We consider the Hamiltonian

$$H = H_0 - \mathbf{E} \cdot \mathbf{M}^{(1)}. \tag{2.54}$$

If conventional perturbation theory is used, the matrix elements, $\langle n' | \mathbf{M}^{(1)} | n \rangle$ are calculated with the eigenfunction of $H_0 = H_h + H_{anh}$; the difficulties encountered with this method have been discussed above, and it is necessary to resort to many-body techniques. This approach leads to the concept of the response function which has been discussed in [Ref. 2.17, Sects. 5.5.2, 3].

If H_{anh} represents the $\Phi^{(3)}$ term of the potential energy, the interaction mechanism involves three phonons as illustrated in Fig. 2.18.

The photon $(\omega, \mathbf{k} \cong 0)$ couples with an infrared TO phonon $(\omega_{TO}, \mathbf{q} \cong 0)$. The anharmonicity then couples the TO phonon with two other phonons $(\omega_{j'}, \mathbf{q}')$ and $(\omega_{j''}, \mathbf{q}'')$; thus, the TO phonon acts only as an intermediate state and overall the $\Phi^{(3)}$ mechanism gives rise to two-phonon processes represented by Fig. 2.18. Momentum is conserved at each vertex and energy is conserved overall. The processes must therefore satisfy the conditions

$$\mathbf{k} = \mathbf{q} = \pm\mathbf{q}' \pm \mathbf{q}'' \cong 0, \quad \text{and} \tag{2.55}$$

$$\omega(\mathbf{k}) = \pm\omega_{j'}(\mathbf{q}') \pm \omega_{j''}(\mathbf{q}''). \tag{2.56}$$

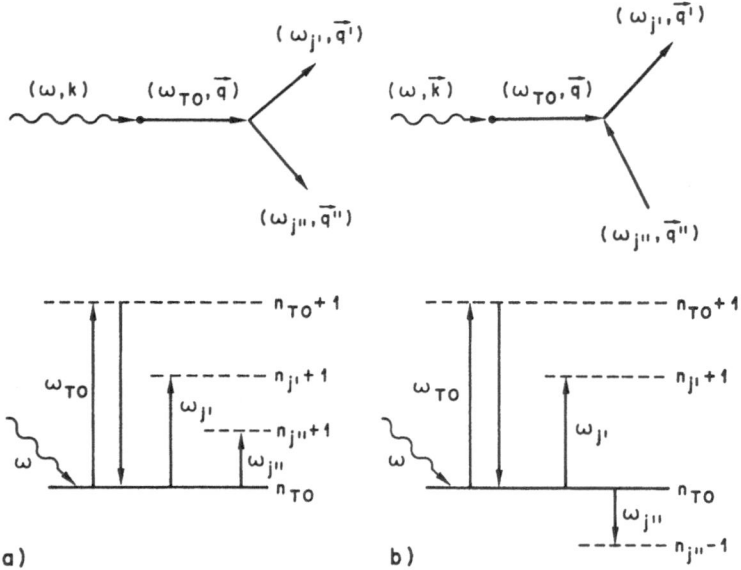

Fig. 2.18a,b. Two-phonon optical processes based on the anharmonic coupling ($\Phi^{(3)}$ mechanism) of the infrared active phonon (ω_{TO}, $q = 0$) with two phonons ($\omega_{j'}, q'$) and $\omega_{j''}, q''$). (a) process leading to a summation band ($\omega_{j'} + \omega_{j''}$); (b) process leading to a difference band $\omega_{j'} - \omega_{j''}$)

The positive and negative signs correspond to phonon creation and annihilation, respectively. Note that the simultaneous annihilation of two phonons can not occur, since it would violate the energy conservation rule (2.56), and there are therefore only two distinct possibilities. Figure 2.18a represents the case where two phonons are created; both signs in (2.55, 56) are positive and it follows that the two vectors q' and q'' have equal magnitude and opposite signs. Such processes give rise to what is known as *summation bands* since

$$\omega(\mathbf{k}) = \omega_{j'}(q') + \omega_{j''}(q''). \tag{2.57}$$

The other possibility is illustrated in Fig. 2.18b; here, the phonon (q', j') is created and the phonon (q'', j'') is annihilated. These processes give rise to a *difference band*, for which

$$\omega(\mathbf{k}) = \omega_{j'}(q') - \omega_{j''}(q''). \tag{2.58}$$

In both cases one has $|q'| = |q''|$ and the branches j' and j'' may either be optical or acoustic.

In physical terms, the interaction process described by $\Phi^{(3)}$ constitutes a mechanism whereby the TO phonon q can decay, i.e. exhibit a finite life time $\tau_{TO}(q)$, as described in [Ref. 2.17, Sect. 5.5.3]. In other words, anharmonic coupling between phonons is responsible for the broadening of the absorption line at $\omega = \omega_{TO}$. The width of the absorption band is given by the damping function $2\Gamma_j(q,\omega,T)$ [Ref. 2.17, Eq. (5.153)]. It depends on temperature and on the frequency ω with which the crystal is probed. The frequency dependence of $\Gamma_j(0,\omega)$ for KBr at 300 K has been shown in [Ref. 2.17, Fig. 5.13]. It is this frequency dependence which gives rise to some of the weak subsidiary bands shown in Fig. 2.5. The effect of anharmonic terms on the infrared absorption of simple ionic crystals has also been investigated by *Born* and *Blackman* using classical theory [2.64, 65].

It should be noted that the mechanism based on the anharmonic interaction $\Phi^{(3)}$ is only operative if there exists an infrared active TO phonon, i.e. a phonon which induces a nonvanishing dipolemoment $M^{(1)}$. This is the case in the alkali-halides but not, for example, in simple covalent crystals such as diamond, silicon and germanium. The weak infrared absorption in diamond and silicon, for example (Figs. 2.9, 23–25), is due to the second- and higher-order dipole moments $M^{(2)}$, $M^{(3)}$, etc. On the other hand, in ionic crystals, both the anharmonic terms $\Phi^{(3)}, \Phi^{(4)}$, etc., as well as the higher-order dipole terms $M^{(2)}$, $M^{(3)}$, etc. may be operative, but the $\Phi^{(3)}$ mechanism is usually dominating.

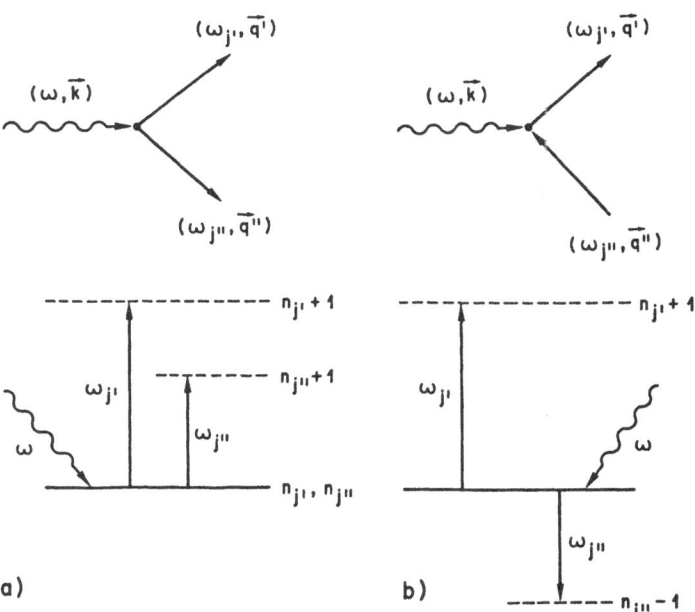

Fig. 2.19a,b. Two-phonon optical processes based on the second-order dipole moment ($M^{(2)}$-mechanism) in which the photon (ω, k) couples directly with two phonons $(\omega_{j'}, q')$ and $(\omega_{j''}, q'')$. (a) Process leading to a summation band $(\omega_{j'} + \omega_{j''})$; (b) process leading to a difference band $(\omega_{j'} - \omega_{j''})$

c) Two-Phonon Processes Due to the $M^{(2)}$ Mechanism

The relevant Hamiltonian is

$$H = H_{\mathrm{h}} - E \cdot M^{(2)}. \tag{2.59}$$

Here the matrix elements $\langle n'|M^{(2)}|n\rangle$ are evaluated with the eigenfunction of the harmonic Hamiltonian H_{h}. The second-order moment $M^{(2)}$ contains the products of two atomic displacements and hence of two normal coordinates; the $M^{(2)}$ mechanism therefore provides the direct coupling of a photon (ω, k) to two phonons $(\omega_{j'}, q')$ and $(\omega_{j''}, q'')$ (Fig. 2.19).

Figure 2.19a illustrates a summation process and Fig. 2.19b a difference process. The overall momentum and energy conditions are the same as (2.55, 56). The contribution of the $M^{(2)}$ mechanism to $\varepsilon_2(\omega)$ will be derived in Sect. 2.3.6.

2.3.2 Relation Between $\varepsilon_2(\omega)$ and Transition Probabilities

Let $U(\omega)$ be the absorption of energy per unit range of ω (i.e., per $2\pi\mathrm{s}^{-1}$) and per unit time. The relation between $U(\omega)$ and the time average absorption of the electromagnetic field per unit volume given by (2.20) is

$$U(\omega) = \frac{V}{2\pi}\overline{W}(\omega) = \frac{V}{16\pi^2}\omega\varepsilon_2(\omega)E_0^2(\omega). \tag{2.60}$$

On the other hand, $U(\omega)$ can be related with the transition probabilities as follows: If the crystal is in the electronic ground state and in the vibrational state $|n\rangle = |n_1, n_2, \ldots\rangle$, then the probability of a transition induced by the light to the vibrational state $|n'\rangle = |n_1', n_2', \ldots\rangle$ is according to the, "golden rule" of perturbation theory [2.66] proportional to the modulus squared of the matrix elements of the perturbation (2.49), $|\langle n'|H'|n\rangle|^2$, that is

$$P_{nn'} = \frac{\pi^2}{h^2}E_0^2(|\omega_{n'n}|)|\langle n'|M|n\rangle|^2. \tag{2.61}$$

In this relation M is the toal dipole moment of the crystal introduced in (2.49, 50) and $\omega_{n'n}$ is defined by (2.63). The absorbed (or emitted) energy is then given by

$$L_{nn'} = w_n P_{nn'}\hbar\omega_{n'n}, \tag{2.62}$$

where w_n is the propability that the crystal is in the state $|n\rangle$ and

$$\hbar\omega_{n'n} = E_{n'} - E_n \tag{2.63}$$

is the energy difference between states $|n'\rangle$ and $|n\rangle$. Note that $\omega_{n'n}$ may be positive or negative. We now select a range of frequencies $d\omega$ in which the variation of the absorption is small, but which contains many transitions of the type (2.62). The absorption of energy in the range $d\omega$ is then given by

$$U(\omega)d\omega = \sum_{nn'} {}_{(\omega,d\omega)} L_{n'n}, \tag{2.64}$$

where the suffix $(\omega, d\omega)$ indicates that the summation is restricted to those transitions for which $\omega_{n'n}$ lies between ω and $\omega + d\omega$. This restriction may also be expressed by introducting the δ-function, namely

$$\sum_{nn'} {}_{(\omega,d\omega)} L_{n'n} = \sum_{nn'} L_{nn'}\delta(\omega - |\omega_{n'n}|)d\omega. \tag{2.65}$$

Since $\omega>0$ the absolute value of $\omega_{n'n}$ has been introduced in the δ-function. Using (2.65) we obtain

$$U(\omega) = \sum_{nn'} L_{nn'}\delta(\omega - |\omega_{n'n}|). \tag{2.66}$$

Substituting (2.61, 62) into (2.66) and noting that for all transitions considered $E_0^2(|\omega_{n'n}) = E_0^2(\omega)$, a comparison of (2.66) with (2.60) gives the following expression for the imaginary part of the dielectric constant

$$\varepsilon_2(\omega) = \frac{4\pi^2}{V\hbar} \sum_{nn'} w_n |\langle n'|M|n\rangle|^2 \frac{\omega_{n'n}}{\omega}\delta(\omega - |\omega_{n'n}|). \tag{2.67}$$

In the general case the matrix elements of M have to be evaluated with the eigenfunctions $|n\rangle$ and $|n'\rangle$ of the anharmonic crystal with $H_0 = H_{\mathrm{h}} + H_{\mathrm{anh}}$ of (2.48) [2.58].

2.3.3 Interaction of a Charged Oscillating Particle with the Radiation Field

Consider a single particle with charge q, mass m and momentum $\boldsymbol{p} = -i\hbar\partial/\partial u$ oscillating around the equilibrium position $\boldsymbol{r}$. its actual position is $\boldsymbol{R} = \boldsymbol{r} + \boldsymbol{u}$. The Hamiltonian is $H = H_0 + H'$, where H_0 refers to the oscillating particle and H' is the perturbation due to the radiation field. For the application in Sect. 2.3.4 it is more convenient to write H' not in the form (2.49) but in terms of the momentum $\boldsymbol{p}$ and the vector potential $\boldsymbol{A}$ of the radiation field. To first order in the radiation field, H' is given by [2.67, 68]

$$H' = -\frac{q}{mc}\boldsymbol{p}\cdot\boldsymbol{A}, \tag{2.68}$$

where $\boldsymbol{A}$ is given by

$$\boldsymbol{A}(\boldsymbol{R},t) = A_0(\omega)\boldsymbol{f}\exp[i(\boldsymbol{k}\cdot\boldsymbol{R} - \omega t)]. \tag{2.69}$$

Here, ω is the angular frequency, A_0 the amplitude and $\boldsymbol{k}$ the wave vector; $\boldsymbol{f}$ is a unit vector specifying the polarization of the radiation. We require the matrix elements $\langle n'|H'|n\rangle$. For a harmonic oscillator the eigenfunctions $|n\rangle$ have been given in [Ref. 2.17, Eq. (2.85)]. Since in the optical region the wavelength of the light is much larger than the spatial spread of the wave function of the oscillating particle, we may evaluate the vector potential at the equilibrium position r of the particle, that is, we may replace $\exp(i\mathbf{k}\cdot\mathbf{R})$ by $\exp(i\mathbf{k}\cdot\mathbf{r})$ in (2.69). In this approximation $\mathbf{A}$ is independent on $\mathbf{u}$ and we have to consider only the matrix elements $\langle n'|\mathbf{p}|n\rangle$, for which we apply the result [2.69].

$$\langle n'|\mathbf{p}|n\rangle = im\omega_{n'n}\langle n'|\mathbf{u}|n\rangle, \tag{2.70}$$

where $\omega_{n'n}$ is given by (2.63), a result which follows also from [Ref. 2.17, Eqs. (2.97, 98)]. For an electromagnetic field such as that associated with a light wave, the electric field is given by [2.67, 68]

$$\mathbf{E} = -\frac{1}{c}\frac{\partial}{\partial t}\mathbf{A} = \frac{i\omega}{c}\mathbf{A}, \tag{2.71}$$

and the condition $\omega = \omega_{n'n}$ required in the subsequent calculation of the transition probability, can be inserted here. Using (2.68–71) we obtain

$$\langle n'|H'|n\rangle = -\langle n'|\mathbf{M}|n\rangle\mathbf{E}, \tag{2.72}$$

where $\mathbf{M} = q\mathbf{u}$ is the electric dipole moment associated with the displacement of the particle. This result can, of course, be obtained directly from (2.49), but (2.68, 70) will be useful in Sect. 2.3.4. It should be noted that the matrix elements $\langle n'|\mathbf{M}|n\rangle$ for $n' \neq n$ are independent on the origin; the addition of a constant vector $\mathbf{r}$ to $\mathbf{u}$ leaves the value unchanged since the wave functions are orthogonal:

$$\langle n'|\mathbf{R}|n\rangle = \langle n'|\mathbf{r} + \mathbf{u}|n\rangle = \langle n'|\mathbf{u}|n\rangle.$$

Using the matrix elements given by [Ref. 2.17, Eq. (2.97)], the selection rules for $\langle n'|\mathbf{M}|n\rangle$ are $n' = n\pm1$ and one obtains

$$|\langle n'|\mathbf{M}|n\rangle|^2 = \begin{cases} \dfrac{q^2\hbar(n+1)}{2m\omega_0} & \text{for } n' = n+1 \\[3mm] \dfrac{q^2\hbar n}{2m\omega_0} & \text{for } n' = n-1 \ , \end{cases} \tag{2.73}$$

where $\omega_0 = \omega_{n'n}$. From (2.67, 73) it is easy to calculate $\varepsilon_2(\omega)$ for a system of N identical harmonic oscillators. Note the for a given value of n the summation over n' in (2.67) gives only the two terms with $n' = n\pm1$, that $\omega_{n+1,n} = \omega_0$, $\omega_{n-1,n} = -\omega_0$, and that $\sum w_n = 1$. Using $V = Nv_a$, where v_a is the volume of the unit cell, one obtains

$$\varepsilon_2(\omega) = \frac{2\pi^2 q^2}{m v_a \omega_0} \delta(\omega - \omega_0). \tag{2.74}$$

We note that $\varepsilon_2(\omega)$ is independent on the mean occupation number $\bar{n} = \sum n w_n$ and hence on temperature and consists of a δ-function centered at the transition frequency ω_0.

2.3.4 Quantum Mechanical Formulation of the Dipole Moment of the Crystal

From the proceeding subsections it is clear that the crucial quantity in the interaction of a system with light is the dipole moment of the system. We now apply the results of the last subsection to the entire system of particles constituting the crystal, with the object of deriving a total dipole moment whose matrix elements refer to transitions between vibrational states only.

Generalizing (2.68) the total Hamiltonian is $H = H_0 + H'$, where H_0 is the Hamiltonian of the crystal in the absence of radiation and H' is given by [2.70]

$$H' = \sum_i \left(\frac{e}{mc}\right) \boldsymbol{p}_i \cdot \boldsymbol{A}_i(\boldsymbol{r}_i, t) - \sum_{l\kappa} \left(\frac{Z_\kappa e}{m_\kappa c}\right) \boldsymbol{p}\binom{l}{\kappa} \cdot \boldsymbol{A}\left(\boldsymbol{R}\binom{l}{\kappa}, t\right). \tag{2.75}$$

The perturbation consists of summations over all the electrons with charge $-e$, mass m, position vectors $\boldsymbol{r}_i$, and over all the nuclei with charges $Z_\kappa e$, masses m_κ and position vectors $\boldsymbol{R}\binom{l}{\kappa} = \boldsymbol{r}\binom{l}{\kappa} + \boldsymbol{u}\binom{l}{\kappa}$. $\boldsymbol{r}\binom{l}{\kappa} = \boldsymbol{r}(l) + \boldsymbol{r}(\kappa)$ is the equilibrium position vector of atom κ in unit cell l and $\boldsymbol{u}\binom{l}{\kappa}$ is its displacement vector (Fig. 2.20).

In the following we assume that each electron i may be assigned to a certain nucleus $\binom{l}{\kappa}$; thus $\boldsymbol{r}_i\binom{l}{\kappa}$ denotes the position vector of electron i belonging to the nucleus $\binom{l}{\kappa}$.

In (2.75) the vector potential $\boldsymbol{A}_i(\boldsymbol{r}_i, t)$ is then written as $\boldsymbol{A}_i(\boldsymbol{r}_i\binom{l}{\kappa}, t)$. Since $\boldsymbol{k}$ is small we may evaluate both vector potentials in (2.75) at the

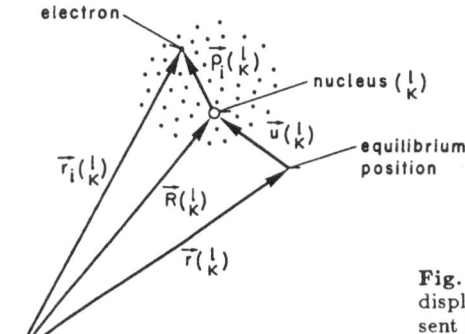

electron

$\vec{P}_i\binom{l}{\kappa}$

nucleus $\binom{l}{\kappa}$

$\vec{u}\binom{l}{\kappa}$

equilibrium position

$\vec{r}_i\binom{l}{\kappa}$

$\vec{R}\binom{l}{\kappa}$

$\vec{r}\binom{l}{\kappa}$

Origin

Fig. 2.20. The dots ($\cdot$) grouped around the displaced position of the nucleus $\binom{l}{\kappa}$ represent the electrons with position vectors $\boldsymbol{r}_i\binom{l}{\kappa}$ assigned to this nucleus. The vector potential is evaluated at the equilibrium position $\boldsymbol{r}\binom{l}{\kappa}$

43

equilibrium positions $r\binom{l}{\kappa}$ as we have done in Sect. 2.3.3. Using (2.69) we obtain

$$H' = -\frac{1}{c}\boldsymbol{A}(\omega,t)\cdot\sum_{l\kappa}\exp\left[\mathrm{i}\boldsymbol{k}\cdot\boldsymbol{r}\binom{l}{\kappa}\right]\left[\left(\frac{Z_\kappa e}{m_\kappa}\right)\boldsymbol{p}\binom{l}{\kappa}\right.$$

$$\left.-\sum_i\left(\frac{e}{m}\right)\boldsymbol{p}_i\binom{l}{\kappa}\right]\ ,\quad\text{where}\tag{2.76}$$

$$\boldsymbol{A}(\omega,t) = A_0(\omega)\mathrm{e}^{-\mathrm{i}\omega t}\boldsymbol{f}.\tag{2.77}$$

In (2.76) the summation over i extends over all electrons belonging to the nucleus $\binom{l}{\kappa}$.

The matrix elements of H' are to be formed with the eigenfunctions $\varXi$ of H_0. $\varXi(r,R)$ is a function of both the electronic coordinates r and the nuclear coordinates R. Adopting the adiabatic approximatin [Ref. 2.17, Sect. 1.3], the initial and final state functions may be written as

$$\begin{aligned}i\rangle &= \varXi_{\mathrm{i}}(r,R) = \chi_e(r,R)\varPsi_{en}(R),\\ f\rangle &= \varXi_{\mathrm{f}}(r,R) = \chi_{e'}(r,R)\varPsi_{e'n'}(R).\end{aligned}\tag{2.78}$$

In (2.78) e and n specify the electronic and nuclear vibrational states, respectively. Note that $|n\rangle$ represents the quantum numbers $|n_1,n_2,\ldots,n_s,\ldots\rangle$ of all the phonons in the state $|n\rangle$. $\chi_e(r,R)$ is the electronic wave function of the crystal in state e for a fixed nuclear configuration R. $\varPsi_{en}(R)$ is the wave function of the vibrational state n, in which the nuclei are moving in an effective potential which is influenced by the electrons in the state e [Ref. 2.17, Sect. 1.3]. From (2.70, 76) we obtain

$$\langle f|H'|i\rangle = -\mathrm{i}\frac{\omega_{\mathrm{fi}}}{c}\boldsymbol{A}(\omega,t)$$

$$\times\left\langle f\left|\sum_{l\kappa}\exp[\mathrm{i}\boldsymbol{k}\cdot\boldsymbol{r}\binom{l}{\kappa}]M_{l\kappa}(R,r)\right|i\right\rangle,\tag{2.79}$$

where

$$M_{l\kappa}(R,r) = Z_\kappa e\boldsymbol{R}\binom{l}{\kappa} - \sum_i e\boldsymbol{r}_i\binom{l}{\kappa},\tag{2.80}$$

is the electric dipole moment at the site $\binom{l}{\kappa}$. Using (2.78) the matrix elements (2.79) can be written in the form

$$\langle f|H'|i\rangle = -\mathrm{i}\frac{\omega_{\mathrm{fi}}}{c}\boldsymbol{A}(\omega,t)\cdot\int\varPsi_{e'n'}^*(R)\langle e'|M(R,\boldsymbol{k})|e\rangle\varPsi_{en}(R)dR,\tag{2.81}$$

where

$$\langle e'|M(R,\boldsymbol{k})|e\rangle = \sum_{l\kappa}\exp[\mathrm{i}\boldsymbol{k}\cdot\boldsymbol{r}\binom{l}{\kappa}]$$

$$\times\int\chi_{e'}^*(r,R)M_{l\kappa}(R,r)\chi_e(r,R)dr.\tag{2.82}$$

Since we are not concerned with electronic transitions it will be understood that $e' = e$ and that we are evaluating the matrix elements for the electronic ground state. Physically this means that we disregard the contribution from the electronic polarisabilities to the dipole moment which arises from transitions between different electronic states. Omitting therefore the subscript e we obtain

$$\langle n'|H'|n\rangle = -i\frac{\omega_{n'n}}{c} A(\omega, t)\cdot\langle n'|M(R, k)|n\rangle, \qquad (2.83)$$

where

$$\langle n'|M(R, k)|n\rangle = \int \Psi_{n'}^*(R)M(R, k)\Psi_n(R)dr, \qquad (2.84)$$

and

$$M(R, k) = \sum_{l\kappa}\exp[i k\cdot r(\begin{smallmatrix}l\\\kappa\end{smallmatrix})] \int \chi^*(r, R)M_{l\kappa}(R, r)\chi(r, R)dr. \qquad (2.85)$$

Using (2.71, 77) is follows from (2.83) that

$$\langle n'|H'|n\rangle = -\langle n'|M(R, k)|n\rangle\cdot E(\omega, t), \qquad (2.86)$$

where

$$E(\omega, t) = E_0(\omega)e^{-i\omega t} f. \qquad (2.87)$$

Equation (2.86) is the relation analogous to (2.72). According to (2.86) the perturbation is determined by the matrix elements of the dipole moment, $\langle n'|M(R, k)|n\rangle$, from an initial vibrational state $|n\rangle$ to a final vibrational state $|n'\rangle$. $M(R, k)$ defined by (2.85) is the "phased" electric dipole moment of the crystal; the effect of the phase retardation is taken into account by multiplying the electric dipole moment of each atom $(\begin{smallmatrix}l\\\kappa\end{smallmatrix})$ by the phase factor $\exp[i k\cdot r(\begin{smallmatrix}l\\\kappa\end{smallmatrix})]$ [2.70]. For sufficiently small systems one often makes the "dipole approximation", which consists in putting $k = 0$. For a system such as a crystal, whose size is large compared to the wavelength, the phase retardation cannot be completely neglected.

In order to discuss the significance of $M(R, k)$ defined by (2.85) we introduce the relative position vectors of the electrons (Fig. 2.20)

$$\varrho_i(\begin{smallmatrix}l\\\kappa\end{smallmatrix}) = r_i(\begin{smallmatrix}l\\\kappa\end{smallmatrix}) - R(\begin{smallmatrix}l\\\kappa\end{smallmatrix}) = r_i(\begin{smallmatrix}l\\\kappa\end{smallmatrix}) - r(\begin{smallmatrix}l\\\kappa\end{smallmatrix}) - u(\begin{smallmatrix}l\\\kappa\end{smallmatrix}), \qquad (2.88)$$

and replace the electronic wave functions $\chi(r, R)$ in (2.85) by $\phi(\varrho, R)$. $M_{l\kappa}(R, r)$ given by (2.80) can then be written in the form

$$M_{l\kappa}(r, R) = z_\kappa er(\begin{smallmatrix}l\\\kappa\end{smallmatrix}) + z_\kappa eu(\begin{smallmatrix}l\\\kappa\end{smallmatrix}) - \sum_i e\varrho_i(\begin{smallmatrix}l\\\kappa\end{smallmatrix}), \qquad (2.89)$$

where $z_\kappa e$ is the ionic charge at the site $(\begin{smallmatrix}l\\\kappa\end{smallmatrix})$, i.e.

$$z_\kappa e = Z_\kappa e - \sum_i e. \qquad (2.90)$$

The first term in (2.89) contributes a permanent dipole moment to $M(R, r)$. The second term is independent of the integration variables in (2.85). The total dipole moment can therefore be written as

$$M(R, \mathbf{k}) = \sum_{l\kappa} \exp[i\mathbf{k}\cdot\mathbf{r}(^l_\kappa)][z_\kappa e\mathbf{r}(^l_\kappa) + z_\kappa e\mathbf{u}(^l_\kappa) - e\sum_i \mathbf{v}_{il\kappa}(R)] \quad (2.91\text{a})$$

where

$$\mathbf{v}_{il\kappa}(R) = \int \phi^*(\varrho, R)\,\varrho_i(^l_\kappa)\phi(\varrho, R)d\varrho, \quad (2.91\text{b})$$

represents the average position of the electron i belonging to the nucleus $(^l_\kappa)$ and depends on the displacement $\mathbf{u}(^l_\kappa)$. Thus $\sum \mathbf{v}_{il\kappa}(R)$ can be considered as the displacement of the "electron cloud" due to the nuclear displacement $\mathbf{u}(^l_\kappa)$.

Apart from the irrelevant permanent ionic dipole moment in (2.91a) which gives a zero contribution in the final result (2.84) if $n' \neq n$, we have thus separated $M(R, \mathbf{k})$ into two contributions, one arising from the ionic displacements, and the other arising from the effect of the electrons grouped around the sites $(^l_\kappa)$. If all these electrons were rigidly attached to there nuclei (rigid-ion model) [Ref. 2.17, Sect. 4.2], the electron configuration and hence the electronic wave functions in (2.91) would be independent on the nuclear positions R; thus $\mathbf{v}_{il\kappa}$ would be independent on R, and the last term in (2.91a) would reduce to a permanent dipole moment.

In more refined models, such as the shell model discussed in [Ref. 2.17, Sect. 4.3] only the inner most core electrons can be regarded as moving rigidly with the nucleus; the wave functions of the outer electrons, the deformable or shell electrons, will depend both on ϱ and on R, and $\mathbf{v}_{il\kappa}(R)$ may be expanded in terms of the displacement vectors $\mathbf{u}(^l_\kappa)$. Thus the dipole moment of the crystal can be expressed formally as a Taylor expansion of similar form as the potential energy [Ref. 2.17, Eq. (3.6)], namely

$$M(R, \mathbf{k}) = M^{(0)} + \sum_{l\kappa} \exp[i\mathbf{k}\cdot\mathbf{r}(^l_\kappa)]e(^l_\kappa)\mathbf{u}(^l_\kappa) + \sum_{l'\kappa'} \mathbf{u}(^l_\kappa)e(^{ll'}_{\kappa\kappa'})\mathbf{u}(^{l'}_{\kappa'}) + \dots$$
$$= M^{(0)} + M^{(1)} + M^{(2)} + \dots \quad (2.92)$$

In the first-order moment $M^{(1)}$ the coefficient $e(^l_\kappa)$ is a second-rank tensor representing the effective charge on the atom. Due to translational symmetry these coefficients are independent of l

$$e(^l_\kappa) = e_\kappa, \quad (2.93)$$

and charge neutrality requires

$$\sum_\kappa e_\kappa = 0. \quad (2.94)$$

Written out explicitly, the α-component of $M^{(1)}$ is

$$M_\alpha^{(1)} = \sum_{l\kappa\beta} \exp[i\boldsymbol{k}\cdot\boldsymbol{r}(^l_\kappa)]e_{\alpha\beta}(^l_\kappa)u_\beta(^l_\kappa). \qquad (2.95)$$

In most cases of interest $e_{\alpha\beta}(^l_\kappa)$ is taken to be isotropic, that is

$$e_{\alpha\beta}(^l_\kappa) = \delta_{\alpha\beta}e_\kappa, \qquad (2.96)$$

and (2.95) reduces to

$$M_\alpha^{(1)} = \sum_{l\kappa} \exp[i\boldsymbol{k}\cdot\boldsymbol{r}(^l_\kappa)]e_\kappa u_\alpha(^l_\kappa). \qquad (2.97)$$

In the rigid ion model $e_\kappa = z_\kappa e$. In the shell model, however, e_κ has two contributions, one is the ionic charge $z_\kappa e$ and the other arises from the electrons represented by the integral in (2.91). Thus, in the shell model, e_κ represents the *Szigeti* charge e_s^* which has been discussed in [Ref. 2.17, Sect. 4.3.1].

The coefficient $e(^{ll'}_{\kappa\kappa'})$ in the second-order moment $M^{(2)}$ of (2.92) is a third rank tensor, and $e(^{ll'}_{\kappa\kappa'})u(^{l'}_{\kappa'})$ is the charge induced on atom $(^l_\kappa)$. The properties of $e(^{ll'}_{\kappa\kappa'})$ are similar to those of the force constants $\Phi(^{ll'}_{\kappa\kappa'})$ discussed in [Ref. 2.17, Sect. 3.1]; thus $e(^{ll'}_{\kappa\kappa'})$ depends only on the relative positions of the unit cells l and l' :

$$e(^{ll'}_{\kappa\kappa'}) = e(^{l-l'}_{\kappa\kappa'}), \quad \text{and} \qquad (2.98)$$

$$\sum_{l'\kappa'} e(^{ll'}_{\kappa\kappa'}) = 0. \qquad (2.99)$$

The α-component of $M^{(2)}$ has the form

$$M_\alpha^{(2)} = \sum_{\beta\gamma} \sum_{\substack{l\kappa \\ l'\kappa'}} \exp[i\boldsymbol{k}\cdot\boldsymbol{r}(^l_\kappa)]e_{\alpha\beta\gamma}(^{ll'}_{\kappa\kappa'})u_\beta(^l_\kappa)u_\gamma(^{l'}_{\kappa'}). \qquad (2.100)$$

The coefficients $e(^{ll'}_{\kappa\kappa'})$ are due entirely to electronic deformations; in the rigid-ion model these coefficients, and those appearing in higher-order terms of (2.92), are all zero. Very little is known about these coefficients. In order to calculate them, it is necessary to introduce a model which includes deformable shells as well as anharmonic interactions. It turns out that the coefficients $e(^{ll'}_{\kappa\kappa'})$ depend on terms which are proportional to the parameters of the shell model and to the anharmonic coupling coefficients in the potential energy [2.71]. Thus in the harmonic approximation $M^{(2)} = 0$ and any calculation of the higher-order moments must be based on a physical model which includes anharmonicity and polarizable ions (Problem 2.4.3).

2.3.5 One-Phonon Absorption

We calculate $\varepsilon_2(\omega)$ given by (2.67) by substituting for M the first-order dipole moment $M^{(1)}$, which according to (2.97) has the form

$$M^{(1)} = \sum_{l\kappa} \exp[i\boldsymbol{k}\cdot\boldsymbol{r}(\tfrac{l}{\kappa})] e_\kappa \boldsymbol{u}(\tfrac{l}{\kappa}). \tag{2.101}$$

We evaluate $\varepsilon_2(\omega)$ on the basis of the eigenvalues and eigenfunctions of the harmonic Hamiltonian. According to [Ref. 2.17, Eqs. (2.104, 110)] the eigenfunctions and eigenvalues are given by

$$|n\rangle = \prod_s |n_s\rangle, \quad \text{and} \tag{2.102}$$

$$E_n = \sum_s (n_s + 1/2)\hbar\omega_s, \tag{2.103}$$

where $s = (\boldsymbol{q}, j)$ denotes the phonon with wave vector $\boldsymbol{q}$ and branch j. From (2.63 and 103) it follows that

$$\hbar\omega_{n'n} = \sum_s (n'_s - n_s)\hbar\omega_s. \tag{2.104}$$

Representing $\boldsymbol{u}(\tfrac{l}{\kappa})$ in (2.101) in terms of creation and annihilation operators [Ref. 2.17, Eq. (3.79)] we write

$$\boldsymbol{u}(\tfrac{l}{\kappa}) = \left(\frac{\hbar}{2Nm_\kappa}\right)^{1/2} \sum_{\boldsymbol{q}j} [\omega_j(\boldsymbol{q})]^{-1/2} e(\kappa|\tfrac{\boldsymbol{q}}{j}) \exp[i\boldsymbol{q}\cdot\boldsymbol{r}(\tfrac{l}{\kappa})][a^+(\tfrac{-\boldsymbol{q}}{j}) + a(\tfrac{\boldsymbol{q}}{j})].$$

$$\tag{2.105}$$

The matrix element $\langle n'|M^{(1)}|n\rangle$ can then be written in the form

$$\langle n'|M^{(1)}|n\rangle = \left(\frac{\hbar}{2N}\right)^{1/2} \sum_\kappa \frac{e_\kappa}{\sqrt{m_\kappa}} \exp[i(\boldsymbol{k}+\boldsymbol{q})\cdot\boldsymbol{r}(\kappa)]$$

$$\times \sum_{\boldsymbol{q}j} \frac{e(\kappa|\tfrac{\boldsymbol{q}}{j})}{\omega_j(\boldsymbol{q})^{1/2}} \langle n'|a^+(\tfrac{-\boldsymbol{q}}{j}) + a(\tfrac{\boldsymbol{q}}{j})|n\rangle \sum_l \exp[i(\boldsymbol{k}+\boldsymbol{q})\cdot\boldsymbol{r}(l)].$$

$$\tag{2.106}$$

Using the results derived in [Ref. 2.17, Sects. 2.2.2, 3] and introducing the notation $s = (-\boldsymbol{q}, j)$ we obtain

$$\langle n'|a^+_{-s}|n\rangle = \langle n'_{-s}|a^+_{-s}|n_s\rangle \prod_{p\neq -s} \langle n'_p|n_p\rangle \quad \text{and}$$

$$\langle n'|a_s|n\rangle = \langle n'_s|a_s|n_s\rangle \prod_{p\neq s} \langle n'_p|n_p\rangle.$$

48

Due to the orthogonality of the eigenfunctions only those matrix elements are different from zero, for which $n'_p = n_p$. We must therefore consider only transitions for which the quantum numbers of the phonons $-s$ or s differ by unity, while the quantum numbers of all the other phonons are the same. These transitions will be referred to as "main transitions" [2.58]. From the normalization condition [Ref. 2.17, Eq. (2.107)] and the relations in [Ref. 2.17, Eqs. (2.124, 125)] we obtain

$$\langle n'|a^+_{-s}|n\rangle = (n_{-s} + 1)^{1/2} \quad \text{for} \quad n'_{-s} = n_{-s} + 1 \tag{2.107}$$

and

$$\langle n'|a_s|n\rangle = n_s^{1/2} \quad \text{for} \quad n'_s = n_s - 1. \tag{2.108}$$

For the main transitions we obtain from (2.104)

$$\hbar\omega_{n'n} = \begin{cases} \hbar\omega_{-s} = \hbar\omega_s & \text{for } n'_{-s} = n_{-s} + 1 \\ -\hbar\omega_s & \text{for } n'_s = n_s - 1. \end{cases} \tag{2.109}$$

Here we have used the relation $\omega_j(-q) = \omega_j(q)$ [Ref. 2.17, Eq. (2.28)].

From (2.109) and the δ-function in (2.67) it follows that

$$\omega(k) = vk = \omega_j(q), \tag{2.110}$$

where v is the velocity of light in the crystal.

The condition (2.110) is illustrated in Fig. 1.1: The dispersion curve of the light, $\omega(k) = vk$, must intersect the dispersion curves $\omega_j(q)$ of the optical phonons. The initial slope of the acoustic branches is the velocity of sound, which is about $10^{-5}v$. It therefore follows, that there is no intersection with the acoustic branches. Moreover, the values of q at the intersections with the optical branches is extremely small, of the order of the ratio of sound velocity to light velocity, and may usually be replaced by zero.

Having realized, that both, k and q, are approximately zero, the matrix elements (2.106) can be written in a simpler form. First we note that according to [Ref. 2.17, Eqs. (2.51, 52)] the sum over l in (2.106) is equal to $N\Delta(k + q)$, where $\Delta(k + q) = 1$ if $k + q = \tau$ and zero otherwise. τ is a vector of the reciprocal lattice which in the present case is clearly zero, hence $q = -k$. For a main transition associated with the absorption of a photon k and the creation of a phonon $(-q, j)$, represented by $a^+\left(\begin{smallmatrix} -q \\ j \end{smallmatrix}\right) = a^+\left(\begin{smallmatrix} k \\ j \end{smallmatrix}\right)$ we obtain from (2.106, 107)

$$|\langle n'|M^{(1)}|n\rangle|^2 = \frac{N\hbar}{2\omega_j}(n_j(k) + 1)|\sum_\kappa e_\kappa \frac{e(\kappa|j)}{\sqrt{m_\kappa}}|^2. \tag{2.111}$$

On the other hand, for a main transition associated with the emission of a photon $-k$ and the annihilation of a phonon (q, j), represented by $a\left(\begin{smallmatrix} q \\ j \end{smallmatrix}\right) = a\left(\begin{smallmatrix} -k \\ j \end{smallmatrix}\right)$, we obtain from (2.106, 108)

$$|\langle n'|M^{(1)}|n\rangle|^2 = \frac{N\hbar}{2\omega_j} n_j(-k) \left| \sum_\kappa e_\kappa \frac{e(\kappa|j)}{\sqrt{m_\kappa}} \right|^2 . \tag{2.112}$$

In (2.111, 112) we have written $\omega_j(\boldsymbol{q} \cong 0) = \omega_j$ and $e(\kappa|^{\boldsymbol{q}\cong 0}_j) = e(\kappa|j)$. The expressions (2.111, 112) are generalizations of the corresponding expression (2.73) for a single charged oscillator. Noting that the summation over n' in (2.67) is equivalent to a summation over the optical phonon branches j and that $\sum w_n n_j(-\boldsymbol{q}) = \sum w_n n_j(\boldsymbol{q}) = \bar{n}_j(\boldsymbol{q})$ is the mean occupation number of the phonon $(\boldsymbol{q}, j)$ as defined in [Ref. 2.17, Eq. (2.126)], we obtain after substitution of (2.109, 111, 112) into (2.67) the result

$$\varepsilon_2(\omega) = \frac{4\pi^2}{v_a} \sum_j \frac{1}{2\omega_j} \left| \sum_\kappa e_\kappa \frac{e(\kappa|j)}{\sqrt{m_\kappa}} \right|^2 \delta(\omega - \omega_j). \tag{2.113}$$

In this exprssion v_a is the volume of the primitive unit cell and the summation over κ extends over the ions in v_a while the summation over j extends over the optical phonons at $\boldsymbol{q} \cong 0$, which at normal experimental conditions are TO phonons (Fig. 2.16).

We note that in the present approximation, $\varepsilon_2(\omega)$ is independent on the mean occupation numbers $\bar{n}_j$ and hence independent on temperature. The observed absorption represented by $\varepsilon_2(\omega)$ is always a *nett* absorption, which is due to the *difference* between the two fundamental processes shown in Fig. 2.17, namely the absorption and emission of a photon. The probability for the absorption of a photon (creation of a phonon) is proportional to $\bar{n}_j + 1$ compared with $\bar{n}_j$ for the emission of a photon (annihilation of a phonon); this is accounted for by the different signs in (2.109) with the result that $\varepsilon_2(\omega)$ is independent on $\bar{n}_j$. As will be shown below, the expression (2.113) is identical with the classical result for harmonic crystals. $\varepsilon_2(\omega)$ consists of a number of δ-functions, one for each optical frequency ω_j. Each δ-function implies conservation of energy between photon and phonon, and each absorption line is infinitely sharp which is a consequence of the harmonic approximation.

It should be noted that not all the optical phonons at $\boldsymbol{q} \cong 0$ necessarily contribute to $\varepsilon_2(\omega)$ but only those for which the sum

$$c_j = \sum_\kappa e_\kappa \frac{e(\kappa|j)}{\sqrt{m_\kappa}} , \tag{2.114}$$

which appears in (2.113), is different from zero. The value of c_j depends on the pattern of ionic displacements of the mode j which is characterized by the vectors $e(\kappa|j)$. If $c_j = 0$, the dipole moment M_j induced by the mode j is also zero and the mode cannot couple to the light since the interaction energy $W_j = -E \cdot M_j^{(1)}$ is zero. The modes for which c_j is different from zero must satisfy certain symmetry requirements and can be predicted by group theory [2.72]. Such modes are called *infrared active*.

As an illustration we apply the results of this section to the alkali-halides. In (2.113) there is only one TO mode which is infrared active, and the sum over κ extends over the two ions in the primitive unit cell. Thus we obtain

$$\varepsilon_2(\omega) = \frac{2\pi^2}{v_a \omega_{\mathrm{TO}}} \left| \sum_\kappa e_\kappa \frac{e(\kappa|\mathrm{TO})}{\sqrt{m_\kappa}} \right|^2 \delta(\omega - \omega_{\mathrm{TO}}). \qquad (2.115)$$

From [Ref. 2.17, Eqs. (2.41, 42] we have

$$e(1|\mathrm{TO}) = -\frac{m_2^{1/2}}{(m_1 + m_2)^{1/2}}, \quad e(2|\mathrm{TO}) = \frac{m_1^{1/2}}{(m_1 + m_2)^{1/2}} \,,$$

and identifying e_κ with the Szigeti charge as discussed in Sect. 2.3.4, that is $e_1 = e_{\mathrm{s}}^*$, $e_2 = -e_{\mathrm{s}}^*$, we obtain

$$\varepsilon_2(\omega) = \frac{2\pi^2 e_{\mathrm{s}}^{*2}}{\mu v_a \omega_{\mathrm{TO}}} \delta(\omega - \omega_{\mathrm{TO}}). \qquad (2.116)$$

In this expression $\mu = m_1 m_2/(m_1 + m_2)$ is the reduced mass. Using the Kramers-Kronig relation (2.36a) we obtain from (2.116) for the real part of the dielectric constant the following expression

$$\varepsilon_1(\omega) = \varepsilon_\infty + \frac{4\pi e_{\mathrm{s}}^{*2}}{\mu v_a} (\omega_{\mathrm{TO}}^2 - \omega^2)^{-1}. \qquad (2.117)$$

This is the classical result which is obtained if in (2.33) the damping γ is neglected and the transverse effective charge e_{T}^* is replaced by the Szigeti charge e_{s}^*. The reason for the different effective charges occurring in (2.33 and 117) is due to the fact that in deriving (2.117) we have neglected the coupling between the electronic and ionic polarizations via the macroscopic polarization and the dipolar contribution to the effective field E^* [2.73]. The classical model leading to (2.117) is treated in Problem 2.4.1.

2.3.6 Two-Phonon Processes Due to the Second-Order Dipole Moment Mechanism

We calculate the contribution of $M^{(2)}$ defined by (2.92, 100) to $\varepsilon_2(\omega)$. As mentioned in Sect. 2.3.1, this is the mechanism responsible for the absorption in Si, Ge and diamond, in which by symmetry $M^{(1)} = 0$. According to (2.67) we have to evaluate

$$\varepsilon_2^{(2)}(\omega) = \frac{4\pi^2}{V\hbar} \sum_{n'n} w_n |\langle n'|M^{(2)}|n\rangle|^2 \frac{\omega_{n'n}}{\omega} \delta(\omega - |\omega_{n'n}|). \qquad (2.118)$$

In the present approximation the matrix elements are calculated with the eigenfunctions of the harmonic Hamiltonian H_{h} in (2.59). From (2.92) we have

$$M^{(2)} = \sum_{l\kappa} \exp[i\mathbf{k}\cdot\mathbf{r}(^l_\kappa)] \sum_{l'\kappa'} \mathbf{u}(^l_\kappa) e(^{ll'}_{\kappa\kappa'}) \mathbf{u}(^{l'}_{\kappa'}). \tag{2.119}$$

Substituting (2.105) into (2.119) it is shown in Appendix C that $M^{(2)}$ can be written in the form

$$M^{(2)} = \frac{\hbar}{2} \sum_{qjj'} \frac{S(^q_{jj'})}{[\omega_j(\mathbf{q})\omega_{j'}(-\mathbf{q})]^{1/2}} \, A(^q_j)A(^{-q}_{j'}), \tag{2.120}$$

where

$$S(^q_{jj'}) = \sum_{\kappa\kappa'}(m_\kappa m'_\kappa)^{-1/2} \, e(\kappa|^q_j)R(^q_{\kappa\kappa'})e(\kappa'|^{-q}_{j'}) \tag{2.121}$$

$$R(^q_{\kappa\kappa'}) = \sum_L e(^L_{\kappa\kappa'})\exp[i\mathbf{q}\cdot\mathbf{r}(^L_{\kappa\kappa'})], \tag{2.122}$$

with

$$\mathbf{r}(^L_{\kappa\kappa'}) = \mathbf{r}(L) + \mathbf{r}(\kappa) - \mathbf{r}(\kappa'); \quad L = l - l', \tag{2.123}$$

and

$$A(^q_j) = a^+(^{-q}_j) + a(^q_j). \tag{2.124}$$

The matrix elements in (2.118) have the form

$$\langle n'|M^{(2)}|n\rangle = \frac{\hbar}{2} \sum_{qjj'} C(^q_{jj'})\langle n'|A(^q_j)A(^{-q}_{j'})|n\rangle, \tag{2.125}$$

where

$$C(^q_{jj'}) = [\omega_j(\mathbf{q})\omega_{j'}(-\mathbf{q})]^{-1/2}S(^q_{jj'}). \tag{2.126}$$

The evaluation of the matrix elements in (2.125) is straigthforward. Writing $s = (\mathbf{q}, j)$, $-s' = (-\mathbf{q}, j')$ we have

$$A_s A_{-s'} = a^+_{-s}a^+_{s'} + a_s a^+_{s'} + a^+_{-s}a_{-s'} + a_s a_{-s'}.$$

As an example we obtain

$$\langle n'|a^+_{-s}a^+_{s'}|n\rangle = (n_{-s} + 1)^{1/2}(n_{s'} + 1)^{1/2}$$

for the double transition

$$n'_{-s} = n_{-s} + 1, \; n'_{s'} = n_{s'} + 1; \quad n'_p = n_p \quad (p \neq -s, s'),$$

and analogous expressions for the other matrix elements. Assuming that $\omega_{j'}(-\mathbf{q}) > \omega_j(\mathbf{q})$, using $\bar{n}_j(-\mathbf{q}) = \bar{n}_j(\mathbf{q})$ and (2.104), and following the same procedure as in Sect. 2.3.5 we obtain from (2.118)

$$\varepsilon_2^{(2)}(\omega)$$

$$= \frac{\pi^2 \hbar}{V} \sum_{qjj'} |C(\substack{q \\ jj'})|^2 \{[\bar{n}_j(q) + \bar{n}_{j'}(q) + 1]\delta(\omega - \omega_j(q) - \omega_{j'}(-q))$$

$$+ [\bar{n}_j(q) - \bar{n}_{j'}(q)]\delta(\omega + \omega_j(q) - \omega_{j'}(-q))\}. \qquad (2.127)$$

It is shown in Appendix D that (2.127) can be written in the compact form

$$\varepsilon_2^{(2)}(\omega, T) = \frac{\hbar}{8\pi} \sum_{jj'} \int_{S_\omega} |C(\substack{q \\ jj'})|^2 \frac{(\bar{n}_j + \frac{1}{2}) \pm (\bar{n}_{j'} + \frac{1}{2})}{|\nabla \Omega_{jj'}(q)|_{\Omega_{jj'}=\omega}} dS_\omega. \qquad (2.128)$$

In this relation the integration in q space extends over the surface of the combined branches (jj') for which

$$\Omega_{jj'}(q) = \omega_{j'}(-q) \pm \omega_j(q) = \omega. \qquad (2.129)$$

The sum over j and j' takes into account that possibly several branch combinations give a contribution to the total absorption at the frequency ω. In the following we discuss the temperature dependence and frequency dependence of $\varepsilon_2^{(2)}$ based on (2.128).

Neglecting the relatively weak temperature dependence of the phonon frequencies, the temperature dependence of $\varepsilon_2^{(2)}$ is essentially given by the factor $[(\bar{n}_j + 1/2) \pm (\bar{n}_{j'} + 1/2)]$. The plus sign in (2.128, 129) describes a two-phonon *sum absorption* leading to a *summation band* (Fig. 2.19a) with

$$\omega = \omega_{j'}(-q) + \omega_j(q), \qquad (2.130a)$$

and a temperature dependence of ε_2 which is essentially controlled by

$$\bar{n}_j + \bar{n}_{j'} + 1 = \frac{1}{2}\left[\operatorname{ctgh}\left(\frac{\hbar\omega_j}{2k_BT}\right) + \operatorname{ctgh}\left(\frac{\hbar\omega_{j'}}{2k_BT}\right)\right]$$

$$\cong \begin{cases} 1 & \text{for } k_BT \ll \hbar\omega_j, \hbar\omega_{j'}, \\ \frac{k_BT}{\hbar}\left(\frac{1}{\omega_j} + \frac{1}{\omega_{j'}}\right) & \text{for } k_BT \gg \hbar\omega_j, \hbar\omega_{j'} \end{cases}. \qquad (2.130b)$$

On the other hand, the minus sign in (2.128, 129) describes a two-phonon *difference absorption* leading to a difference band (Fig. 2.19b) with

$$\omega = \omega_{j'}(-q) - \omega_j(q), \qquad (2.131a)$$

and an associated temperature dependence given by

$$\bar{n}_j - \bar{n}_{j'} = \begin{cases} \exp[-\hbar\omega_j/k_BT] - \exp[-\hbar\omega_{j'}/k_BT] & \text{for } k_BT \ll \hbar\omega_j, \hbar\omega_{j'} \\ \dfrac{k_BT}{\hbar}\left(\dfrac{1}{\omega_j} - \dfrac{1}{\omega_{j'}}\right) & \text{for } k_BT \gg \hbar\omega_j, \hbar\omega_{j'}. \end{cases} \quad (2.131\mathrm{b})$$

The relations (2.130b, 2.131b) show the extrem difference in the temperature dependence between summation and difference bands. At low temperatures the absorption of a summation band becomes constant while that of a difference band tends to zero. The fact that the absorption of the difference bands vanishes at low temperatures is the reason for the "super-transparency" of many crystals at low temperaturs and at low frequencies (Fig. 2.25).

From (2.128) it follows that the main structures in $\varepsilon_2^{(2)}(\omega)$ will appear in the neighbourhood of those frequencies for which $\nabla\Omega_{jj'}(\boldsymbol{q}) = 0$. This corresponds to the frequencies at which the slope of the combined branches (jj') becomes zero, and these points are known as the *critical points* (c.p.) [Ref. 2.17, Sect. 3.5.9].

According to (2.129) the condition for a c.p. is satisfied if either

$$\nabla\omega_{j'}(-\boldsymbol{q}) = \nabla\omega_j(\boldsymbol{q}) = 0 \quad \text{or} \qquad (2.132)$$

$$\nabla\omega_{j'}(-\boldsymbol{q}) = \pm\nabla\omega_j(\boldsymbol{q}),$$

with the plus sign for a difference band and the minus sign for a summation band. The three cases represented by (2.132) are illustrated in Fig. 2.21 [2.74].

The relation (2.128) shows a close analogy to the expression for the density of states. Since the phonon dispersion is usually small in regions of $\boldsymbol{q}$ space with large values of $\boldsymbol{q}$, i.e. large density of states, $C(^{\boldsymbol{q}}_{jj'})$ is only weakly dependent on $\boldsymbol{q}$. If this $\boldsymbol{q}$-dependence is neglected (2.128) can be written in the form

$$\varepsilon_2^{(2)}(\omega) \cong \frac{h}{8\pi} \sum_{jj'} |C_{jj'}|^2 g_{jj'}(\omega)[(\bar{n}_j + \tfrac{1}{2}) \pm (\bar{n}_{j'} + \tfrac{1}{2})], \qquad (2.133)$$

where

$$g_{jj'}(\omega) = \int \frac{dS_\omega}{|\nabla\Omega_{jj'}(\boldsymbol{q})|_{\Omega_{jj'}=\omega}} \ , \qquad (2.134)$$

is the *two-phonon density of states*; $g_{jj'}(\omega)$ is a generalization of the one-phonon density of states $g_j(\omega)$ defined in [Ref. 2.17, Eq. (3.84)]. In this approximation $\varepsilon_2^{(2)}(\omega)$ reflects the two-phonon density of states.

The non-vanishing coupling parameters $C(^{\boldsymbol{q}}_{jj'})$ at the c.p.'s can be predicted by group theory . These symmetry selection rules have been worked out for crystals of the zinc blende and diamond structure by *Birman* [2.75].

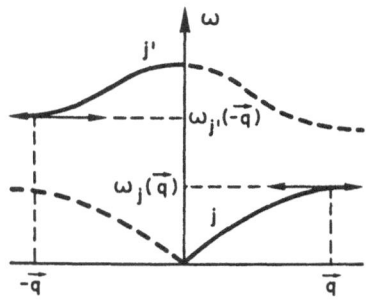

(a) $\nabla \omega_{j'}(-\vec{q}) = \nabla \omega_j(\vec{q}) = 0$

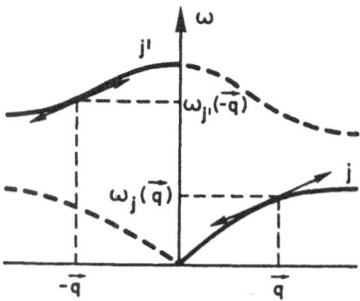

(b) $\nabla \omega_{j'}(-\vec{q}) = \nabla \omega_j(\vec{q})$

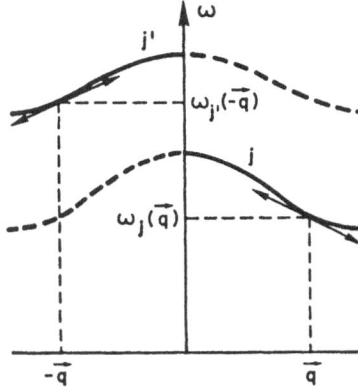

(c) $\nabla \omega_{j'}(-\vec{q}) = -\nabla \omega_j(\vec{q})$

Fig. 2.21a–c. Critical points of the two-phonon density of states

An analysis of the experimental absorption spectra of Si, Ge and diamond within the region of their two- and three-phonon absorption based on the theoretical aspects discussed above has been carried out by *Bilz* et al. [2.76]. Figure 2.22 illustrates the analysis for the case of Si. The experimental data originate from *Collins* and *Fan* [2.77], and from *Johnson* [2.78], while the data of the combined branches were taken from the neutron-scattering measurements of *Dolling* [2.79] and *Ghose* [2.80]. Note the near-coincidence of the most pronounced absorption maxima with the flat regions of the combination branches, such as the TO + TA, LO + TA combinations. Based on an approximate calculation of the two-phonon density of states it is possible to calculate in simple cases not only the position but also the form of the absorption bands [2.76]. The absorption coefficient of Si as a function of frequency and temperature is shown in Figs. 2.23–26 [2.81].

The absorptions shown in Fig. 2.23 are due to three-phonon summation processes, while those shown in Fig. 2.24 are mainly due to two-phonon summation processes. Note that at low temperatures the intensities of these

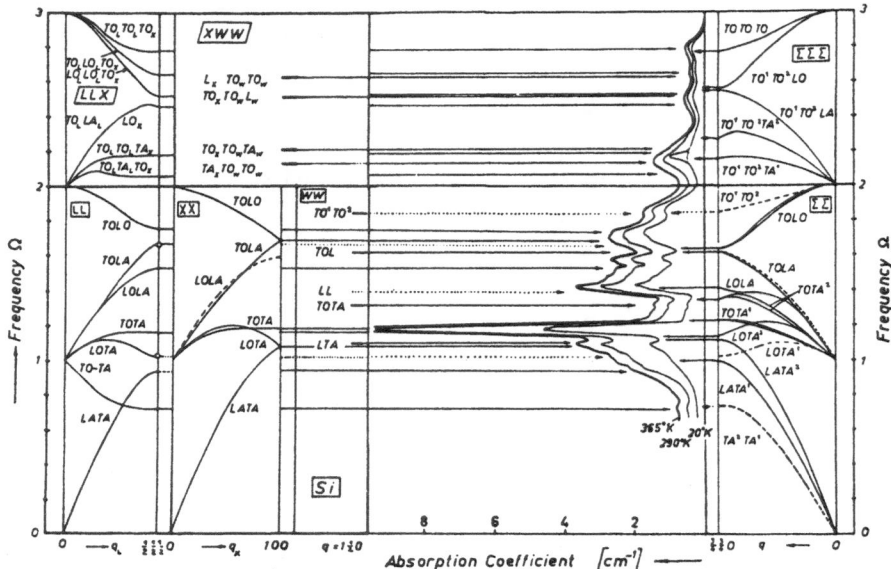

Fig. 2.22. Absorption coefficient for silicon and dispersion curves of the two- and three-phonon combination branches. (———): symmetry alowed combinations; (- - - - -): forbidden combinations; (•••): forbidden at the endpoint of·this direction

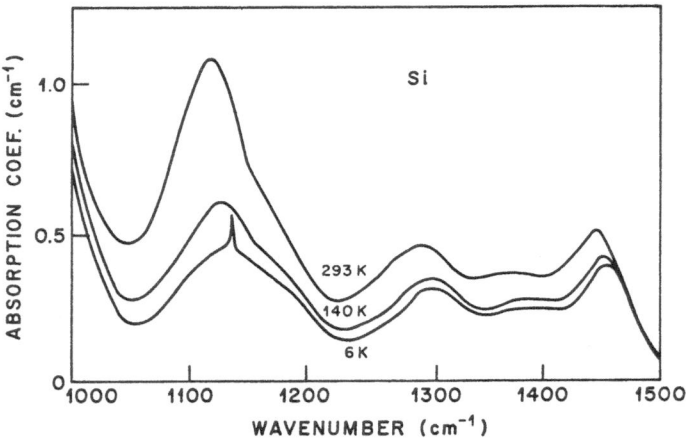

Fig. 2.23. Absorption spectrum in the three-phonon excitation region of silicon

summation processes approach a constant value. The far-infrared absorptions shown in Fig. 2.25 originate from two-phonon difference processes and show a dramatic decrease with decreasing temperature. Figure 2.26 shows the temperature dependence of the difference absorption intensity of the peak at $152\,\mathrm{cm}^{-1}$. This peak is assigned to the $\mathrm{LO}(\Sigma) - \mathrm{TA}(\Sigma)$ combina-

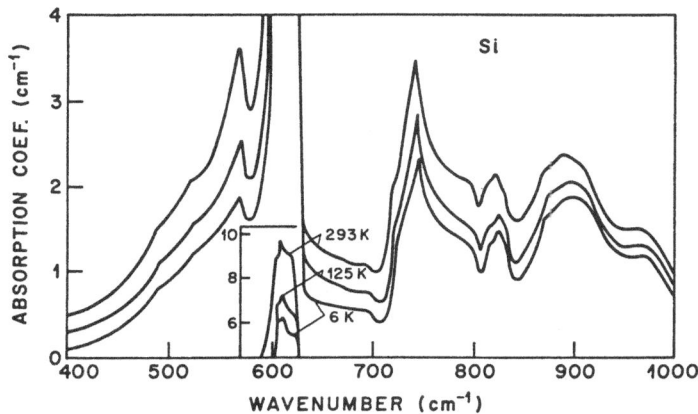

Fig. 2.24. Absorption spectrum in the two-phonon excitation region in silicon. The inset shows intense peaks

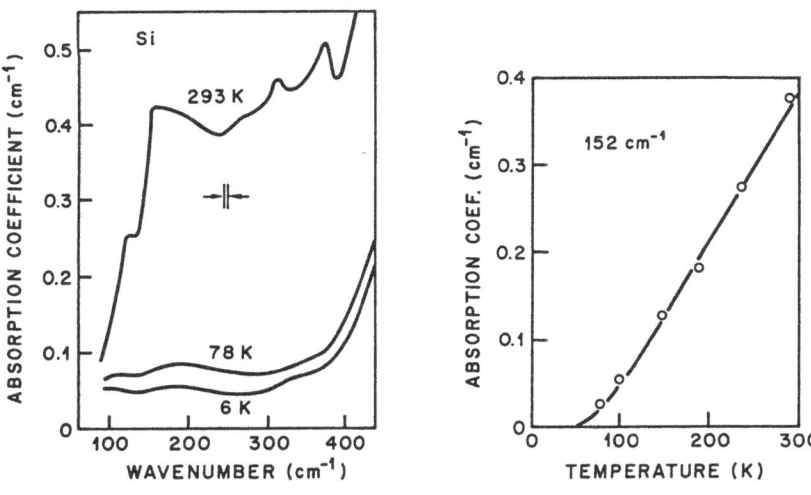

Fig. 2.25   **Fig. 2.26**

Fig. 2.25. Absorption spectrum by the two-phonon difference process in silicon

Fig. 2.26. Temperature dependence of the difference absorption intensity of silicon. (○) are measured at $152\,\mathrm{cm}^{-1}$ (Fig. 2.25) and (——) is the calculated relative intensity according to the assignment of the absorption structure. See text for details

tion; the frequency of the $LO(\Sigma)$ phonon is about $370\,\mathrm{cm}^{-1}$ and that of the $TA(\Sigma)$ phonon about $220\,\mathrm{cm}^{-1}$. The solid line in Fig. 2.26 shows the calculated relative intensity according to this assignment. According to (2.133) this intensity is given by

$$I = \alpha(\bar{n}_j - \bar{n}_{j'}),$$

where α is a constant which is proportional to the two-phonon density of states and to the square of the second-order dipole moment.

Figure 2.9 shows the observed [2.82] and calculated [2.83] infrared absorption $\varepsilon_2(\omega)$ of diamond. The calculated spectrum is based on a nonlinear extension of the bond-charge model [Ref. 2.17, Sect. 4.4] and contains only one adjustable parameter.

2.3.7 Two-Phonon Processes Due to Anharmonic Coupling

The formal relations have already been discussed in [Ref. 2.17, Sect. 5.5]. The dieletric constant is given by [Ref. 2.17, Eqs. (5.136, 139)], namely

$$\varepsilon(\omega) = \varepsilon_\infty + \sum_j S_j R_j(0, \omega), \tag{2.135}$$

where the response function is defined by

$$R_j(0, \omega) = [\omega_j^2(0) - \omega^2 + 2\omega_j(0)\Sigma_j(0, \omega)]^{-1}. \tag{2.136}$$

The complex self-energy of the mode $\binom{0}{j}$ is given by

$$\Sigma_j(0, \omega) = \Delta_j(0, \omega) - i\Gamma_j(0, \omega), \tag{2.137}$$

where $\Delta_j(0, \omega)$ is the change of the frequency $\omega_j(0)$ due to anharmonicity and $\Gamma_j(0, \omega)$ is the damping of the mode $\binom{0}{j}$. For a definition of these quantities the reader is referred to [Ref. 2.17, Sects. 5.5.2,3].

In the following we confine ourselves to simple ionic crystals such as the alkali-halides with an infrared active mode of frequency ω_{TO} and write $\omega_j(0) = \omega_{TO}$, $\Delta_j(0, \omega) = \Delta(\omega)$ $\Gamma_j(0, \omega) = \Gamma(\omega)$, and $S_j = S$. Since we assume that $\Delta(\omega)$ is small compared with ω_{TO} we obtain

$$\varepsilon(\omega) = \varepsilon_\infty + \frac{S}{\Omega_{TO}^2(\omega) - \omega^2 - 2i\omega_{TO}\Gamma(\omega)}, \tag{2.138}$$

wehre $\Omega_{TO}(\omega) = \omega_{TO} + \Delta(\omega)$ is the renormalized eigenfrequency of the TO-mode and $\Gamma(\omega)$ is its damping. Both these quantities depend on the frequency ω with which the crystal is probed [Ref. 2.17, Fig. 5.13]. The reciprocal of $\Gamma(\omega)$ is the lifetime of the TO phonon which interacts with the light and decays into two other phonons, as illustrated in Fig. 2.18. From (2.138) we obtain for the imaginary part of the dielectric constant

$$\varepsilon_2(\omega) = \frac{2S\omega_{TO}\Gamma(\omega)}{(\Omega_{TO}^2 - \omega^2)^2 + 4\omega_{TO}^2\Gamma^2(\omega)}. \tag{2.139}$$

The main absorption peak is in the vicinity of the frequency for which $\Omega_{TO}(\omega) - \omega = 0$. For frequencies ω sufficiently different from $\Omega_{TO}(\omega)$, the second term in the denominator can often be neglected, and in this case $\varepsilon_2(\omega)$ is proportional to $\Gamma(\omega)$. For this reason we discuss in more detail the frequency dependence of Γ. According to [Ref. 2.17, Eq. (5.153)], $\Gamma(\omega)$ is given by

$$\Gamma(\omega) = \frac{18\pi}{\hbar^2} \sum_{q_1} \sum_{j_1 j_2} \left| V \begin{pmatrix} 0 & -q_1 & q_1 \\ \text{TO} & j_1 & j_2 \end{pmatrix} \right|^2 K(\omega).$$

Here we have used [Ref. 2.17, Eqs. (5.153, 154)] and restricted ourselves to the first Brillouin zone, i.e. $\tau = 0$. The coefficients $V(\cdots)$ are the third-order anharmonic coupling coefficients defined in [Ref. 2.17, Appendix P] and $K(\omega)$ is defined in [Ref. 2.17, Eq. 5.153]. In order to illustrate the similarities with the results for the $M^{(2)}$ coupling we change the notation by writing $-q_1 \to q, j_1 \to j, j_2 \to j'$, and using corresponding changes in the coefficients $V(\cdots)$, we obtain

$$\Gamma(\omega) = \frac{18\pi}{\hbar^2} \sum_{q} \sum_{jj'} |V(\substack{q \\ 0jj'})|^2 K(\omega), \quad \text{where} \tag{2.140}$$

$$K(\omega) = (\bar{n}_j + \bar{n}_{j'} + 1)[\delta(\omega - \omega_j - \omega_{j'}) - \delta(\omega + \omega_j + \omega_{j'})] + (\bar{n}_j - \bar{n}_{j'})[\delta(\omega + \omega_j - \omega_{j'}) - \delta(\omega - \omega_j + \omega_{j'})]. \tag{2.141}$$

Following the same procedure as in Sect. 2.3.6 it can be shown that $\Gamma(\omega)$ can be written in the form

$$\Gamma(\omega) \sim \sum_{jj'} \int_{S_\omega} |V(\substack{q \\ 0jj'})|^2 \frac{(\bar{n}_j + 1/2) \pm (\bar{n}_{j'} + 1/2)}{|\nabla \Omega_{jj'}(q)|_{\Omega_{jj'} = \omega}} dS_\omega \ , \tag{2.142}$$

where $\Omega_{jj'} = \omega_{j'} \pm \omega_j$. Neglecting the q-dependence of $V(\substack{q \\ jj'})$ and of $\bar{n}_j, \ \bar{n}_{j'}$ as discussed in Sect. 2.3.6, and introducing the two-phonon density of states $g_{jj'}(\omega)$ defined by (2.134) we obtain

$$\Gamma(\omega) \sim \sum_{jj'} |V_{0jj'}|^2 g_{jj'}(\omega)[(\bar{n}_j + 1/2) \pm (\bar{n}_{j'} + 1/2)]. \tag{2.143}$$

The frequency dependence of $\varepsilon_2(\omega)$ is obtained if (2.142) or (2.143) is substituted in (2.139). If ω is sufficiently different from Ω_{TO} we neglect the second term in the denominator of (2.139) and obtain with (2.142) the following expression for $\varepsilon_2(\omega)$:

$$\varepsilon_2(\omega) \sim \frac{S \omega_{\text{TO}}}{(\Omega_{\text{TO}}^2 - \omega^2)^2} \sum_{jj'} \int_{S_\omega} |V(\substack{q \\ 0jj'})|^2 \frac{(\bar{n}_j + 1/2) \pm (\bar{n}_{j'} + 1/2)}{|\nabla \Omega_{jj'}(q)|_{\Omega_{jj'} = \omega}} dS_\omega \tag{2.144}$$

This equation resembles (2.128) obtained for the absorption due to second-order electric dipole moment. All the conclusions concerning the temperature dependence of the summation and difference bands as well as the structures of $\varepsilon_2(\omega)$ originating from the critical points derived in Sect. 2.3.6 also apply to (2.144) for the case of anharmonic coupling. The summation and difference processes are illustrated in Fig. 2.18. The non-vanishing coefficients $V\binom{q}{0jj'}$ at the c.p.'s can be predicted by group theoretical methods [2.84].

If (2.143) is substituted into (2.139) and if ω is sufficiently different from $\Omega_{TO}(\omega)$, we obtain an approximate exprssion for $\varepsilon_2(\omega)$ which resembles the expression (2.133) for $M^{(2)}$ coupling:

$$\varepsilon_2(\omega) \sim \frac{S\omega_{TO}}{(\Omega_{TO}^2 - \omega^2)^2} \sum_{jj'} |V_{0jj'}|^2 g_{jj'}(\omega)[\bar{n}_j + 1/2) \pm (\bar{n}_{j'} + 1/2)]. \quad (2.145)$$

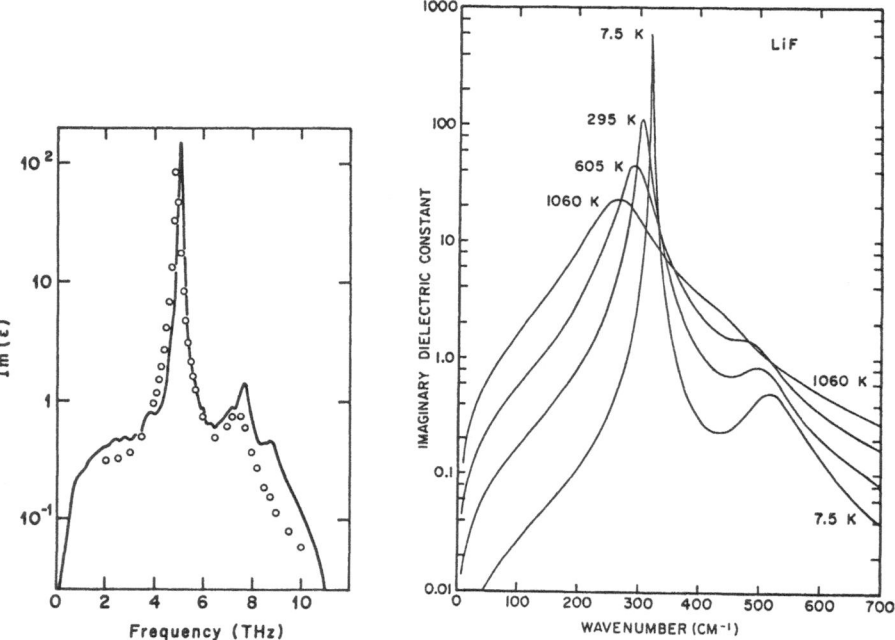

Fig. 2.27. Imaginary part $\varepsilon_2(\omega)$ of the dielectric constant $\varepsilon(\omega)$ of NaCl. (ooo) experimental points; (——) calculation

Fig. 2.28. Imaginary part $\varepsilon_2(\omega)$ of the dielectric constant $\varepsilon(\omega)$ of LiF as obtained from the observed reflectivity shown in Fig. 2.6

Figure 2.27 illustrates $\varepsilon_2(\omega)$ of NaCl at 300 K. The experimental points are from *Genzel* et al. [2.85], while the calculated curve has been obtained by *Cowley* [2.86] based on the general exprssion (2.139) and a shell-model calculation to provide frequencies and eigenvectors. The structures at both sides of the main absorption band arise from summation and difference bands.

Figure 2.28 shows the frequency and temperature dependence of ε_2 for LiF as obtained from *Jaspers* et al. [2.16a]. These data have been obtained on the basis of the observed reflectivity shown in Fig. 2.6. We note that with decreasing temperature the absorption strongly decreases at the low-frequency side of the main absorption band which is due to the strong temperatur dependence of the difference bands.

Finally, it should be remembered that although the $\varPhi^{(3)}$ mechanism is usually dominating in ionic crystals, the $M^{(2)}$ mechanism will, in general, also contribute to $\varepsilon(\omega)$, and that in addition there are cross-terms arising from a coupling of the two mechanisms [2.58]. Since (2.133 and 144) are nearly similar, it is impossible to separate experimentally the contributions of the anharmonicity from that of the second-order dipole moment.

2.4 Problems

2.4.1 Dielectric Constant of Diatomic Cubic Crystals

The equations of motion for the TO mode at $q \cong 0$ in an alkali-halide-type crystal are

$$m_1 \ddot{u}_1 = R_0(u_2 - u_1) - m_1 \gamma \dot{u}_1 + e_s^* E^*$$
$$m_2 \ddot{u}_2 = -R_0(u_2 - u_1) - m_2 \gamma \dot{u}_2 - e_s^* E^* \,, \tag{2.4.1}$$

where u_1 and u_2 are the displacements of the cation and anion, respectively, R_0 is the short-range force constant, γ is a phenomenological damping constant, e_s^* is the Szigeti charge and E^* the effective field.

a) Introducing the relative displacement $w = u_1 - u_2$ and the transverse effective charge $e_T^* = e_s^*(\varepsilon_\infty + 2)/3$, show that (2.4.1) can be written in the form

$$\mu \ddot{w} + R_0 w + \mu \gamma \dot{w} = \frac{3}{\varepsilon_\infty + 2} e_T^* E^* , \tag{2.4.2}$$

where $\mu = m_1 m_2/(m_1 + m_2)$ is the reduced mass.

b) From $E^* = E + (4\pi/3)P$ and $P = v_a^{-1}(e_s^* w + \alpha E^*)$, where P is the polarization, E the macroscopic field, $\alpha = \alpha_1 + \alpha_2$ the electronic polarizability and v_a the volume of the primitive unit cell, show that E^* and P can be written in the form

$$E^* = \frac{4\pi}{3v_a} e_T^* w + \frac{\varepsilon_\infty + 2}{3} E; \quad P = \frac{1}{v_a} e_T^* w + \frac{\varepsilon_\infty - 1}{4\pi} E. \tag{2.4.3}$$

Hint: Use the Clausius-Mosotti relation $(\varepsilon_\infty - 1)/(\varepsilon_\infty + 2) = 4\pi/3v_a$ [Ref. 2.17, Eq. (4.103)].

c) From (2.4.2 and 3) and the relation $D = \varepsilon E = E + 4\pi P$ show that

$$\varepsilon(\omega) = \varepsilon_1 + i\varepsilon_2 = \varepsilon_\infty + \frac{4\pi e_T^{*\,2}}{\mu v_a}(\omega_{TO}^2 - \omega^2 - i\gamma\omega)^{-1}$$

$$= \varepsilon_\infty \frac{\omega_{LO}^2 - \omega^2 - i\gamma\omega}{\omega_{TO}^2 - \omega^2 - i\gamma\omega}, \quad \text{where} \qquad (2.4.4)$$

$$\omega_{TO}^2 = \frac{1}{\mu}\left(R_0 - \frac{4\pi e_T^{*\,2}}{v_a(\varepsilon_\infty + 2)}\right) \quad \text{is the TO frequency.}$$

The last form in (2.4.4) has been obtained by using $4\pi e_T^{*\,2}/\mu v_a = (\varepsilon_0 - \varepsilon_\infty)\omega_{TO}^2$ and $\varepsilon_0/\varepsilon_\infty = \omega_{LO}^2/\omega_{TO}^2$.

d) The energy of light absorbed by the system per unit time (power absorption) is given by

$$W = \langle \text{Re}\{Ee_T^*\}\cdot\text{Re}\{\dot{w}\}\rangle_t = \frac{1}{2}e_T^*\text{Re}\{E\dot{w}^*\},$$

where Re means the real part and $\langle\cdots\rangle_t$ the time average. Show that W is proportional to the frequency dependent conductivity $\sigma(\omega) = (\omega/4\pi)\varepsilon_2(\omega)$. Note that $\sigma(\omega)$ is the conductivity due to the oscillating charges of the ions.

e) In a less consistent treatment it is assumed that the ionic polarization P_i and the electronic polarization P_e are independent. The corresponding relations are

$$\mu\ddot{w} + R_0 w + \mu\gamma\dot{w} = e_s^*\left(E + \frac{4\pi}{3}P_i\right),$$

$$P_i = \frac{1}{v_a}e_s^* w, \quad P_e = \frac{\alpha}{v_a}\left(E + \frac{4\pi}{3}P_e\right).$$

Using the Clausius-Mosotti relation show that

$$\varepsilon(\omega) = \varepsilon_\infty + \frac{4\pi e_s^{*\,2}}{\mu v_a}(\omega_{TO}^2 - \omega^2 - i\gamma\omega)^{-1},$$

where

$$\omega_{TO}^2 = \frac{1}{\mu}\left(R_0 - \frac{4\pi}{3v_a}e_s^{*\,2}\right).$$

f) From the expression

$$\varepsilon_2(\omega) = \frac{S\gamma\omega}{(\omega_{TO}^2 - \omega^2)^2 + \gamma^2\omega^2},$$

show that in the limit $\gamma \to 0$ one obtains

$$\varepsilon_2(\omega) = \frac{\pi}{2} \frac{S}{\omega_{TO}} \delta(\omega - \omega_{TO}).$$

Hint: Use the representation of the δ-function

$$\lim_{s \to 0} \frac{s}{x^2 + s^2} = \pi \delta(x),$$

and the general relation

$$\delta[g(\omega)] = \sum_n \frac{1}{|g'(\omega_n)|} \delta(\omega - \omega_n),$$

where $g(\omega_n) = 0$ and $g'(\omega_n) \neq 0$. Put $g(\omega) = \omega_{TO}^2 - \omega^2$ and note that $\delta(\omega + \omega_{TO})$ can be disregarded since $\omega > 0$.

2.4.2 Conductivity Sum Rule for Lattice Vibrations

Using $\sigma(\omega) = (\omega/4\pi)\varepsilon_2(\omega)$ and (2.30) for $\varepsilon_2(\omega)$ with $S_j = 4\pi\varrho_j\omega_j^2$, show that

$$\int_0^{\omega_c} \sigma(\omega) d\omega \cong \frac{\pi}{2} \sum_j \varrho_j \omega_j^2,$$

where ω_c is a frequency large compared with the lattice frequencies ω_j but small compared with the electronic transition frequencies.

Hint: Use

$$\int_0^{t_{cj}} \frac{t_j^{1/2} dt_j}{t_j^2 + (k_j^2 - 2)t_j + 1} = \frac{\pi}{k_j} - \frac{k_j}{t_{cj}^{1/2}},$$

where $t_j = \omega^2/\omega_j^2$, $t_{cj} = \omega_c^2/\omega_j^2$, and $k_j = \gamma_j/\omega_j$.

2.4.3 Second-Order Dipole Moment

Consider a cation and a nearest anion of a diatomic cubic crystal as shown in Fig. 2.29. For simplicity, we assume that only the anion is polarizable. The cation has mass m_1, displacement u_1 and charge ze, while the anion is composed of a core with mass m_2, displacement u_2, charge xe, and a shell with mass $m_s \cong 0$, displacement v, charge ye [Ref. 2.17, Sect. 4.3.1].

f and g are the harmonic and cubic anharmonic force constants, respectively, which couple the cation with the shell of the anion; k is the harmonic core-shell force constant.

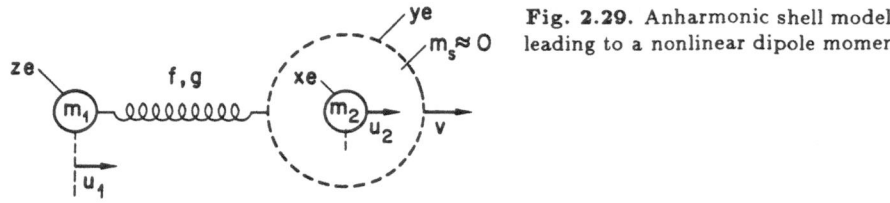

Fig. 2.29. Anharmonic shell model leading to a nonlinear dipole moment

a) Neglecting internal field effects and assuming a potential energy of the form

$$\phi = \frac{1}{2}k(u_2 - v)^2 + \frac{1}{2}f(u_1 - v)^2 + \frac{1}{3!}g(u_1 - v)^3, \qquad 1)$$

for each ion pair, show that the dipole moment induced by the $q \cong 0$ TO mode is given by

$$M(w) \cong q_1 w + q_2 w^2, \quad \text{where} \qquad 2)$$

$$q_1 = e\left(x + y\frac{k}{k+f}\right); \quad q_2 = \frac{1}{2}gye\frac{k^2}{(k+f)^3} \; ; \quad w = u_1 - u_2 \; . \qquad 3)$$

Note that from 2) and 3) it follows that for $k = \infty$ (rigid-ion model) as well as for $g = 0$ (harmonic approximation) the second-order moment vanishes.

Hint: $M = e(zu_1 + xu_2 + yv)$. From 1) calculate the force $m_s \ddot{v} = -\partial\phi/\partial v \cong 0$ and eliminate v. Use the electrical neutrality condition $x + y + z = 0$.

b) Using (2.67) and $w = (\hbar/2\mu\omega_{TO})^{1/2}(a^+ + a)$, where a^+ and a are creation and annihilation operators, respectively [Ref. 2.17, Problem 5.6.2], show that the contribution from the second-order dipole moment $M^{(2)} = q_2 w^2$ to $\varepsilon_2(\omega)$ is given by

$$\varepsilon_2^{(2)}(\omega) = \hbar\frac{2\pi^2 q_2^2}{v_a(\mu\omega_{TO})^2}(2\overline{n}_{TO} + 1)\delta(\omega - 2\omega_{TO}).$$

Here $v_a = V/N$ is the volume of the primitive unit cell.

64

3. Raman Spectroscopy

In this chapter we consider the scattering of photons with energies of the order of 2 to 4 eV by optical phonons. When the spectrum of the radiation scattered by the crystal illuminated with monochromatic light of frequency ω_L is analyzed, it is found that it consists of a very strong line at the frequency ω_L, as well as of a series of much weaker lines with frequencies $\omega_L \pm \omega_j(\boldsymbol{q})$, where $\omega_j(\boldsymbol{q})$ are optical phonon frequencies (Fig. 3.1). The strong line centered at ω_L is due to *elastic scattering* of photons and is known as *Rayleigh scattering*. The series of weak lines at $\omega_L \pm \omega_j(\boldsymbol{q})$ originates from *inelastic scattering* of photons by phonons and constitutes the *Raman spectrum*. The Raman bands at frequencies $\omega_L - \omega_j(\boldsymbol{q})$ are called *Stokes lines*, those at frequencies $\omega_L + \omega_j(\boldsymbol{q})$ are known as *anti-Stokes* lines. The intensities of the anti-Stokes lines are usually considerably weaker than those of the Stokes lines.

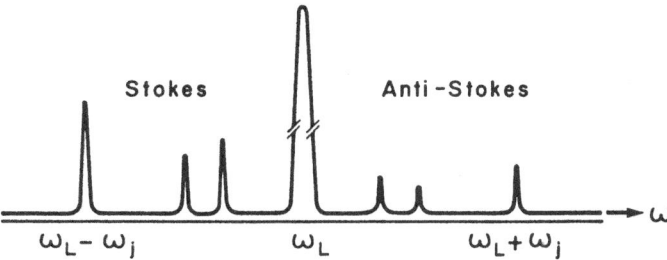

Fig. 3.1. Stokes and anti-Stokes Raman spectrum (schematic). The strong line at ω_L is due to Rayleigh scattering

In *first-order* Raman scattering only optical phonons with $\boldsymbol{q} \cong 0$ are involved. This is a consequence of momentum conservation (Sect. 3.3.1). Since the selection rules in infrared and Raman scattering are different, the two techniques provide complementary information.

The chapter starts with a discussion of the experimental techniques including Raman scattering apparatus and a discussion of some important scattering geometries. We then give a discussion of the classical theory of the Raman effect on the basis of the radiation emitted by an oscillating dipole. The intensity of the light is proportional to $|\bar{\boldsymbol{M}}|^2$, where $\boldsymbol{M} = \alpha \boldsymbol{E}$ is the dipole moment induced by the electric field $\boldsymbol{E}$ of the light, and α is the electronic polarizability tensor with components α_{xx}, α_{xy}, α_{xz}, etc. The components $\alpha_{\varrho\sigma}$ depend on the normal coordinates of the vibrating system; this is illustrated by considering the diatomic molecule [1]. Section 3.2.2 is devoted to a discussion of the properties of the polarizability tensor; its dependence on the normal coordinates of molecules and crystals is discussed in Sect. 3.2.3, 4. The quantum theory of the Raman effect is introduced by a qualitative discussion of the scattering of photons and phonons, including first- and second-order scattering (Sect. 3.3.1). In Sect. 3.3.2 the quantum mechanical calculation of the induced dipole moment and of the intensity of the scattered light on the basis of standard time-dependent perturbation theory and the correspondence principle is discussed. The actual calculation of the Raman intensities is based on Placzek's polarizability theory (Sect. 3.3.3). The chapter is terminated with a discussion of *polaritons* observed by Raman scattering and by several illustrating problems.

3.1 Experimental Techniques

3.1.1 Raman-Scattering Apparatus

The principle type of instrumentation is illustrated in Fig. 3.2. The monochromatic and polarized light of the laser (1) HeNe:$\lambda_L = 6328$ Å; Ar : $\lambda_L = 5145$ Å, 4880 Å, 4579 Å, etc.) passes through an interference filter or a small grating monochrometer (2) that rejects spurious lines and background from the laser sources. The light beam then enters the polarization rotator (3) and is focused by the lens (4) onto the sample (5). Light scatterd from the sample is focused by the lens (6) onto the entrance slit (9) of the double-grating spectrometer (10). It is advantageous if the laser beam is parallel to the entrance slit. After the lens (6) the scattered light passes through the analyzer (7) and the polarization scrambler (8). The latter effectively depolarizes the light so that unpolarized light enters the monochromator; this is necessary if intensities of differently polarized light are compared, because the transmission of the spectrometer depends on the polarization of the light. The double-grating spectrometer acts as a tunable filter of extremely high contrast; its tandem mode of operation is required to prevent the internally scattered intense Rayleigh light from overpowering the weak Raman lines. Light leaving the final exit slit (11) of the double-grating monochromator is focused onto the cathode of the photomultiplier (12), whose output is processed with "photon counting" electronics which includes the preamplifier (13) and the discriminator (14). The discriminator output pulses can then be converted to an analog signal with a count-rate meter (15) and

---

[1]   In crystals α should be replaced by the elctronic susceptibility χ_∞ but in the following we use the symbol α. $\chi_\infty = (\varepsilon_\infty - 1)/4\pi$, ε_∞ being the optical dielectric constant. A change $\Delta\alpha$ produced by a normal mode is equivalent to a change $\Delta\chi_\infty = \Delta\varepsilon_\infty$

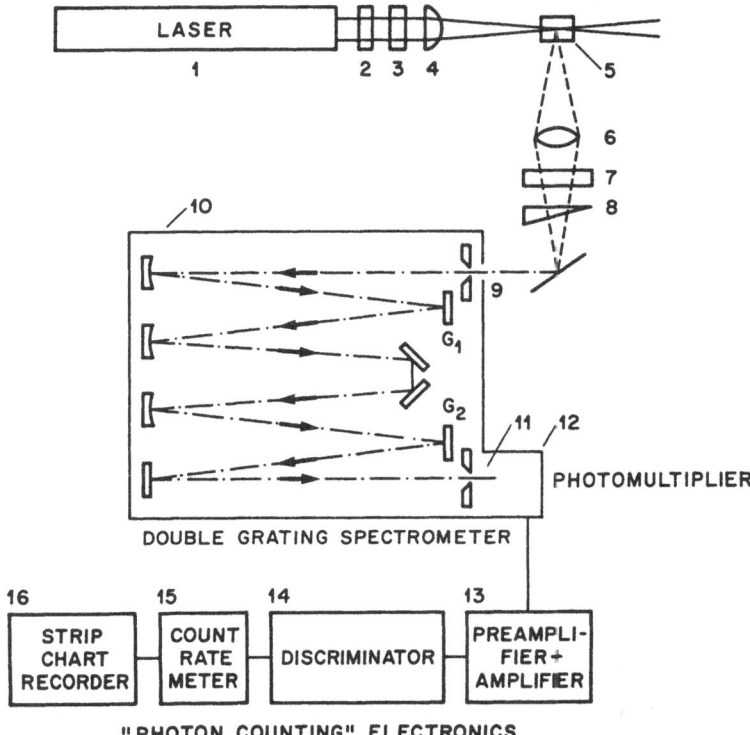

LASER

1   2 3 4   5

6

7

8

10

9

G_1

G_2

11 12

PHOTOMULTIPLIER

DOUBLE GRATING SPECTROMETER

16   15   14   13

| STRIP CHART RECORDER | COUNT RATE METER | DISCRIMINATOR | PREAMPLI-FIER + AMPLIFIER |

"PHOTON COUNTING" ELECTRONICS

Fig. 3.2. Raman scattering apparatus (*see text*)

recorded on a strip-chart recorder (16) as shown, or else stored in the memory of a multiscaler whose memory adress is swept synchronously with the spectrometer.

Instrumentation of the type shown in Fig. 3.2 has been applied to the study of a wide variety of excitations in crystals such as optical phonons, color centers, magnons, electronic transitions, plasmons, etc. [3.1–11]. For many purposes the 90°scattering geometry shown in Fig. 3.2 can not be used, for instance, for opaque samples or for the study of polaritons. We therefore discuss in the following subsection some other important scattering geometries.

3.1.2 Scattering Configurations

Figure 3.3 shows two possible backscattering configurations which are used for opaque samples. In both arrangements the incident beam is normal to the surface of the sample. In Fig. 3.3a the mirrors M_1 and M_2 direct the beam to the highly polished surface of the sample S. The cylindrical lens L_1

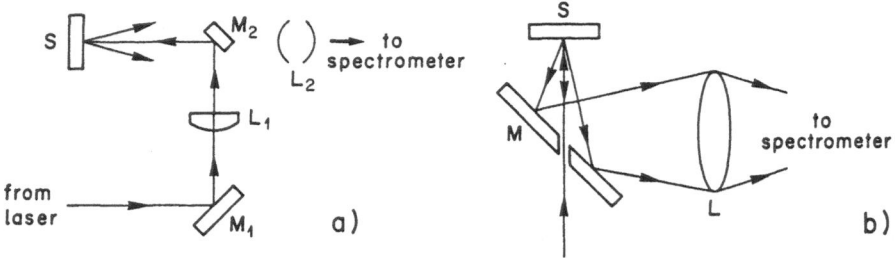

Fig. 3.3.a,b. Backscattering configurations for opaque samples (*see text*)

forms a line image of the laser light on the sample surface. The collecting lens L_2 focuses the backscattered Raman light in the usual way on the entrance slit of the monochromator. Although the small mirror M_2 stops some Raman light entering the monochromator, it also blocks the direct reflections from the crystal surface and cryostate windows. In Fig. 3.3b the incident and reflected laser beams pass through the hole in the mirror M. The radiation scattered by the sample S is reflected from the mirror M, passes through the lens L and enters the monochromator. The reflected beam from the sample passes back through the hole in mirror M.

A scattering arrangement for near-forward directions at small angles θ away from the direction of the incident laser beam is shown in Fig. 3.4 [3.10]. Such configurations are used for the observations of polaritons (Sects. 2.2.7 and 3.4.5). The direct beam passes through the crystal and is stopped by the diaphragm D, while part of the scattered light passes through the circular aperture in the diaphragm and is collected by the lens L and focused onto the entrance slit of the monochromator. More sophisticated arrangements for observing scattered light from polaritons, especially in anisotropic crystals, have been described in [3.10].

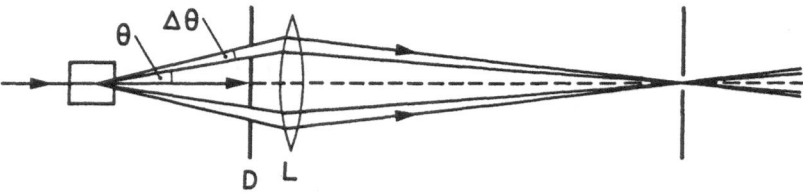

Fig. 3.4. Near forward scattering configuration used for the observation of polaritons

Figure 3.5 illustrates some possible scattering configurations. To describe a particular experiment of this type, we follow the convention of

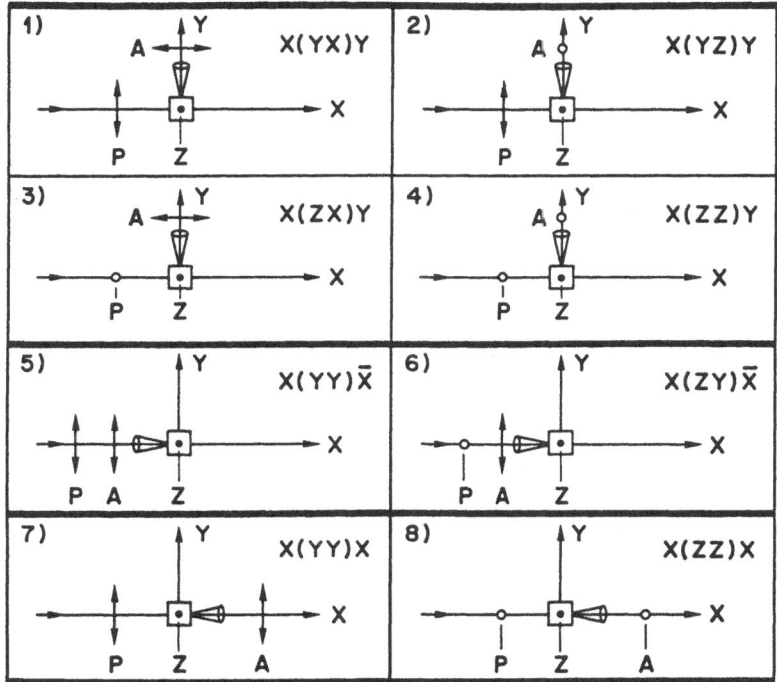

Fig. 3.5. Some possible scattering configurations (*see text*)

Damen et al. [3.12]and use for the case of right-angle scattering the simple notation $X(ZZ)Y$ (configuration 4 in Fig. 3.5) to indicate an experiment in which light propagating in the X direction and polarized in the Z direction is scattered into the Y direction with Z polarization. Similarly, $Z(XY)X$ signifies incident radiation propagating in the Z direction, X polarized, which is scattered into the X direction with Y polarization, etc. The backscattering configuration 6 in Fig. 3.5 is denoted by $X(ZY)\overline{X}$ and indicates that the incident light propagates in the positive X direction with Z polarization which is scattered in the negative X direction $(\overline{X})$ with Y polarization. By choosing different settings of polarizer and analyzer it is possible to observe the different components of the polarizability tensor α. In configuration 1 of Fig. 3.5, the component α_{YX} is observed and in configuration 4 the component α_{ZZ} , etc. This information is of considerable importance for the assignement of the observed phonons to the different symmetry species of the space group of the crystal [3.7,13]. Figure 3.6 shows the Raman spectrum (Stokes lines) of calcite, $CaCO_3$, for different scattering configurations. The numbers in paranthesis indicate the frequencies in cm^{-1} of the Raman active phonon modes [3.14].

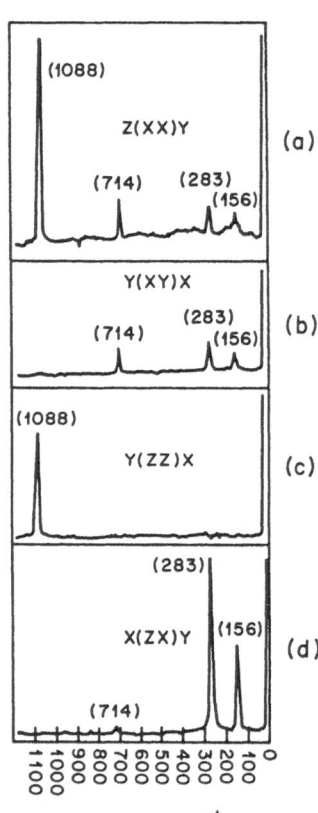

Fig. 3.6a–d. Raman spectrum of calcite [3.14]

(1088)

Z(XX)Y (a)

(714) (283)
(156)

Y(XY)X (b)
(283)
(714) (156)

(1088)

Y(ZZ)X (c)

(283)

X(ZX)Y (156) (d)

(714)

FREQUENCY (cm^{-1})

3.2 Classical Theory

3.2.1 Basic Model

The classical theory of Raman scattering is based on the idea that the electromagnetic field of the incident light induces in the system a time dependent dipole moment $M(t) = \sum e_i r_i(t)$. It is shown in textbooks of electromagnetic theory [3.15] that an accelerated charge e_i with acceleration $\ddot{r}_i(t)$ radiates electromagnetic waves with the result that the intensity of the emitted radiation is proportional to $|\ddot{M}(t)|^2$. Let $E = E_0 \cos \omega_L t$ be the electric vector of the incident light. Since ω_L is much larger than the vibrational frequencies of the atoms, only the electrons but not the atoms can respond to the light field E. For the dipole moment M induced by E we write

$$M = \alpha E + \frac{1}{2} \beta E^2 + \dots . \tag{3.1}$$

Here α is the electronic polarizability and β is a third-rank tensor, called the hyperpolarizability, which gives rise to the *Hyper-Raman effect*. Although this phenomenon has been observed after the introduction of lasers [3.16, 17], we shall not discuss it here and confine ourselves to the normal Raman effect governed by the linear term in (3.1).

In general, the direction of M does not coincide with the direction of E, that is, α is a second-rank tensor with components $\alpha_{\varrho\sigma}$ (Sect. 3.2.2). For the time being we assume, however, that M is parallel to E; this is the case for isotropic systems or if E is parallel to the direction of one of the axis of high symmetry of the system.

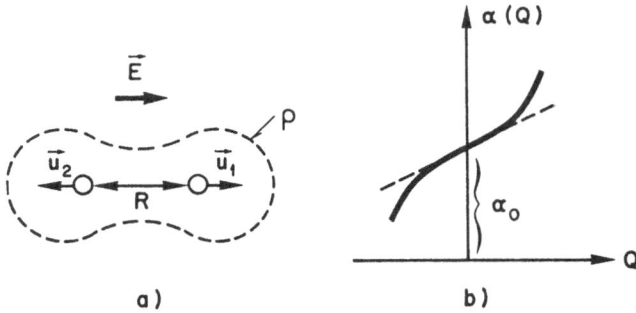

Fig. 3.7. (a) During the vibration of a diatomic molecule, the electronic charge distribution ϱ and hence the electronic polarizability $\alpha(\varrho)$ change. (b) Electronic polarizability α as a funciton of the normal coordinate Q of a diatomic molecule (schematic)

The electronic polarizability α depends on the electric charge distribution ϱ of the system: $\alpha = \alpha(\varrho)$. If the atomic configuration changes during the vibration, ϱ and hence α will also change. This is illustrated in Fig. 3.7a und b for the vibration of a diatomic molecule (Problem 3.4.1). We assume that E is parallel to the axis of the molecule. During the vibration, ϱ and α will change: α is larger than the equilibrium value α_0 in one half period and smaller in the other. For sufficiently small displacements of the nuclei from their equilibrium positions, α will change linearly with the normal coordinate $Q = \sqrt{\mu}(u_2 - u_1)$ (μ: reduced mass), as indicated in Fig. 3.7b. Expanding α in a Taylor's series we obtain

$$\alpha = \alpha_0 + \left(\frac{\partial\alpha}{\partial Q}\right)_0 Q + \frac{1}{2}\left(\frac{\partial^2\alpha}{\partial Q^2}\right)_0 Q^2 + \dots . \tag{3.2}$$

The *first-order Raman effect* is determined by the term linear in Q, the *second-order Raman effect* by the term quadratic in Q, etc. In the following we confine ourselves to first-order Raman scattering. If the molecule vibrates with the frequency ω_s we have $Q = Q_0 \cos \omega_s t$ and obtain

$$\alpha(t) = \alpha_0 + \left(\frac{\partial \alpha}{\partial Q}\right)_0 Q_0 \cos \omega_s t. \tag{3.3}$$

Substituting (3.3) into (3.1) gives

$$M(t) = \alpha_0 E_0 \cos \omega_L t + \left(\frac{\partial \alpha}{\partial Q}\right)_0 Q_0 E_0 \cos \omega_L t \cos \omega_s t.$$

From this, using well-known trigonometric formulae, we obtain

$$M(t) = a \cos \omega_L t + b[\cos (\omega_L - \omega_s)t + \cos (\omega_L + \omega_s)t], \tag{3.4}$$

where

$$a = \alpha_0 E_0, \quad b = \frac{1}{2}\left(\frac{\partial \alpha}{\partial Q}\right)_0 Q_0 E_0. \tag{3.5}$$

Equation (3.4) shows that the induced dipole moment M vibrates not only with the frequency ω_L of the incident light, but also with the frequencies $\omega_L \pm \omega_s$. These latter frequencies arise from the modulation of the electronic polarizability α by the vibration of the atoms.

The classical radiation theory of an oscillating dipole is based on the description of the electromagnetic field produced by an accelerated charge employing Maxwell's equation. The intensity of radiation emitted by the dipole moment $M(t)$ into the solid angle $d\Omega = \sin \vartheta \, d\vartheta \, d\phi$ (Fig. 3.8) is given by [3.15]

$$dI(t) = \frac{d\Omega}{4\pi c^3} \sin^2 \vartheta |\ddot{M}(t)|^2, \tag{3.6a}$$

and per unit solid angle $d\Omega$ by

$$I(t) = \frac{1}{4\pi c^3} \sin^2 \vartheta |\ddot{M}(t)|^2 = A|\ddot{M}(t)|^2. \tag{3.6b}$$

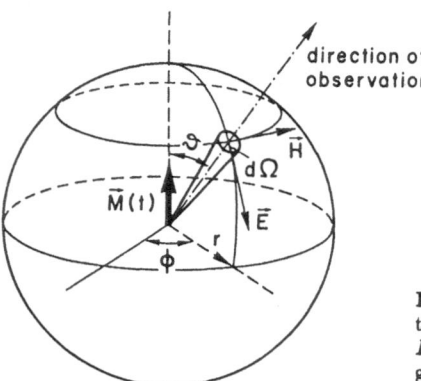

Fig. 3.8. Polarization of the radiation field emitted by an oscillating electric dipole $M(t)$. E and H are the field vectors of the radiation propagating in the direction of observation

From (3.6) it follows that $M(t)$ does not emit radiation parallel to his axis ($\vartheta = 0$). Integrating (3.6a) over ϑ and ϕ gives the intensity emitted into the solid angle $\Omega = 4\pi$,

$$\hat{I}(t) = \frac{2}{3c^3}|\ddot{M}(t)|^2. \tag{3.7}$$

By use of (3.4–6), the intensity of the scattered light per unit solid angle is now given by

$$I(t) = AE_0^2[k_0^2\cos^2 \omega_L t + k_1^2\cos^2 (\omega_L - \omega_s)t$$
$$+ k_2^2\cos^2 (\omega_L + \omega_s)t] + \text{cross terms}, \tag{3.8}$$

where

$$k_0^2 = \alpha_0^2 \omega_L^4, \tag{3.9}$$

$$k_1^2 = \frac{1}{4}\left(\frac{\partial\alpha}{\partial Q}\right)_0^2 Q_0^2(\omega_L - \omega_s)^4, \tag{3.10}$$

$$k_2^2 = \frac{1}{4}\left(\frac{\partial\alpha}{\partial Q}\right)_0^2 Q_0^2(\omega_L + \omega_s)^4. \tag{3.11}$$

The cross terms in (3.8) can be neglected since the power they irradiate averages to zero on time intervals of sufficient length. This can easily be verified by calculating the time average

$$I = \lim_{\tau\to\infty} \frac{1}{\tau}\int_0^\tau I(t)dt$$

$$= \tfrac{1}{2}AE_0^2(k_0^2 + k_1^2 + k_2^2). \tag{3.12}$$

From (3.8) we expect that the scattered light will have peaks at the frequencies ω_L and $\omega_L\pm\omega_s$. This can directly be verified by calculating the frequency dependence of the scattered light, the power spectrum, which is obtained by forming the square of the Fourier transform of $M(t)$. Using (3.6b) the power spectrum is defined by [3.18].

$$P(\omega) = A \lim_{\tau\to\infty} \frac{2}{\tau}\left|\int_{-\tau/2}^{\tau/2} \ddot{M}(t)e^{-i\omega t}dt\right|^2. \tag{3.13}$$

The calculation of (3.13) is outlined in Appendix E with the result

$$P(\omega) = \pi AE_0^2\{k_0\delta(\omega - \omega_L) + k_1^2\delta[\omega - (\omega_L - \omega_s)]$$
$$+ k_2^2\delta[\omega - (\omega_L + \omega_s)]\}. \tag{3.14}$$

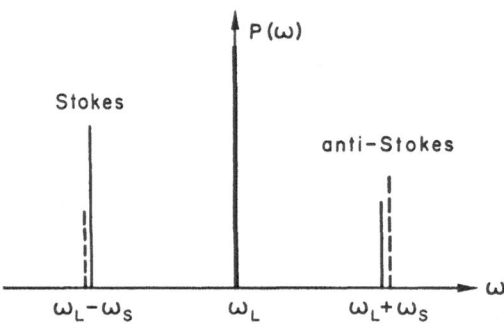

From (3.12, 14) we obtain at once that

$$I = \frac{1}{2\pi} \int\limits_0^\infty P(\omega)d\omega. \tag{3.15}$$

The power spectrum given by (3.14) is illustrated in Fig. 3.9. The first term in (3.14) is the power scattered per unit solid angle at the frequency ω_L of the incident radiation and is known as *elastic scattering* or *Rayleigh scattering*. The second and the third terms represent the *inelastic* or *Raman scattering* consisting of the radiation scattered at the Stokes frequency $\omega_L - \omega_s$ and at the anti-Stokes frequency $\omega_L + \omega_s$, respectively.

The classical theory thus correctly predicts the occurance of the Stokes and anti-Stokes lines, but leads to an incorrect ratio of their intensities. From (3.10–12) it follows that the ratio of the intensities of the Stokes and anti-Stokes lines should be

$$\frac{I_{\text{Stokes}}}{I_{\text{anti-Stokes}}} = \frac{(\omega_L - \omega_s)^4}{(\omega_L + \omega_s)^4}, \tag{3.16}$$

which will certainly be less than unity, whereas experimentally it is found that the Stokes lines are more intense than the anti-Stokes lines. This inconsistency is eliminated in the quantum theory of the Raman effect (Sect. 3.3.4).

3.2.2 The Polarizability Tensor

We have already mentioned in Sect. 3.2.1 that, in general, the direction of the induced dipole moment M does not coincide with the direction of the electric field E. The relation $M = \alpha E$ is a vectorial relation and the polarizability is thus a second-order tensor. Neglecting the quadratic term in (3.1) this equation written in full has the form

$$M_x = \alpha_{xx}E_x + \alpha_{xy}E_y + \alpha_{xz}E_z$$
$$M_y = \alpha_{yx}E_x + \alpha_{yy}E_y + \alpha_{yz}E_z$$
$$M_z = \alpha_{zx}E_x + \alpha_{zy}E_y + \alpha_{zz}E_z, \tag{3.17a}$$

or

$$M_\varrho = \sum_\sigma \alpha_{\varrho\sigma}E_\sigma. \tag{3.17b}$$

Here the components of E and M are referred to the system (x, y, z) which we assume to coincide with the laboratory system (X, Y, Z). In matrix notation we may write

$$\begin{pmatrix} M_x \\ M_y \\ M_z \end{pmatrix} = \begin{pmatrix} \alpha_{xx} & \alpha_{xy} & \alpha_{xz} \\ \alpha_{yx} & \alpha_{yy} & \alpha_{yz} \\ \alpha_{zx} & \alpha_{zy} & \alpha_{zz} \end{pmatrix} \begin{pmatrix} E_x \\ E_y \\ E_z \end{pmatrix}. \tag{3.18}$$

By considering the energy of the polarized system it may be shown that α is a symetrical tensor [3.19], that is

$$\alpha^T = \alpha, \quad \text{or}$$
$$\alpha_{\varrho\sigma} = \alpha_{\sigma\varrho}. \tag{3.19}$$

It can further be shown that there exists always a coordinate system with axes (x', y', z') such that the relation between M and E, when referred to these axes, assumes the simple form [3.19]

$$\begin{pmatrix} M'_{x'} \\ M'_{y'} \\ M'_{z'} \end{pmatrix} = \begin{pmatrix} \alpha'_{x'x'} & 0 & 0 \\ 0 & \alpha'_{y'y'} & 0 \\ 0 & 0 & \alpha'_{z'z'} \end{pmatrix} \begin{pmatrix} E'_{x'} \\ E'_{y'} \\ E'_{z'} \end{pmatrix} \tag{3.20}$$

or

$$M' = \alpha' E', \tag{3.21}$$

where α' is a diagonal matrix. Such axes are called *principle axis of polarizability* (Fig. 3.10). It is easy to find the principal axes of polarizability for a symmetrical system (molecule or crystal), since they must coincide with the symmetry axes present and be perpendicular to any plane of symmetry.

Let R be the transformation matrix from the system (x, y, z) to the principal axes system (x', y', z'). Then

$$M' = RM, \quad E' = RE, \tag{3.22}$$

where E' and M' are the vectors in the principal axes system. R is an orthogonal matrix, that is $R^{-1} = R^T$ and its elements are the direction cosines of the system (x', y', z') with respect to the system (x, y, z). From (3.1, 21, 22) we obtain

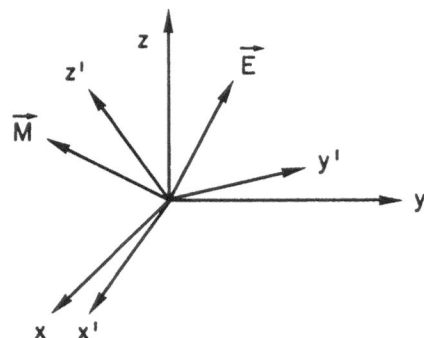

Fig. 3.10. Coordinate system (x, y, z) (identical with the laboratory system (X, Y, Z)) and principal axes system of polarizability, (x', y', z'). $\boldsymbol{M}$ is the dipole moment induced by the electric field $\boldsymbol{E}$ of the light

$$\boldsymbol{M'} = R\boldsymbol{M} = R\alpha\boldsymbol{E} = \alpha'\boldsymbol{E'} = \alpha'R\boldsymbol{E}, \quad \text{or}$$
$$\alpha = R^T\alpha'R. \tag{3.23}$$

It is convenient to define the system (x, y, z) (or (X, Y, Z)) in such a way that it coincides with the *principal axes of the equilibrium configuration* of the molecule or crystal. For the equilibrium configuration the components of α are

$$\alpha_{\varrho\sigma} = \delta_{\varrho\sigma}\alpha_{\varrho\sigma}^{(0)}, \quad \text{that is}$$

$$\alpha = \begin{pmatrix} \alpha_{xx}^{(0)} & 0 & 0 \\ 0 & \alpha_{yy}^{(0)} & 0 \\ 0 & 0 & \alpha_{zz}^{(0)} \end{pmatrix}. \tag{3.24}$$

If as a result of thermal fluctuation the system is in a distorted configuration there will be a new principal-axes system, the system (x', y', z') shown in Fig. 3.10, which in general will not coincide with (x, y, z). The polarizability will again be diagonal if referred to (x', y', z') but not necessarily if referred to (x, y, z). In the latter case we may expand each component $\alpha_{\varrho\sigma}$ in terms of the normal coordinates Q_s as in (3.2) and obtain

$$\alpha_{\varrho\sigma} = \alpha_{\varrho\sigma}^{(0)} + \sum_s \alpha_{\varrho\sigma,s}Q_s + \frac{1}{2}\sum_{s's''}\alpha_{\varrho\sigma,s's''}Q_{s'}Q_{s''} + \ldots \,. \tag{3.25}$$

For a given normal coordinate Q_s we may define changes in the polarizability components

$$\Delta\alpha_{\varrho\sigma,s} = \alpha_{\varrho\sigma,s}Q_s = \left(\frac{\partial\alpha_{\varrho\sigma}}{\partial Q_s}\right)_0 Q_s \tag{3.26}$$

and a matrix with elements

$$\alpha_{\varrho\sigma,s} = \left(\frac{\partial\alpha_{\varrho\sigma}}{\partial Q_s}\right)_0, \quad \text{namely} \tag{3.27}$$

$$\delta\alpha^{(s)} = \begin{pmatrix} \alpha_{xx,s} & \alpha_{xy,s} & \alpha_{xz,s} \\ \alpha_{yx,s} & \alpha_{yy,s} & \alpha_{yz,s} \\ \alpha_{zx,s} & \alpha_{zy,s} & \alpha_{zz,s} \end{pmatrix}. \tag{3.28}$$

Substitution of (3.25) into (3.17b) and using

$$E_\sigma = E_{0\sigma}\cos\omega_L t, \tag{3.29a}$$

$$Q_s = Q_0\cos\omega_s t, \quad \text{we obtain} \tag{3.29b}$$

$$M_\varrho(t) = a_\varrho\cos\omega_L t \\ + \sum_s b_{s\varrho}[\cos(\omega_L - \omega_s)t + \cos(\omega_L + \omega_s)t], \tag{3.30}$$

where

$$a_\varrho = \sum_\sigma \alpha_{\varrho\sigma}^{(0)} E_{0\sigma} \tag{3.31a}$$

$$b_{s\varrho} = \tfrac{1}{2}Q_{s0}\sum_\sigma \alpha_{\varrho\sigma,s}E_{0\sigma}. \tag{3.31b}$$

The expression (3.30) is a generalization of the corresponding expression (3.4) for the diatomic molecule. The first term in (3.30) is responsible for Rayleigh scattering while the other terms give rise to Raman scattering. According to Sect. 3.2.1 we therefore expect Raman lines at the frequencies $\omega_L\pm\omega_s$. From (3.31b) it is, however, evident that the normal mode Q_s will appear in the Raman spectrum only if at least one of the six components $\alpha_{\varrho\sigma,s}$ of the matrix $\delta\alpha^{(s)}$ is different from zero. If this is the case, the normal mode Q_s is Raman active. Whether or not a normal mode is Raman active depends on the symmetry of the equilibrium configuration and of the symmetry of the normal modes Q_s. Examples of Raman active and Raman inactive modes are discussed in the next subsection.

Finally we consider the radiation emitted per unit solid angle in the direction X; in (3.6) ϑ is then the angle between M and the x-axis, and $\sin^2\vartheta|\ddot{M}(t)|^2 = \ddot{M}_y^2(t) + \ddot{M}_z^2(t)$, thus from (3.6b) we obtain

$$I(t) = \frac{1}{4\pi c^3}[\ddot{M}_y^2(t) + \ddot{M}_z^2(t)]. \tag{3.32}$$

Substituting (3.30) into (3.32), forming the time average according to (3.12) and observing that all cross terms vanish one obtains

$$I = \frac{1}{8\pi c^3}\left\{A_x\omega_L^4 + \sum_s B_{sx}[(\omega_L - \omega_s)^4 + (\omega_L + \omega_s)^4]\right\}, \tag{3.33}$$

where

$$A_x = a_y^2 + a_z^2, \quad B_{sx} = b_{sy}^2 + b_{sz}^2$$

with a_ϱ and $b_{s\varrho}$ given by (3.31a, b). Using (3.26, 31) and assuming that the incident electric field $\boldsymbol{E}$ of the light is polarized in the direction z, we obtain for the intensity of the Stokes line of the normal mode s

$$I_s(z) = \frac{1}{4c^4}(\omega_L - \omega_s)^4 (\Delta\alpha_{yz,s0}^2 + \Delta\alpha_{zz,s0}^2) I_0, \qquad (3.34)$$

where $\Delta\alpha_{\varrho\sigma,s0} = \alpha_{\varrho\sigma,s} Q_{s0}$ is the amplitude of the change of $\alpha_{\varrho\sigma}$ due to the mode s and

$$I_0 = \frac{c}{8\pi} E_{0z}^2 \qquad (3.35)$$

is the intensity of the incident light. If the scattered light is observed with the analyzer in the direction y, only $\ddot{M}_y(t)$ is observed with the result

$$I_s(zy) = \frac{1}{4c^4}(\omega_L - \omega_s)^4 \Delta\alpha_{yz,s0}^2 I_0. \qquad (3.36a)$$

If the polarizer is in the direction ϱ and the analyzer in the direction σ we obtain with $\alpha_{\varrho\sigma} = \alpha_{\sigma\varrho}$

$$I_s(\varrho\sigma) = \frac{1}{4c^4}(\omega_L - \omega_s)^4 \Delta\alpha_{\varrho\sigma,s0}^2 I_0, \qquad (3.36b)$$

where

$$\Delta\alpha_{\varrho\sigma,s0} = \alpha_{\varrho\sigma,s} Q_{s0}. \qquad (3.37)$$

3.2.3 Raman Active and Raman Inactive Modes: Simple Molecules

If the structure and symmetry of the vibrating system is known, it is possible to predict the number of Raman active modes for each symmetry species of the symmetry group under consideration. This is possible by applying group theoretical methods which are described elsewhere [3.7, 13]. We shall not treat group theory here but illustrate Raman active and inactive modes of simple molecules and crystals by using elementary symmetry arguments [3.20].

We have seen in Chap. 2 that a mode is infrared active only if it induces a change in the dipole moment. In Sect. 3.2.2 we came to the conclusion that a mode will be Raman active only if it changes at least one of the six independent components of the polarizability tensor $\alpha_{\varrho\sigma}$.

Consider first the diatomic molecule shown in Fig. 3.11. There is just one normal mode $Q_1 = \sqrt{\mu}(u_1 - u_2)$ (μ: reduced mass), which obviously does not change the symmetry of the molecule, and hence the principal axes (x, y, z) of the equilibrium configuration remain principal axes during the whole of the vibration and coincide with the principle axes (x', y', z') of

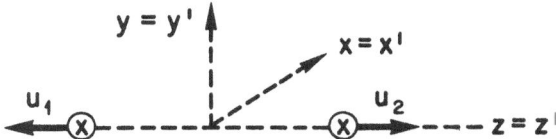

Fig. 3.11. In a diatomic molecule the principle axes system (x, y, z) of the equilibrium configuration remains principle axes system during the whole of the vibration

a displaced configuration. For an arbitrary configuration of the electric field E the components of the induced dipole moment M are therefore given by (3.20), namely $M_\varrho = \alpha_{\varrho\varrho} E_\varrho$, all components $\alpha_{\varrho\sigma}$ being zero for $\varrho \neq \sigma$. The matrix $\delta\alpha^{(1)}$ defined by (3.28) is accordingly given by

$$\delta\alpha^{(1)} = \begin{pmatrix} \alpha_{xx,1} & 0 & 0 \\ 0 & \alpha_{yy,1} & 0 \\ 0 & 0 & \alpha_{zz,1} \end{pmatrix} = \begin{pmatrix} a & 0 & 0 \\ 0 & a & 0 \\ 0 & 0 & c \end{pmatrix}.$$

Due to symmetry $\alpha_{xx,1} = \alpha_{yy,1} \neq \alpha_{zz,1}$. According to Sect. 3.2.1 and Fig. 3.7b this mode is certainly Raman active since the charge distribution and hence α will be different for the oppositely displaced positions. In Fig. 3.12 we reproduce the general behaviour of one of the components $\alpha_{\varrho\varrho}(Q_1)$ (Curve I).

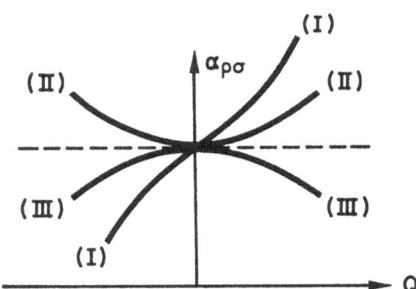

Fig. 3.12. Polarizability component $\alpha_{\varrho\sigma}$ as a function of the normal coordinates Q_s (schematic). Curve *(I)*: Raman active modes; curves *(II)* and *(III)*: Raman inactive modes [3.20]

Next we consider the normal modes of the XY_2 molecule, such as H_2O (Fig. 3.13) [3.20]. The directions of the principal axes (x, y, z) of the equilibrium configuration are fixed by the symmetry elements (two-fold axes and mirror planes) of the molecule. Since the vibrations Q_1 and Q_2 do not disturb the symmetry of the equilibrium configuration, the axes (x, y, z) remain principle axes in the displaced positions, that is, $\alpha_{\varrho\sigma,s} = 0$ for $\sigma \neq \varrho$ and $s = 1, 2$. However, $\alpha_{\varrho\varrho,s} \neq 0$ for $s = 1, 2$ since the charge distribution and hence $\alpha_{\varrho\varrho}$ are different in opposite phases of the vibrations (curve I in Fig. 3.12). The matrixes $\delta\alpha^{(s)}$ for $s = 1, 2$ have therefore the following form

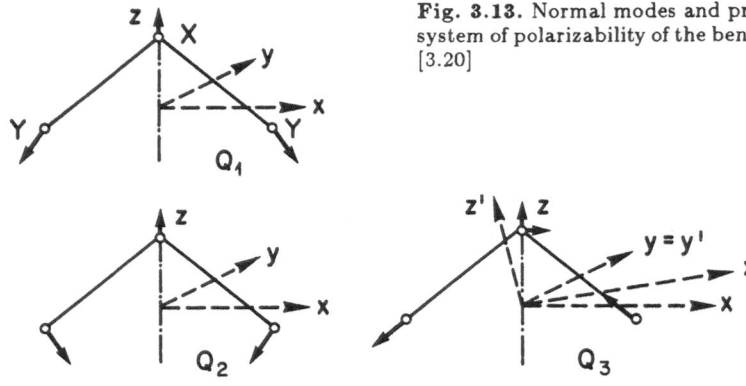

$$\delta\alpha^{(s)} = \begin{pmatrix} \alpha_{xx,s} & 0 & 0 \\ 0 & \alpha_{yy,s} & 0 \\ 0 & 0 & \alpha_{zz,s} \end{pmatrix} = \begin{pmatrix} a & 0 & 0 \\ 0 & b & 0 \\ 0 & 0 & c \end{pmatrix}.$$

For the vibration Q_3, however, we have $\alpha_{\varrho\varrho,3} = 0$, since the configurations of phase and anti-phase have the same bond lengths and bond angle and therefore the same charge distribution; $\alpha_{\varrho\varrho}$ has therefore the same value in opposite phases (Curve II or III in Fig. 3.12). On the other hand, the x and the z axes (in the plane of the molecule) no longer remain principle axes of α during the whole of the vibration Q_3; in a displaced configuration of Q_3 the principle axes are rather $x', y' = y$ and z'. Therefore, $\alpha_{xz,3} \neq 0$ and the normal mode Q_3 is Raman active even though only one of the components $\alpha_{\varrho\sigma,3}$ is different from zero; the corresponding matrix $\delta\alpha^{(3)}$ has the form

$$\delta\alpha^{(3)} = \begin{pmatrix} 0 & 0 & \alpha_{xz,3} \\ 0 & 0 & 0 \\ \alpha_{zx,3} & 0 & 0 \end{pmatrix} = \begin{pmatrix} 0 & 0 & d \\ 0 & 0 & 0 \\ d & 0 & 0 \end{pmatrix}.$$

While in the bent XY_2 molecule all three normal modes are Raman active, this is not the case for the normal modes of the linear XY_2 molecule, such as CO_2 as illustrated in Fig. 3.14 [3.20]. From Fig. 3.14 it is obvious that for all three vibrations the principle axes (x', y', z') for the displaced positions are identical with the principle axes (x, y, z) of the equilibrium configuration, that is, $\alpha_{\varrho\sigma,s} = 0$ for $\varrho \neq \sigma$ and $s = 1, 2, 3$. From the above considerations it is also clear that Q_1 is Raman acitve and that $\delta\alpha^{(1)}$ has the same form as for the diatomic molecule. For Q_2 and Q_3 we have in addition $\alpha_{\varrho\varrho,s} = 0$, since opposite phases of vibration have the same polarizability (Curve II and III in Fig. 3.12). Thus the modes Q_2 and Q_3 are not Raman active and

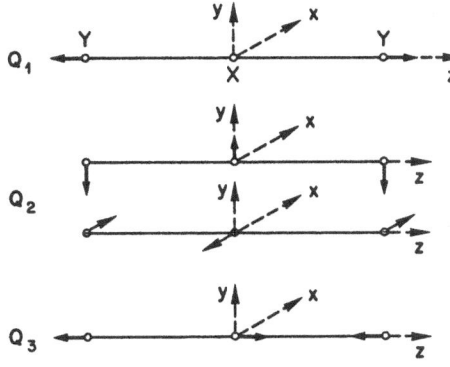

$$\delta\alpha^{(2,3)} = \begin{pmatrix} 0 & 0 & 0 \\ 0 & 0 & 0 \\ 0 & 0 & 0 \end{pmatrix}.$$

Figure 3.14 shows that in the equilibrium configuration the central atom X is located at a center of inversion i. If we apply the symmetry operation i to the pattern of atomic displacements of the mode Q_1 we obtain exactly the same mode, that is

$$iQ_1 = +Q_1.$$

On the other hand, if we apply the inversion operation on Q_2 or Q_3 all displacement arrows have opposite directions, that is

$$iQ_2 = -Q_2, \quad iQ_3 = -Q_3.$$

The *parity of the mode* Q_1 is therefore different from the parity of the modes Q_2 and Q_3. Modes for which $iQ_s = +Q_s$ have even parity and are called "g modes" (German: gerade), while modes for which $iQ_{s'} = -Q_{s'}$ have odd parity and are called "u modes" (German: ungerade). In general, some of the g modes are Raman active, while u modes are Raman inactive but often infrared active. The distinction between g and u modes can only be made if the molecule possesses a center of inversion.

3.2.4 Raman Active and Raman Inactive Modes: Simple Crystals

Figures 3.15a and b shows the NaCl and the CsCl structure, respectively. In these structures each ion is located at a center of symmetry. In first-order Raman scattering only optical modes with $q \cong 0$ are involved (Sect. 3.3.1). One of the tryply degenerate $q = 0$ TO modes is shown by the arrows. It is obvious that the application of the inversion operation i to these normal

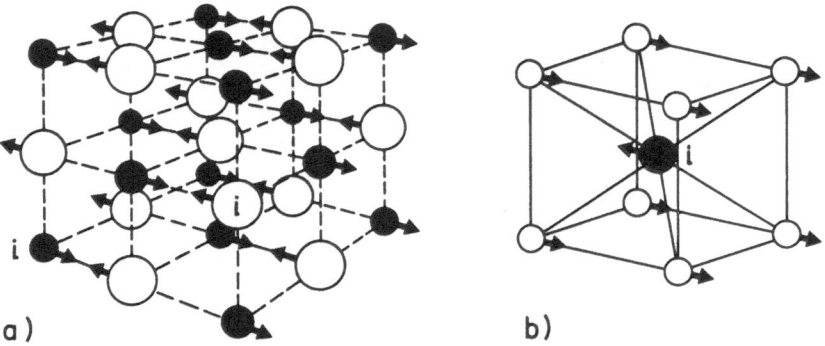

Fig. 3.15. In crystals with NaCl-structure (a), and CsCl-structure (b) each ion is located at a center of inversion i, and the TO modes at $q = 0$ (indicated by the arrows) are not Raman active

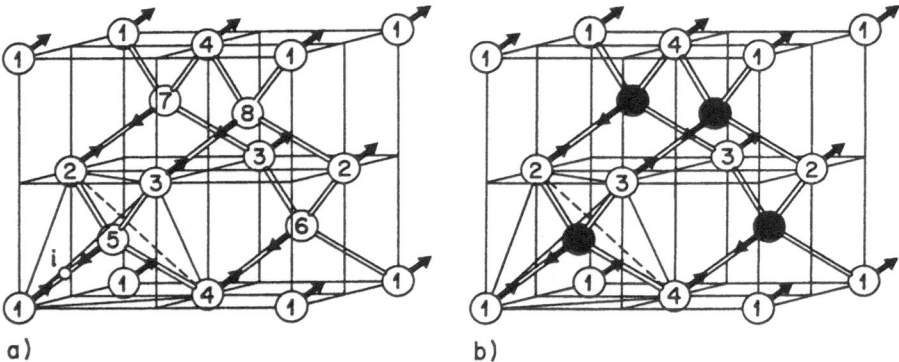

Fig. 3.16. In crystals with the diamond structure (a), and with the zinc blende structure (b), the TO modes at $q = 0$ (indicated by the arrows) are Raman active

modes Q yields the opposite pattern of ionic displacements: $iQ = -Q$. We are therefore dealing with odd-parity or u modes similar to the Q_2 and Q_3 modes of the linear XY_2 molecule (Fig. 3.14). Hence, crystals with NaCl or CsCl structure do not exhibit a first-order Raman effect.

Next we consider the diamond structure shown in Fig. 3.16a. Carbon, silicon, germanium and gray tin crystillize in the diamond structure. The diamond structure is composed of two fcc lattices displaced from each other by one-quarter of a body diagonal. The structure possesses a center of inversion symmetry i, which however, is *not* located at the atoms, but at the mid-point of each line connecting nearest-neighbour atoms. The triply degenerate $q = 0$ TO mode is shown by the arrows; it can be pictured as a relative motion of the two interpenetrating fcc lattices against each other.

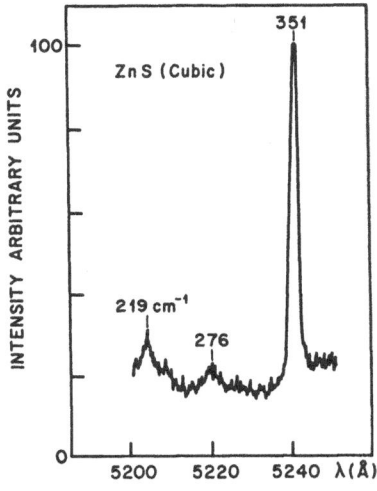

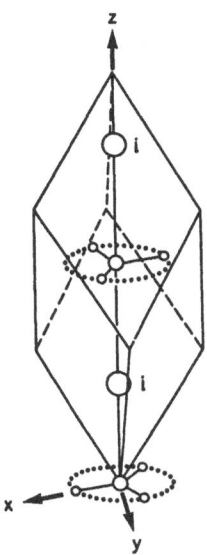

Fig. 3.17. Raman spectrum of cubic zinc blende with 5145 Å excitation at room temperature [3.22a] (*see text*)

Fig. 3.18. Unit cell of calcite, $CaCO_3$ [3.22b]

Application of the inversion operation i on Q clearly gives the identical mode: $iQ = Q$. The TO mode is therefore a g mode and resembles the normal oscillation of a homonuclear diatomic molecule (Fig. 3.11) and as such is Raman active but not infrared active.

The cubic ZnS or zinc blende structure results from the diamond structure when Zn atoms are placed on one fcc lattice and S atoms on the other fcc lattice, as in Fig. 3.16b. The ZnS structure does not have inversion symmetry and therefore the distinction between g and u modes is not possible. The arrows in Fig. 3.16b indicate the optical mode at $q = 0$; it resembles the normal oscillation of a heteronucleus diatomic molecule (Figs. 3.11, 12) and as such is expected to be Raman active. Since ZnS is a polar crystal, the macroscopic electric field associated with the vibration removes the degeneracy of the optical mode yielding a doubly degenerate TO mode and a LO mode [Ref. 3.21, Sect. 4.2]. In ZnS $\omega_{TO} = 276\,\mathrm{cm}^{-1}$ and $\omega_{LO} = 351\,\mathrm{cm}^{-1}$. Both, the TO and LO mode are Raman active, but the intensity of the TO mode is much weaker than that of the LO mode (Fig. 3.17) [3.22a]. The TO mode is also infrared active. The line at $219\,\mathrm{cm}^{-1}$ observed in Fig. 3.17 is probably due to a two-phonon process.

Finally we consider some of the $q = 0$ normal modes of the calcite structure, $CaCO_3$. The unit cell shown in Fig. 3.18 is an elongated rhombohedron and contains two formula units, that is, two calcium ions at (1/4, 1/4, 1/4), (3/4, 3/4, 3/4) and two planar carbonate ions at (0, 0, 0) and (1/2, 1/2, 1/2) [3.22b].

The calcium ions are located at a center of symmetry i, but not the carbonate ions. The latter are well defined complexes with relatively strong intramolecular bonds, and it is therefore possible to distinguish between internal and external modes [Ref. 3.21, Sect. 4.6]. In the internal modes only the atoms within the carbonate ions are displaced, while external modes are motions in which the carbonate ions (considered as rigid units) vibrate against the calcium ions. Figure 3.19a shows four internal modes and Fig. 3.19b four external modes [3.22b]. The $q = 0$ modes of calcite, having point symmetry D_{3d}, can be classified according to the irreducible representations of D_{3d}, namely, A_{1g}, A_{2g}, E_g and A_{1u}, A_{2u} and E_u [3.14, 20, 22b]. The totally symmetric A_{1g} mode in Fig. 3.19a is observed as a strong Raman line at $1088\,\mathrm{cm}^{-1}$ (Fig. 3.6); the corresponding u mode of species A_{1u} is Raman and infrared inactive. The symmetry coordinate of species E_g gives rise to a Raman line at $714\,\mathrm{cm}^{-1}$ (Fig. 3.6), while the corresponding E_u mode is Raman inactive but infrared active and observed at $712\,\mathrm{cm}^{-1}$. From the four external symmetry modes shown in Fig. 3.19b the two E_g modes are observed at 156 and $283\,\mathrm{cm}^{-1}$ (Fig. 3.6). The E_u mode corrsponding to the E_g mode at $156\,\mathrm{cm}^1$ is a pure translational mode with zero frequency, while

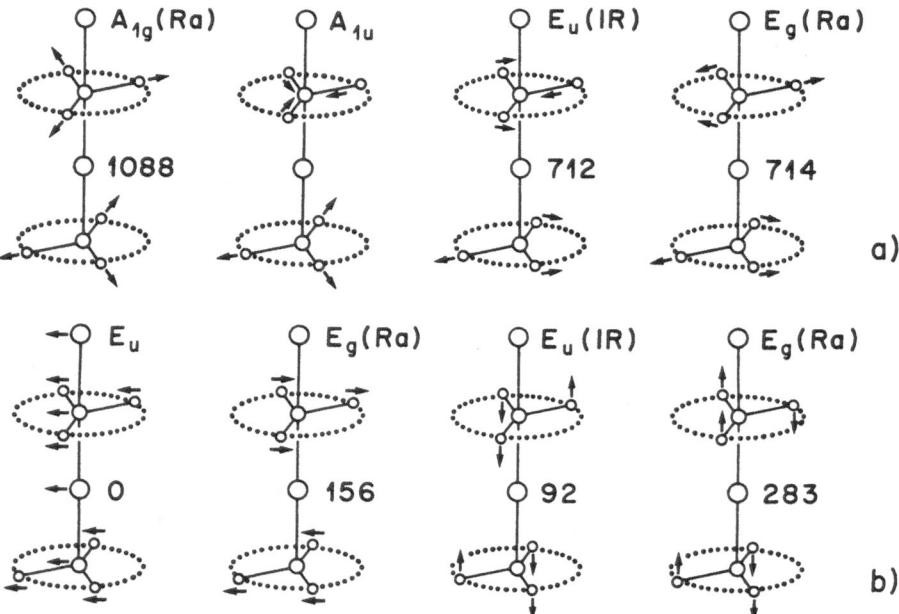

Fig. 3.19. (a) Four internal $q = 0$ symmetry modes of CaCO$_3$: the A_{1g} and E_g modes are Raman active, the E_u mode is infrared active. (b) Four external $q = 0$ symmetry modes of CaCO$_3$; the E_g modes are Raman active. The first E_u mode (with $\tilde{\nu} = 0$) corresponds to a pure translation [3.22b]

the E_u mode corresponding to the E_g mode at $283\,\mathrm{cm}^1$ is a low-frequency infrared active mode at $92\,\mathrm{cm}^{-1}$. Note that the g modes shown in Fig. 3.19 are symmetric with respect to the inversion operation i at the calcium ions, while the u modes are antsymmetric with respect to i. It should also be mentioned that the symmetry modes of species E_g and E_u shown in Fig. 3.19 are only approximate normal modes because they couple with other symmetry modes of the same species, not all of them are shown in the figures.

3.3 Quantum Theory

3.3.1 Qualitative Discussion

The complete quantum theory of light scattering is rather complex and will be outlined in Sects. 3.3.2–4. In this subsection we give an introductionary qualitative discussion in terms of discrete energy levels and of photons.

In terms of the corpuscular theory of light Rayleigh scattering corresponds to an *elastic collision* process between the photon and the crystal, whereas Raman scattering corresponds to an *inelastic collision* in which the photon either loses one or more quanta of vibrational energy (Stokes lines) or acquires one or more such quanta (anti-Stokes lines). In *first-order scattering* only one phonon is involved; this corresponds to the terms linear in the normal coordinates Q_s in (3.25). In second-order scattering two phonons are involved, corresponding to the terms proportional to $Q_{s'}Q_{s''}$ in (3.25), or to anharmonic coupling of a phonon which is active in first-order Raman or Brillouin scattering with two other phonons.

a) First-Order Scattering

Figure 3.20 shows the transitions for Rayleigh scattering and for first-order Stokes and anti-Stokes scattering. Let (ω_L, k_L) be the incident photon of the Laser with frequency ω_L and wavevector k_L, (ω_{sc}, k_{sc}) the scattered photon, and (ω_j, q) the optical phonon $s = (q, j)$ involved in the scattering process. Energy and momentum are conserved between initial and final states of the system. For Rayleigh scattering we have

$$\omega_L = \omega_{sc}, \qquad\qquad\qquad (3.38)$$

$$k_L = k_{sc}. \qquad\qquad\qquad (3.39)$$

For Raman scattering the conservation of energy and momentum are

$$\omega_L = \omega_{sc} \pm \omega_j(q), \qquad\qquad\qquad (3.40)$$

$$k_L = k_{sc} \pm q. \qquad\qquad\qquad (3.41)$$

In the Stokes process a phonon $\omega_j(q)$ is created ($+$sign) (Fig. 3.20b), while in the anti-Stokes process the phonon (ω_j, q) is annihilated ($-$sign) (Fig. 3.20c).

85

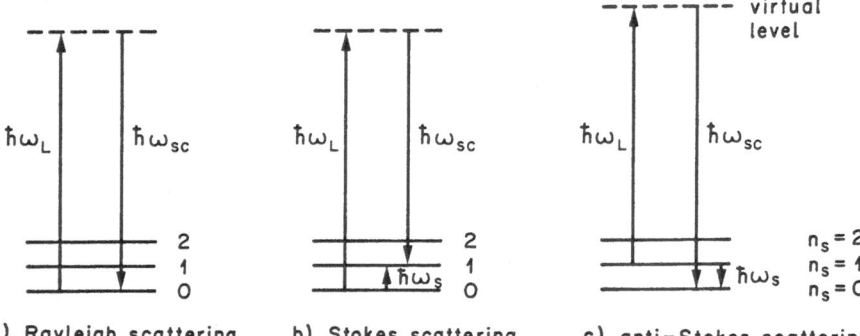

Fig. 3.20. Transitions for (a) Rayleigh scattering, (b) first-order Stokes scattering, and (c) first-order anti-Stokes scattering

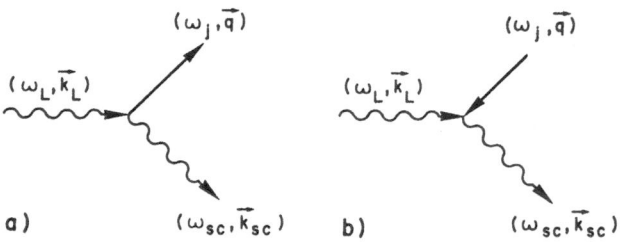

Fig. 3.21. (a) Stokes process, (b) anti-Stokes process

The two processes are shown schematically in Fig. 3.21. Since $\omega_L \gg \omega_s = \omega_j(\boldsymbol{q})$ it follows from (3.40) that $\omega_L \cong \omega_{sc}$. Furthermore, the experiments are generally carried out at frequencies where there is essentially no dispersion of the refractive index n, that is in the transparent region of the crystal. Since $\boldsymbol{k}_L$ and $\boldsymbol{k}_{sc}$ are the wave vectors within the crystal, we have $k_L = 2\pi/\lambda_L$, $k_{sc} = 2\pi/\lambda_{sc}$, where $\lambda_L = \lambda_v/n(\omega_L)$ and $\lambda_{sc} = \lambda_v/n(\omega_{sc})$ (λ_v : wavelength in vacuum). From $c = \nu\lambda_v$ we obtain $k_L = n(\omega_L)\omega_L/c$, $k_{sc} = n(\omega_{sc})\omega_{sc}/c$ and since $\omega_L \cong \omega_{sc}$, it follows that $k_L \cong k_{sc}$. In additioin, λ_L and λ_{sc} are much larger than the lattice parameter a, and hence k_L and k_{sc} are much smaller than π/a, the magnitude of the wavevector at the zone boundary. Therefore, according to (3.41) it follows that $q \ll \pi/a$, *that is, in first-order Raman scattering only $\boldsymbol{q} \cong 0$ optical modes can be excited.*

A deeper insight into the scattering of light can be obtained by considering a reasonable model of the elementary peturbation process that involves a photon and a crystal. If a photon of energy $\hbar\omega_L$ in the visible or uv region of the spectrum approaches a crystal, it sets up a perturbation of its electronic wave functions, because only electrons are light enough to follow the fast-changing electric field of the photon. The wave functions of the perturbed system then acquire a mixed character and become linear combinations of all possible wave functions of the unperturbed crystal with time-dependent coefficients, see (3.51–55). We can formally regard the crystal as having attained a non-stationary energy level of higher energy by considering the perturbation as belonging to the crystal. This formal description has nothing to do with the concept of the energy levels r (Fig. 3.20) used to describe the absorption process. For this reason we speak of a *virtual* level in order to indicate that it is introduced into the discussion only for a modellistic description of the perturbation process. Such virtual levels are indicated in Fig. 3.20. In classical language a virtual level corresponds to a *forced oscillation* of the electrons with the frequency ω_L of the incident light.

After the photon has realized that the system has no stationary states of energy $\hbar\omega_L$, it leaves the unstable situation in which it has been trapped. In terms of the same modellistic representation we can formally consider the photon as being emitted by the perturbed crystal which jumps back to one of its stationary states. If it returnes to its initial state this gives rise to Rayleigh scattering in which the frequency of the photon remains the same; this is also called *elastic scattering* and all that can happen to the photon is a change in the direction of propagation, see (3.39). With a much lower probability, however, the photon can loose part of its energy in the interaction process and thus leaves the system with a lower energy $\hbar\omega_{sc}$ (Stokes process). Since the crystal must return to a stationary state, the difference $\hbar\omega_L - \hbar\omega_{sc}$ must correspond to a phonon energy; this process is associated with a creation of a phonon of energy $\hbar\omega_s$ (Fig. 3.20b). In the same way, the photon can leave the crystal with a higher energy if by chance it finds the system in an excited vibrational state, and the system jumps after the interaction process to the ground level. This corresponds to the anti-Stokes process which is associated with an annihilation of a phonon (Fig 3.20c). The Stokes and anti-Stokes processes are also referred too as *inelastic scattering*.

It is important to note that the "absorption" of the photon $\hbar\omega_L$ and the emission of the photon $\hbar\omega_{sc}$ are truly simultaneous events and cannot be separated in time one from the other. In our modellistic description, however, we have established a sequential occurrence of events in which the photon is first "absorbed" and then emitted from the virtual level. These transitions might equally well be reversed, that is, the emission of the photon precedes the "absorption" process.

As the frequency of the incident radiation approaches an electronic transition frequency (electronic level r in Fig. 3.20), the intensity of the

Raman bands is strongly enhanced. Raman spectra obtained with exciting frequencies close to absorption bands are called *resonance Raman spectra;* this terminology emphasizes the fact that in the quantum-mechanical description a resonant dominator occurs, see (3.69, 70). The resonance Raman effect (except for full resonance) is not a sequence of absorption and emission processes as is, for instance, fluorescence. Fluorescence and resonance Raman effect should therefore not be confused. In the case of fluorescence the incident photon is truly absorbed by the molecule, which reaches an excited stationary state r (Fig. 3.20) with a well defined lifetime. After some time the molecule returns to a lower energy state and, if this is different from the initial state, irradiates light of frequency lower than the exciting frequency. In contrast to the normal and resonance Raman effect, the two steps in fluorescence are really sequential in time. In fact, fluorescence can be quenched by adding to the system a species that can take away the excitation energy from the excited system in the time intervall between absorption and emission. This is not possible in the case of the Raman effect, since the annihilation of the incident photon and the creation of the scattered photon occur simultaneously. The fact that the scattered intensity is much greater when ω_L is close to an absorption frequency simply reflects a greater efficiency of the perturbation in these conditions.

The picture developed in this subsection and illustrated in Fig. 3.20 immediately accounts for the higher intensity of the Stokes lines compared with the anti-Stokes lines, because the population of the ground vibrational level (Fig. 3.20b) is much greater, for optical phonons and not too high temperatures, than the population of excited vibrational levels (3.20c). Thus the chance the incident photon finds the system in an excited vibrational level is much smaller than for the ground state. Since the ratio of the two populations, according to Bose-Einstein statistics [Ref. 3.21, Eq. (2.126)] is proportional to $\exp(\hbar\omega_s/k_\mathrm{B}T)$, the ratio of the intensities of a Stokes line to a corresponding anti-Stokes line is expected to be proportional to

$$\left(\frac{\omega_L - \omega_s}{\omega_L + \omega_s}\right)^4 \exp\left(\frac{\hbar\omega_s}{k_\mathrm{B}T}\right),$$

and this ratio is considerably larger than unity, in contrast to (3.16) for the classical case. This will be verified on the basis of Placezk's theory in Sect. 3.3.4.

b) Second-Order Scattering

In second-order scattering processes, the incident photon (ω_L, k_L) excites the crystal from an initial electronic and vibritional state to a virtual state. The crystal then emits a scattered photon $(\omega_\mathrm{sc}, k_\mathrm{sc})$ and makes a transition from the virtual state to a final electronic and vibrational state which differs from the initial state by two vibrational quanta. In these scattering processes, the phonons involved may either be optical or acoustical, or a combination of optical and acoustical. If two optical modes are involved we are dealing with second-order Raman scattering, and if two acoustical modes are involved with second-order Brillouin scattering (Chap. 4). The changes in the vibrational quantum numbers that can occur between the initial and final states in second-order Raman scattering processes are $n_{j'} \rightarrow n_{j'} \pm 1$ and $n_{j''} \rightarrow n_{j''} \pm 1$ for combination frequencies $\omega_{j'} \pm \omega_{j''}$, and $n_j \rightarrow n_j \pm 2$ for overtone frequencies $2\omega_j$. The conservation of energy and momentum between initial and final states are

$$\omega_L = \omega_{sc} \pm \omega_{j'}(\boldsymbol{q'}) \pm \omega_{j''}(\boldsymbol{q''}), \qquad (3.42)$$

$$\boldsymbol{k}_L = \boldsymbol{k}_{sc} \pm \boldsymbol{q}_{j'} \pm \boldsymbol{q}_{j''}. \qquad (3.43)$$

As in first-order scattering $k_L \cong k_{sc} \ll \pi/a$. Figs. 3.22, 23 illustrate a second-order Stokes scattering process. A second-order Stokes process $(\omega_{sc} < \omega_L)$ involves the creation of two phonons $(\omega_{j'}, \boldsymbol{q'})$ and $(\omega_{j''}, \boldsymbol{q''})$ (Figs. 3.22, 23), or the creation of one phonon and the annihilation of a second, lower frequency phonon. On the other hand, a second-order anti-Stokes process $(\omega_{sc} > \omega_L)$ involves the annihilation of two phonons or the annihilation of one phonon and the creation of a second, lower frequency phonon.

As in second-order infrared absorption spectra (Sect. 2.3), the second-order Raman spectra involve contributions from pairs of phonons from throughout the first Brillouin zone; this follows directly from (3.43). Also, there are two types of mechanisms which can give rise to second-order scattering processes: (a) A mechanism involving the second-order change in the electronic polarizability which is produced by the vibrational modes (third term in (3.25)). In this mechanism the radiation is directly scattered by

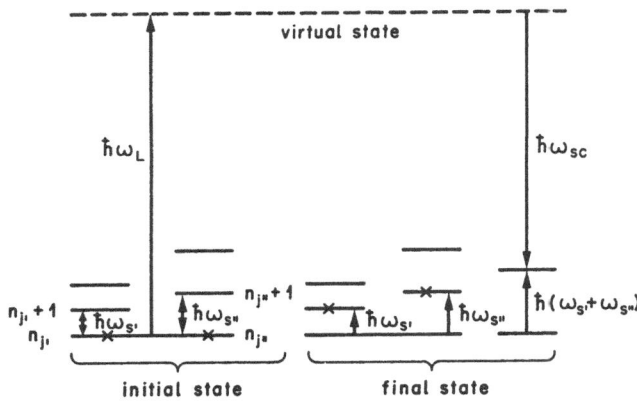

Fig. 3.22. Transitions for a second-order Stokes process in which two phonons $\omega_{s'}$ and $\omega_{s''}$ are created

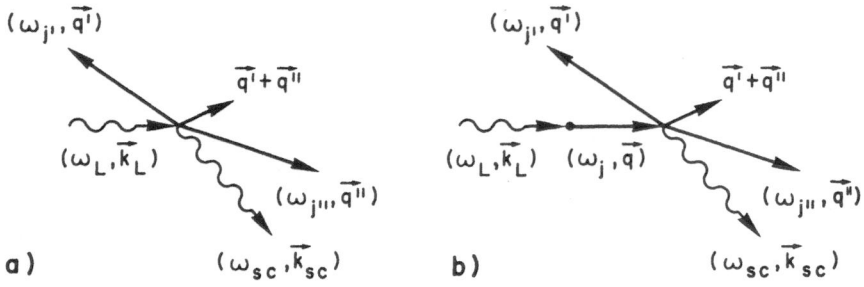

Fig. 3.23. (a) Second-order Stokes process involving the second-order change of the electronic polarizability. (b) Second-order Stokes process involving the anharmonic coupling mechanism

the two phonons $Q_{s'}$ and $Q_{s''}$ (Fig. 3.23a) and the intensity of the scattered radiation depends on the magnitudes of the coefficients $\alpha_{\rho\sigma,s's''}$ in (3.25). (b) A mechanism involving the anharmonic coupling of the vibrational modes active in first-order Raman or Brillouin scattering with two other phonons. In this mechanism, the active phonon serves as an intermediate state and the net result is that the radiation is indirectly scattered by the two other phonons. (Fig. 3.23b). As discussed in Sect. 2.3.1 this interaction process constitutes a mechanism whereby the Raman (or Brillouin) active phonon $(\boldsymbol{q}, j)$ decays, i.e. exhibits a finite life time $\tau_j(\boldsymbol{q})$, which gives rise to a width $2\Gamma_j(\boldsymbol{q}, \omega, T)$ of the Raman (or Brillouin) line. The intensity of the scattered radiation depends on the magnitudes of the coefficients $\alpha_{\rho\sigma,s}$ in (3.25) and of the third-order anharmonic coupling coefficients $V_{0j'j''}$ introduced in Sect. 2.3.7. For NaCl and CsCl type crystals which do not exhibit first-order Raman spectra (Sect. 3.2.4), $\alpha_{\rho\sigma,s}$ represents only the first-order coefficients for the TA and LA modes, $\alpha_{\rho\sigma,\mathrm{TA}}$ and $\alpha_{\rho\sigma,\mathrm{LA}}$; the corresponding coefficients $\alpha_{\rho\sigma,\mathrm{TO}}$ and $\alpha_{\rho\sigma,\mathrm{LO}}$ for optical modes are equal to zero. For diamond and ZnS type crystals which do exhibit first-order Raman spectra (Sect. 3.2.4), the coefficients $\alpha_{\rho\sigma,s}$ represent the first-order coefficients of both acoustical and optical modes.

Independent of the specific mechanism involved, the second-order Raman spectra may be expected to exhibit structure arising from structure in the two-phonon density of states $g_{j'j''}(\omega)$ of the pairs of phonons $(\omega_{j'}, \boldsymbol{q'})$, $(\omega_{j''}, \boldsymbol{q''})$ which are associated with critical points in the dispersion curves. The situation is very similar to the discussion given for the second-order infrared absorption in Sects. 2.3.6, 7. When the major contributions come from the phonons with wave vectors near the Brillouin zone boundary, which are orders of magnitude larger than the wave vectors of photons, the latter can be neglected in (3.43) which then reduces to $\pm\boldsymbol{q'} \pm\boldsymbol{q''} = 0$. Figure 3.24 shows the second-order Raman spectrum of NaF [3.23]. The calculated spectra are based on the deformation dipole model [Ref. 3.21, Sect. 4.3]. The E_{g}

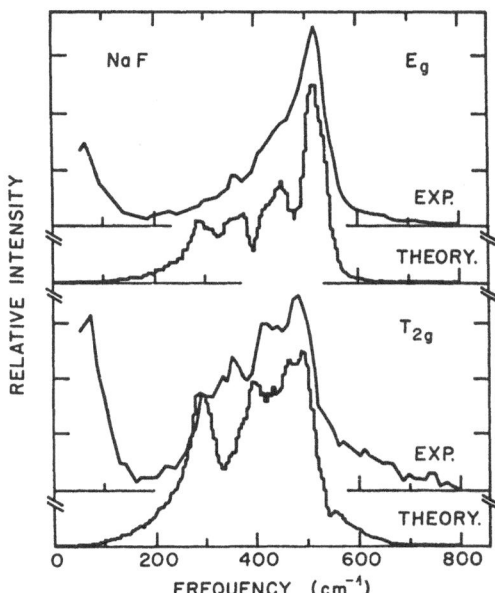

Fig. 3.24. Comparison between the-oretical and measured second-order Stokes spectra of NaF at room tem-perature. The measured spectra be-low $200\,\text{cm}^{-1}$ are attributed to dif-ference bands and not considered in the theoretical calculation [3.23]

spectrum is measured in the configuration $a(ba)c$, where a, b, c refer to the crystal axes (110), $(\bar{1}, 1, 0)$ $(0, 0, 1)$, while the T_{2g} spectrum is observed in the configuration $X(YX)Z$.

3.3.2 The Intensity of the Scattered Light

A relatively simple derivation of the intensity of the scattered light is ob-tained on the basis of the correspondence principle [3.24]. According to this principle the dipole moment induced by the incident light is calculated quantum mechanically, but the radiation emitted by this dipole is calculated classically. Thus, the system (crystal or molecule) is treated quantum me-chanically whereas the radiation field is described classically. The induced dipole moment is calculated using standard time-dependent perturbation theory [3.25]. Our treatment closly follows *Placzek*'s paper [3.26][2]. In this subsection we outline the main steps for the calculation of the intensity of the scattered light; details are derived in Appendix F.

Consider a system composed of particles (nuclei and electrons) obeying the time dependent Schrödinger equation

$$H_0 \Psi^{(0)}(q, t) = i\hbar \frac{\partial}{\partial t} \Psi^{(0)}(q, t), \tag{3.44}$$

---

[2] In the following we essentially use Placzek's notation and definition of the matrix elements; although it differs from our conventions it allows a direct comparison with this important paper and other literature in this field

where q stands for all the coordinates of the particles [3]. The general solution of (3.44) is

$$\Psi^{(0)}(q,t) = \sum_r a_r \Xi_r(q) e^{-i\omega_r t}, \tag{3.45}$$

where $E_r = \hbar\omega_r$ and $\Xi_r(q)$ are the eigenvalues and eigenfunctions of the time independent Schrödinger equation

$$H_0 \Xi_r(q) = E_r \Xi_r(q). \tag{3.46}$$

If the system is in the state k [4] ($a_r = 0$ for $r \neq k$, $a_k = 1$), it follows from (3.45) that

$$\Psi_k^{(0)}(q,t) = \Xi_k(q) e^{-i\omega_k t}. \tag{3.47}$$

We now suppose that this system is perturbed by a light wave whose wavelength is large compared with the interatomic distances. This is certainly the case for light in the visible and uv but not for X-rays. Furthermore we assume that the frequency ω_L of the incident light is different from all the transition frequencies $\omega_{rk} = \omega_r - \omega_k$ of the system (no resonance). For the electric field of the light we write

$$\boldsymbol{E} = \boldsymbol{A} e^{-i\omega_L t} + \boldsymbol{A}^* e^{i\omega_L t}, \tag{3.48}$$

where $\boldsymbol{A}$ is a complex amplitude. The perturbing interaction energy of the light with the system is given by

$$H' = -\boldsymbol{E} \cdot \boldsymbol{M}, \tag{3.49}$$

where $\boldsymbol{M} = \sum e_j \boldsymbol{r}_j$ is the electric dipole moment of the system with charges e_j at positions $\boldsymbol{r}_j$. The Schrödinger equation of the peturbed system is

$$(H_0 - \boldsymbol{E} \cdot \boldsymbol{M}) \Psi(q,t) = i\hbar \frac{\partial}{\partial t} \Psi(q,t). \tag{3.50}$$

If the unperturbed system is in the state k described by (3.47) we make the following Ansatz for the perturbed solution of (3.50):

$$\Psi(q,t) = \Psi_k^{(0)}(q,t) + \Psi_k^{(1)}(q,t) = \Psi_k(q,t). \tag{3.51}$$

It is shown in Appendix F that if only terms linear in $\boldsymbol{E}$ are retained

$$\Psi_k^{(1)} = \Xi_k^+ \exp[-i(\omega_k + \omega_L)t] + \Xi_k^- \exp[-i(\omega_k - \omega_L)t], \tag{3.52}$$

---

[3] q should not be confused with the magnitude of the wave vector $\boldsymbol{q}$

[4] $k = (e, n)$ denotes collectively the electronic quantum number e and the nuclear (vibrational) quantum number n

where

$$\Xi_k^+ = \frac{1}{\hbar} \sum_r \frac{(A \cdot M_{kr})}{\omega_{rk} + \omega_L} \Xi_r, \tag{3.53a}$$

$$\Xi_k^- = \frac{1}{\hbar} \sum_r \frac{(A^* \cdot M_{kr})}{\omega_{rk} + \omega_L} \Xi_r, \quad \text{and} \tag{3.53b}$$

$$M_{kr} = M_{rk}^* = \int \Xi_r^* M \Xi_k dq, \tag{3.54}$$

$$\omega_{rk} = \omega_r - \omega_k. \tag{3.55}$$

For the matrix elements of the perturbed system,

$$M_{km}^{(1)}(t) = \int \Psi_m^*(q, t) M \Psi_k(q, t) dq, \tag{3.56}$$

we obtain from (3.51–53)

$$M_{km}^{(1)}(t) = M_{km} \exp(-i\omega_{km} t) + C_{km} \exp[-i(\omega_{km} + \omega_L)t] + D_{km} \exp[-i(\omega_{km} - \omega_L)t], \tag{3.57}$$

where

$$C_{km} = \frac{1}{\hbar} \sum_r \left(\frac{(A \cdot M_{kr}) M_{rm}}{\omega_{rk} - \omega_L} + \frac{M_{kr}(A \cdot M_{rm})}{\omega_{rm} + \omega_L} \right), \tag{3.58}$$

$$D_{km} = \frac{1}{\hbar} \sum_r \left(\frac{(A^* \cdot M_{kr}) M_{rm}}{\omega_{rk} + \omega_L} + \frac{M_{kr}(A^* \cdot M_{rm})}{\omega_{rm} - \omega_L} \right). \tag{3.59}$$

Since $D_{km} = C_{mk}^*$ we obtain for $k = m$

$$M_{kk}^{(1)} = M_{kk} + C_{kk} e^{-i\omega_L t} + C_{kk}^* e^{i\omega_L t}, \tag{3.60}$$

where

$$C_{kk} = \frac{1}{\hbar} \sum_r \left(\frac{(A \cdot M_{kr}) M_{rk}}{\omega_{rk} - \omega_L} + \frac{M_{kr}(A \cdot M_{rk})}{\omega_{rk} + \omega_L} \right). \tag{3.61}$$

Note that $M_{kk}^{(1)}(t)$ is real; it is the expectation value of the dipole moment in the state k and has the same time dependence as that of the incident radiation. According to the correspondence principle the intensity of the radiation emitted by the dipole moment $M_{kk}^{(1)}(t)$ is given by the classical expression. Using (3.7, 12) we obtain

$$\hat{I}_{kk} = \frac{2}{3c^3} \langle |\ddot{M}_{kk}^{(1)}(t)|^2 \rangle_t = \frac{2}{3c^3} \lim_{\tau \to \infty} \frac{1}{\tau} \int_0^\tau |\ddot{M}_{kk}^{(1)}(t)|^2 dt. \tag{3.62}$$

From (3.60) we find

$$\hat{I}_{kk} = \frac{4}{3c^3}\omega_L^4 |C_{kk}|^2. \tag{3.63}$$

$M_{kk}^{(1)}(t)$ is the dipole moment responsible for Rayleigh scattering whose intensity is given by (3.63).

In contrast to $M_{kk}^{(1)}(t)$ it should be noted that $M_{km}^{(1)}(t)$, as given by (3.57), is complex. In order to relate the individual terms of (3.57) with the classical radiation of a real dipole, the correspondence principle must be used. According to *Klein* [3.24] this can be formulated as follows: The radiation emitted by the component $M_{km}\exp(-i\omega_{km}t)$, for example, ($k$: initial, m: final state) is zero if $\omega_{km}<0$ and is equal to the classical radiation emitted by the real dipole

$$M_{k\leftrightarrow m} = M_{km}\exp(-i\omega_{km}t) + M_{km}^*\exp(i\omega_{km}t),$$

if $\omega_{km} = \omega_k - \omega_m >0$. In order to apply this prescription to (3.57) we first form the real dipole

$$\begin{aligned}
M_{k\leftrightarrow m}^{(1)} &= M_{km}^{(1)} + M_{km}^{(1)*} = M_{km}^{(1)} + M_{mk}^{(1)} \\
&= M_{km}\exp(-i\omega_{km}t) + M_{km}^*\exp(i\omega_{km}t) \\
&\quad + C_{km}\exp[-i(\omega_{km} + \omega_L)t] + C_{km}^*\exp[i(\omega_{km} + \omega_L)t] \\
&\quad + D_{km}\exp[-i(\omega_{km} - \omega_L)t] + D_{km}^*\exp[i(\omega_{km} - \omega_L)t].
\end{aligned} \tag{3.64}$$

In analogy to (3.62) the intensity of the scattered light is given by

$$\tilde{I}_{km} = \frac{2}{3c^3}\langle |\ddot{M}_{k\leftrightarrow m}^{(1)}(t)|^2 \rangle t. \tag{3.65}$$

The evaluation of the time average gives [note that the cross terms vanish as in (3.12)]:

$$\begin{aligned}
\tilde{I}_{km} = \frac{4}{3c^3}\Big[&\omega_{km}^4 |M_{km}|^2 + (\omega_{km} + \omega_L)^4 |C_{km}|^2 \\
&+ (\omega_{km} - \omega_L)^4 |D_{km}|^2 \Big].
\end{aligned} \tag{3.66}$$

According to the correspondence principle there will be radiation only if in the first term of (3.66) $\omega_{km}>0$, in the second term $\omega_{km} + \omega_L >0$, and in the third term $\omega_{km} - \omega_L >0$.

In the first term of (3.66) the energy of the initial state is larger than the energy of the final state: $E_k > E_m$ ($E_k = \hbar\omega_k$, $E_m = \hbar\omega_m$). Furthermore, the frequency of the light, ω_L, does not appear. This term therefore describes *spontaneous emission* associated with the transition $k \to m$ (Fig. 3.25a). In

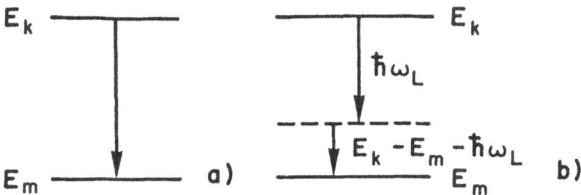

Fig. 3.25. (a) Spontaneous emission associated with the transition $k \to m$, first term in (3.66). (b) Induced emission of two quanta associated with the transition $k \to m$, third term of (3.66)

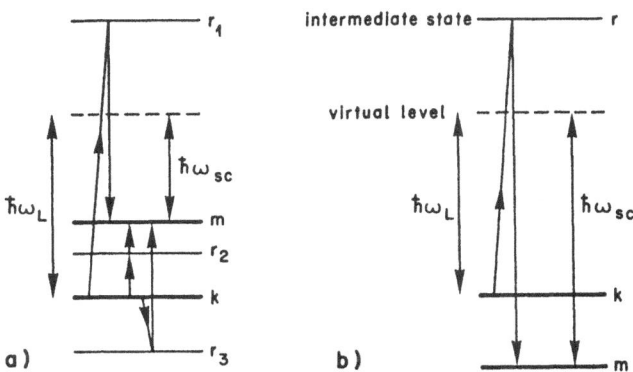

Fig. 3.26. (a) Stokes transition $k \to m$, (b) anti-Stokes transition $k \to m$. $r_1, r_2 r_3, \ldots$ are intermediate levels involved in the transitions. These transitions are described by the second term in (3.66)

the third term of (3.66) we have $E_k > E_m + \hbar\omega_L$. The initial state k is therefore an excited state. This term describes an *induced emission* of two quanta with energies $\hbar\omega_L$ and $E_k - E_m - \hbar\omega_L$ associated with the transition $k \to m$ (Fig. 3.25b). This kind of scattering is only observable if it is possible to populate excited states to a considerable degree, for instance, with optical pumping techniques using high-power lasers.

It is the second term in (3.66) which describes the *normal Raman effect,* and the following discussion is devoted to this term. Here we have $E_m < E_k + \hbar\omega_L$. The energy of the final state m can be larger or smaller than the energy of the initial state k. In the first case the energy of the scattered radiation is smaller than $\hbar\omega_L$ (Stokes line), in the latter it is larger than $\hbar\omega_L$ (anti-Stokes line). These two cases are shown in Fig. 3.26. The quantity C_{km} in (3.66) is defined by (3.58); it is a sum involving the transition moments from the initial state k to all states r of the unperturbed system, and the transition moments from all states r to the final state m . As discussed in Sect. 3.3.1, this does not mean that such transitions, in fact, occur in the scattering process. The summation arises purely as a consequence of

95

the mathematical treatment of the perturbation problem, in which a wave function of the perturbed system is expressed in terms of the full set of its unperturbed wave functions. Note also that the expression for C_{km} involves the product of the transition moments M_{kr} and M_{rm}, and not the individual transition moments themselves. Since these quantities may be either positive or negative, it is possible for terms belonging to different states r either to reinforce one another or the reverse. They may indeed cancel out one another completely and so cause C_{km} to vanish. The Raman line is then forbidden. This is the case for light scattered by a harmonic oscillator (Problem 3.4.3). It is not necessary that an intermediate state r is excited by an absorption process starting from the ground state. Indeed, in general, the summation (3.58) includes not only states r which lie above the initial state k but also states r (if any) which lie below it (Fig. 3.26a). In such cases, the idea of actually attaining the state r by absorption of the incident light is obviously absurd. It should also be emphasized that the transition moment M_{km} which determines the probability for a spontaneous transition between the initial and final states does not enter at all in the expression for C_{km}. For this reason, the intensities for spontaneous emission (or absorption) and for Raman scattering are in no direct relation, and also the selection rules are completely different.

According to the preceeding discussion and (3.66) the intensity for Raman scattering into the solid angle $\Omega = 4\pi$ is given by

$$\hat{I}_{km} = \frac{4}{3c^3}(\omega_L + \omega_{km})^4 |C_{km}|^2. \tag{3.67}$$

For $k = m$ (3.67) reduces to (3.63) which gives the intensity of Rayleigh scattering. The components of C_{km} can be written in the form

$$(C_\varrho)_{km} = \sum_\sigma (c_{\varrho\sigma})_{km} A_\sigma, \quad \text{where} \tag{3.68}$$

$$(c_{\varrho\sigma})_{km} = \frac{1}{\hbar} \sum_r \left(\frac{(M_\sigma)_{kr}(M_\varrho)_{rm}}{\omega_{rk} - \omega_L} + \frac{(M_\varrho)_{kr}(M_\sigma)_{rm}}{\omega_{rm} + \omega_L} \right) \tag{3.69}$$

is the *scattering tensor*. In general, it is a complex and non-symmetric tensor. For $k = m$ we obtain

$$(c_{\varrho\sigma})_{kk} = \frac{1}{\hbar} \sum_r \left(\frac{(M_\sigma)_{kr}(M_\varrho)_{rk}}{\omega_{rk} - \omega_L} + \frac{(M_\varrho)_{kr}(M_\sigma)_{rk}}{\omega_{rk} + \omega_L} \right). \tag{3.70}$$

From (3.54) it follows that

$$(c_{\varrho\sigma})_{kk} = (c_{\sigma\varrho})_{kk}^*. \tag{3.71}$$

Therefore, if $(c_{\varrho\sigma})_{kk}$ is real, then it is also symmetric. This is not only the case if $\omega_L = 0$ (perturbation by a static electric field), but also if the

Hamiltonian is real. Disregarding in (3.60) the permanent dipole moment M_{kk} and using $(c_{\varrho\sigma})_{kk} = (c_{\sigma\varrho})_{kk}$ as well as (3.48) we obtain from (3.60)

$$(M_\varrho)_{kk}^{(1)} = \sum_\sigma (c_{\varrho\sigma})_{kk} E_\sigma = \sum_\sigma \alpha_{\varrho\sigma}^{(k)} E_\sigma. \tag{3.72}$$

The real quantity $\alpha_{\varrho\sigma}^{(k)}$ is called the *electronic polarizability* of the state k. Substituting (3.68) into (3.67) we find

$$\hat{I}_{km} = \frac{4}{3c^3} (\omega_L + \omega_{km})^4 \sum_\varrho |\sum_\sigma (c_{\varrho\sigma})_{km} A_\sigma|^2.$$

Using the notation $A = |A|e$, where e is the unit vector in the polarization direction of the incident light and

$$I_0 = \frac{c}{2\pi} |A|^2 \tag{3.73}$$

is the intensity of the incident light, we obtain

$$\hat{I}_{km} = Q_{km} I_0, \quad \text{where} \tag{3.74}$$

$$Q_{km} = \frac{8\pi}{3c^4} (\omega_L + \omega_{km})^4 \sum_\varrho |\sum_\sigma (c_{\varrho\sigma})_{km} e_\sigma|^2 \tag{3.75}$$

is the *cross section* for the Raman transition $k \rightarrow m$. The dimension of Q_{km} is cm^2; $\hat{I}_{km}$ has the dimension of erg/s, while I_0 has the dimension of erg/s cm^2.

If the incident light is polarized in the direction σ and the scattered light is observed with the analyzer in the dirction ϱ, one finds for the scattered radiation emitted per unit solid angle $d\Omega$

$$I_{km}(\varrho\sigma) = \frac{1}{c^4} (\omega_L + \omega_{km})^4 |(c_{\varrho\sigma})_{km}|^2 I_0. \tag{3.76}$$

This relation should be compared with the classical intensity given by (3.36b). If in the latter we replace $\Delta\alpha_{\varrho\sigma,s0}^2$ by $4|(c_{\varrho\sigma})_{km}|^2$ we obtain the quantum mechanical intensity. In the next subsection it is shown that in Placzek's approximation $c_{\varrho\sigma}$ is replaced by the electronic polarizability $\alpha_{\varrho\sigma}$.

3.3.3 Placzek's Approximation

The direct evaluation of the scattering formulae (3.74–76) with $(c_{\varrho\sigma})_{km}$ given by (3.69) is only possible for simple systems, such as the harmonic oscillator (Problem 3.4.3), the free electron and simple atoms. For molecules and crystals, however, the complexity of the energy levels and the incomplete knowledge of the excited states r occuring in (3.69) prevents a direct evaluation. Using *Placzek*'s approximation [3.26] it is, however, possible to obtain the general result in a more direct way.

Let us consider only scattering processes in which the electronic state is not changed, that is, initial and final states are identical and correspond to the electronic ground state. In this case only the vibrational states are changed and energy conservation is given by (3.40) (Fig. 3.20). Since the scattering of light is due to the electrons of the system, the transfer of the energy of the light to the nuclei, or vice versa, is only possible through the coupling between the motion of the nuclei and the electrons, and the Raman effect is a direct consequence of this coupling. In order to calculate the intensity of the scattered light it is therefore appropriate to consider the radiation scattered by a system with fixed nuclei, and then to study how this radiation is modified by the motion of the nuclei.

We suppose that the electronic ground state is non-degenerate. If the nuclei are in fixed positions such a system will scatter only Rayleigh radiation. The intensity of the scattered light is then determined by the polarizability tensor $(c_{\varrho\sigma})_{00} = \alpha_{\varrho\sigma}^{(0)} = \alpha_{\varrho\sigma}$ defined in (3.72); here $k = 0$ denotes the electronic ground state. We remember that the electronic polarizability is real and symmetric. Since in (3.70) for $(c_{\varrho\sigma})_{00}$ both the eigenfrequencies ω_{r0} as well as the eigenfunctions $\mathcal{E}_r$ (which are used to calculate M_{0r} defined by (3.54)) of a system with fixed nuclei depend on the positions of the nuclei, also the components of the electronic polarizability, $\alpha_{\varrho\sigma}$, will be a function of the nuclear configuration $R : \alpha_{\varrho\sigma} = \alpha_{\varrho\sigma}(R)$. *We now assume that the intensity of the light scattered by a system with vibrating nuclei is the same for each nuclear configuration R as the intensity of the light scattered by the system with the nuclei in the fixed configuration R.* Placzek [3.26] has shown that this assumption is justified if the following conditions are satisfied:

1. The electronic ground state must be non-degenerate.[5]
2. The adiabatic approximation must be valid.
3. The exciting frequency ω_L must be less than any electronic transition frequency of the system, although much larger than any vibrational frequency.

The derivation of this result is given in Appendix F. We now apply Placzek's approximation to a system with vibarting nuclei by considering a transition from the state $k = (0n)$ to the state $m = (0n')$, where 0 denotes the electronic ground state and n, n' vibrational states. On the basis of Placzek's approximation we expect that the matrix elements $(c_{\varrho\sigma})_{0n,0n'}$ are simply given by the matrix elements $(\alpha_{\varrho\sigma}(R))_{0n,0n'}$. In Appendix F it is shown that this is indeed the case, that is

$$
\begin{aligned}
(c_{\varrho\sigma})_{0n,0n'} &= \int \psi_{0n'}^*(R)\alpha_{\varrho\sigma}(R)\psi_{0n}(R)dR \\
&= (\alpha_{\varrho\sigma}(R))_{0n,0n'}.
\end{aligned}
\tag{3.77}
$$

---

[5] A detailed discussion of this condition is given in [Ref. 3.26, p. 269].

Here $\psi_{0n}(R)$ is the vibrational wave function of the electronic ground state 0 and vibrational state n, see (F.10). Substituting (3.77) into (3.76) and omitting the subscript 0 gives the following expression for the intensity of the scattered light associated with the vibrational transition $n \rightarrow n'$:

$$I_{nn'}(\varrho\sigma) = \frac{1}{c^4}(\omega_L + \omega_{nn'})^4 |(\alpha_{\varrho\sigma})_{nn'}|^2 I_0. \tag{3.78}$$

In (3.78) $n = |\ldots n_s \ldots\rangle$ and $n' = |\ldots n'_s \ldots\rangle$ specify the occupation numbers of the two vibrational states and $\omega_{nn'}$ is defined by

$$\hbar\omega_{nn'} = E_n - E_{n'} = \sum_s (n_s - n'_s)\hbar\omega_s. \tag{3.79}$$

3.3.4 Raman Intensities Based on Placzek's Theory

If the incident light is polarized in the direction σ and the scattered light is observed with the analyzer in the direction ϱ, then according to (3.78) one finds for the intensity of the scattered radiation per unit soild angle

$$I_{nn'} = \mathrm{const}(\omega_L + \omega_{nn'})^4 |(\alpha_{\varrho\sigma})_{nn'}|^2 I_0. \tag{3.80}$$

Since the polarizability $\alpha_{\varrho\sigma} = \alpha_{\varrho\sigma}(R)$ depends on the configuration R of the nuclei, we can expand $\alpha_{\varrho\sigma}$ in terms of the normal coordinates Q_s of the system as in (3.25)

$$\alpha_{\varrho\sigma} = \alpha_{\varrho\sigma}^{(0)} + \sum_s \alpha_{\varrho\sigma,s} Q_s + \frac{1}{2}\sum_{s,s'}\alpha_{\varrho\sigma,ss'} Q_s Q_{s'} + \cdots. \tag{3.81}$$

For the matrix elements of $\alpha_{\varrho\sigma}$ associated with a transition $n_s \rightarrow n'_s$ (all other quantum numbers are supposed to be unchanged) one obtains in a first approximation (neglecting terms proportional to $Q_s Q_{s'}$ in (3.81))

$$(\alpha_{\varrho\sigma})_{n_s n'_s} = \alpha_{\varrho\sigma}^{(0)} \delta_{n_s n'_s} + \alpha_{\varrho\sigma,s}(Q_s)_{n_s n'_s}. \tag{3.82}$$

For Rayleigh scattering $n_s = n'_s$ and

$$(\alpha_{\varrho\sigma})_{n_s n_s} = \alpha_{\varrho\sigma}^{(0)}. \tag{3.83}$$

Since according to (3.79) $\omega_{nn} = 0$, the intensity of Rayleigh scattering is given by

$$I_{nn} = \mathrm{const}\,\omega_L^4 |\alpha_{\varrho\sigma}^{(0)}|^2 I_0. \tag{3.84}$$

We note that in this approximation the intensity of Rayleigh scattering is independent on temperature. If in (3.82) second-order terms were included the intensity of Rayleigh scattering becomes temperature dependent.

For first-order Raman scattering $n'_s = n_s \pm 1$. In the case of Stokes scattering $(n'_s = n_s + 1)$ one obtains [Ref. 3.21, Sect. 2.2.3]

$$(\alpha_{\varrho\sigma})_{n_s,n_s+1} = \alpha_{\varrho\sigma,s}(Q_s)_{n_s,n_s+1}$$

$$= \left(\frac{\hbar}{2\omega_s}\right)^{1/2}\alpha_{\varrho\sigma,s}(n_s + 1)^{1/2}, \tag{3.85}$$

and for an anti-Stokes line

$$(\alpha_{\varrho\sigma})_{n_s,n_s-1} = \left(\frac{\hbar}{2\omega_s}\right)^{1/2}\alpha_{\varrho\sigma,s}n_s^{1/2}. \tag{3.86}$$

The mean intensities are obtained from (3.80, 85, 86) and forming the thermal averages. For a Stokes line we have $\omega_{nn'} = -\omega_s$, and for an anti-Stokes $\omega_{nn'} = \omega_s$ according to (3.79). We then obtain

$$\bar{I}_{\text{Stokes}} = \text{const}\,(\omega_L - \omega_s)^4\,\frac{\hbar}{2\omega_s}\alpha^2_{\varrho\sigma,s}(1 + \bar{n}_s), \quad \text{and} \tag{3.87}$$

$$\bar{I}_{\text{anti-Stokes}} = \text{const}\,(\omega_L + \omega_s)^4\,\frac{\hbar}{2\omega_s}\alpha^2_{\varrho\sigma,s}\bar{n}_s. \tag{3.88}$$

Here

$$\bar{n}_s = [\exp(\hbar\omega_s/k_{\text{B}}T) - 1]^{-1} \tag{3.89}$$

is the mean occupation number of the phonon s.

The expressions (3.87, 88) show that the intensities of both the Stokes and the anti-Stokes lines increase with increasing temperatur [Ref. 3.21, Fig. 2.10]. This is understandable because with increasing tempeature the vibrational amplitudes and therefore the scattering will increase. From (3.87-89) we obtain for the intensity ratio of Stokes and anti-Stokes scattering

$$\frac{I_{\text{Stokes}}}{I_{\text{anti-Stokes}}} = \left(\frac{\omega_L - \omega_s}{\omega_L + \omega_s}\right)^4\exp(\hbar\omega_s/k_{\text{B}}T). \tag{3.90}$$

From this expression we expect that the intensity of a Stokes line is larger than the intensity of the corresponding anti-Stokes line, in agreement with experiments but in contradiction with the classical expression (3.16).

The intensities of the second-order Raman lines are found by forming the matrix elements of the products $Q_sQ_{s'}$ occurring in (3.81). These are

Table 3.1. Temperature dependent factors in one- and two-phonon Raman and infrared spectra

Process	Stokes	Raman	Anti-Stokes	Infrared Absorption
One-phonon	$1 + \bar{n}_s$		$\bar{n}_s$	indept. of T
Two-phonon				
overtone	$2 + 3\bar{n}_s + \overline{n_s^2}$		$\overline{n_s^2} - \bar{n}_s$	$1 + 2\bar{n}_s$
summation	$1 + \bar{n}_s + \bar{n}_{s'} + \bar{n}_s\bar{n}_{s'}$		$\bar{n}_s\bar{n}_{s'}$	$1 + \bar{n}_s + \bar{n}_{s'}$
difference	$\bar{n}_{s'} + \bar{n}_s\bar{n}_{s'}$		$\bar{n}_s + \bar{n}_s\bar{n}_{s'}$	$\bar{n}_{s'} - \bar{n}_s$

given in terms of n_s and $n_{s'}$, and the intensity of the scattered radiation is proportional to the square of the result. Then the thermal averaging process is carried out as for first-order scattering. The accompanying table shows the results for first- and second-order Raman scattering, together with those for one- and two-phonon infrared absorption.

3.3.5 Raman Scattering by Phonon-Polaritons

Polaritons are quasi-particles consisting of a photon coupled with a long-wavelength polar (or infrared active) phonon, with an exciton or with a plasmon. We shall confine ourselves to phonon-polaritons which are coupled phonon-photon excitations. An excellent treatment of experimental and theoretical aspects of light scattering by phonon-polaritons is found in [3.10].

In Sect. 2.2.7 we have discussed in some detail phonon-polaritons. Figure 2.13b shows the dispersion of an infrared active TO phonon and the associated LO phonon near $q \cong 0$, and equation (2.41) is the dispersion relation for such polaritons for the case of vanishing damping.

Polaritons can be observed by means of Raman scattering. The reason is, that if near-forward scattering is used (Fig. 3.4), excitations with very small wave vectors q can be observed, and this is a necessary condition for observing polaritons (Fig. 2.13b). For a Stokes line we obtain from (3.41)

$$q = (k_L^2 + k_{sc}^2 - 2k_L k_{sc} \cos 2\theta)^{1/2}, \tag{3.91}$$

where 2θ is the angle between k_L and k_{sc} (Fig. 3.27). In Sect. 3.3.1 we have expressed k_L and k_{sc} in the form

$$k_L = \omega_L n(\omega_L)/c, \quad k_{sc} = \omega_{sc} n(\omega_{sc})/c, \tag{3.92}$$

and since $\omega_L \cong \omega_{sc}$ it follows that

$$k_L \cong k_{sc}. \tag{3.93}$$

Substituting (3.92, 93) into (3.91) one finds

$$q \cong 2\frac{\omega_L n(\omega_L)}{c} \sin\theta. \tag{3.94}$$

This relation shows that q can be changed by varying the angle between the detected Stokes radiation and the incident laser beam. If $2\theta = \frac{\pi}{2}$ then $q \cong \sqrt{2}k_L$; this corresponds to a value which is already large in the q-scale of Fig. 2.13b, and the corresponding frequency is close to ω_{TO}. By decreasing the angle 2θ, the magnitude of q will decrease; the smallest possible value of q for $\theta = 0$ is according to (3.91) given by

$$q_{min} = k_L - k_{sc} \cong \frac{n}{c}(\omega_L - \omega_{sc}). \tag{3.95}$$

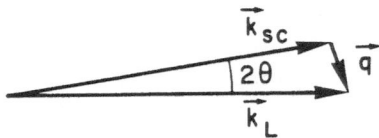

Fig. 3.27. Momentum conservation for near-forward Stokes scattering. For the observation of polaritons the angle 2θ ranges between $0°$ and about $6°$

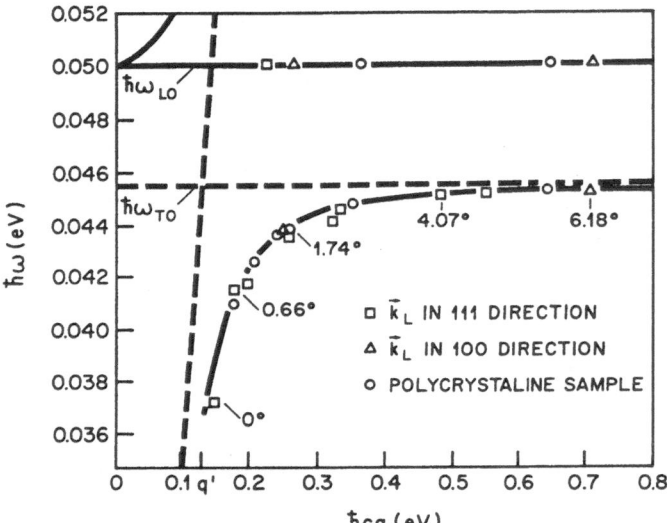

Fig. 3.28. A plot of the observed energies and wave vectors of the polaritons and the LO phonons of GaP; the theoretical dispersion curves are shown by (——). The dispersion curves for the uncoupled photons and phonons are shown by (- - -). Some of the experimental angles 2θ are indicated next to the data points [3.27]

The frequency corresponding to q_{min} will be considerably smaller as ω_{TO} (Fig. 2.13b). Experiments with very small values of q are performed using the near-forward scattering configuration shown in Fig. 3.4.

Figure 3.28 shows the experimental and theoretical phonon-polariton dispersion of GaP [3.27]. The theoretical dispersion is based on (2.41), namely

$$q^2 = \frac{\varepsilon_\infty}{c^2}\omega^2\frac{\omega_{LO}^2 - \omega^2}{\omega_{TO}^2 - \omega^2} = \frac{\omega^2}{c^2}\left(\varepsilon_\infty + \frac{\varepsilon_0 - \varepsilon_\infty}{1 - \omega^2/\omega_{TO}^2}\right). \tag{3.96}$$

Finally, it should be mentioned that for nonpolar (or infrared-inactive) optical modes, the dispersion $\omega(q)$ is essentially independent on q near $q \cong 0$.

Although according to (3.94) q depends on ω_L and θ, the observed Raman frequencies of nonpolar optical modes will therefore be essentially independent on ω_L and θ.

3.4 Problems

3.4.1 Polarizability $\alpha(R)$ of Diatomic Molecules

Consider a diatomic "ionic molecule" such as NaCl held together by ionic binding. Let e, α_1 and $-e, \alpha_2$ be the charges and polarizabilities of the cation and anion, respectively, and R the internuclear distance. Show that if the electric field E of the light is parallel to the axis, the polarizability of the molecule is given by

$$\alpha_{\parallel}(R) = \frac{\alpha_1 + \alpha_2 + 4\alpha_1\alpha_2/R^3}{1 - 4\alpha_1\alpha_2/R^6}, \tag{3.4.1}$$

and if E is perpendicular to the axis, the polarizability is given by

$$\alpha_{\perp}(R) = \frac{\alpha_1 + \alpha_2 - 2\alpha_1\alpha_2/R^3}{1 - \alpha_1\alpha_2/R^6}. \tag{3.4.2}$$

Plot $\alpha_{\parallel}(R)$ for NaCl, assuming $\alpha_1 = 0.41\,\text{Å}^3$, $\alpha_2 = 2.96\,\text{Å}^3$; the equilibrium distance is $r_e = 2.51\,\text{Å}$. Note that the Raman intensities are proportional to $(\partial\alpha_{\parallel}/\partial R)^2_{r_e}$ and $(\partial\alpha_{\perp}/\partial R)^2_{r_e}$. The relations (3.4.1, 2) are reasonably accurate for large distances R, but less accurate for $R \cong r_e$ due to the inhomogeneity of the electric fields [3.29]. Similar expressions are also valid for identical atoms held together by van der Waals binding. Note that in the present calculation we have completely neglected the change of the polarizabilities due to the overlap of the ions.

Hint: The electronic polarizations of the cation and anion are $p_1 = \alpha_1 E_1$ and $p_2 = \alpha_2 E_2$, where

$$E_1 = \frac{e}{R^2} + \frac{2}{R^3}p_2 + E, \tag{3.4.3}$$

$$E_2 = \frac{e}{R^2} + \frac{2}{R^3}p_1 + E, \tag{3.4.4}$$

if E is parallel to the axis. Calculate $p = p_1 + p_2$. Result: $p = \alpha_{\parallel}(e/R^2 + E)$. If E is perpendicular to the axis one finds

$$E_1 = -\frac{p_2}{R^3} + E, \tag{3.4.5}$$

$$E_2 = -\frac{p_1}{R^3} + E. \tag{3.4.6}$$

3.4.2 Decomposition of the Scattering Tensor into Its Isotropic, Symmetric, and Antisymmetric Parts

For $c_{\varrho\sigma}$ defined by (3.69) we write [3.26]

$$c_{\varrho\sigma} = c^0 \delta_{\varrho\sigma} + c_{\varrho\sigma}^s + c_{\varrho\sigma}^a, \tag{3.4.7}$$

where the isotropic part is defined by

$$c^0 = \frac{1}{3} \sum_{\varrho} c_{\varrho\varrho}, \tag{3.4.8}$$

the symmetric part by

$$c_{\varrho\sigma}^s = \frac{1}{2}(c_{\varrho\sigma} + c_{\sigma\varrho}) - c^0 \delta_{\varrho\sigma}, \tag{3.4.9}$$

and the antisymmetric part by

$$c_{\varrho\sigma}^a = \frac{1}{2}(c_{\varrho\sigma} - c_{\sigma\varrho}). \tag{3.4.10}$$

Using (3.69) show that

$$(c^0)_{km} = \frac{1}{3\hbar} \sum_{r} A(kmr) \sum_{\sigma} (M_\sigma)_{kr}(M_\sigma)_{rm} \tag{3.4.11}$$

$$(c^s)_{km} = \frac{1}{2\hbar} \sum_{r} A(kmr) [(M_\sigma)_{kr}(M_\varrho)_{rm} + (M_\varrho)_{kr}(M_\sigma)_{rm}]$$
$$- (c^0)_{km} \delta_{\varrho\sigma} \tag{3.4.12}$$

$$(c^a)_{km} = \frac{1}{2\hbar} \sum_{r} B(kmr) [(M_\sigma)_{kr}(M_\varrho)_{rm} - (M_\varrho)_{kr}(M_\sigma)_{rm}], \tag{3.4.13}$$

where

$$A(kmr) = \frac{\omega_{rk} + \omega_{rm}}{(\omega_{rk} - \omega_L)(\omega_{rm} + \omega_L)}, \tag{3.4.14}$$

$$B(kmr) = \frac{2\omega_L + \omega_{km}}{(\omega_{rk} - \omega_L)(\omega_{rm} + \omega_L)}. \tag{3.4.15}$$

The radiation scattered by c^0 is known as *trace scattering*. Note that the trace of c^s vanishes: $\sum_{\varrho} c_{\varrho\varrho}^s = 0$. Since quadrupole radiation is characterized by a tensor with vanishing trace, the scattering described by c^s is called *quadrupole scattering*. The scattering caused by c^a is analoguous to the radiation emitted by a magnetic dipol, and this part is therefore called *magnetic dipole scattering*. The intensity of the scattered radiation is determined by

$$|(c_{\varrho\sigma})_{km}|^2 = |\delta_{\varrho\sigma}(c^0)_{km} + (c_{\varrho\sigma}^s)_{km} + (c_{\varrho\sigma}^a)_{km}|^2. \tag{3.4.16}$$

3.4.3 Scattering by a Charged Oscillating Particle

Consider a linear harmonic oscillator with mass m, charge e and force constant f, vibrating in the direction x. According to [Ref. 3.21, Eq. (2.84)] the eigenvalues are given by $E_n = \hbar\omega_0(n + 1/2)$, where $\omega_0 = (f/m)^{1/2}$.

a) Show that the intensity of the Rayleigh scattering is given by [3.26]

$$I_{nn} = \text{cst} \cdot \omega_L^4 \left(\frac{e^2/m}{\omega_L^2 - \omega_0^2}\right)^2 I_0, \tag{3.4.17}$$

which is independent of n and equal to the classical radiation emitted by a charged particle executing a forced oscillation governed by

$$m\ddot{x} + fx = eE_{0x}\cos\omega_L t. \tag{3.4.18}$$

b) Show that there is no Raman scattering for this harmonic oscillator. It should be mentioned here, that Raman scattering exists for an anharmonic oscillator, however. This does not mean that Raman scattering by molecules and crystals is due to anharmonic motion of the nuclei and that the intensity of the Raman lines is directly related to anharmonicity. This is only the case for that part of the scattered radiation which originates from the scattering by the nuclei[6]. The essential part of the scattering by molecules and crystals is, however, due to the scattering by the electrons, and the Raman effect is a consequence of the coupling between the motion of the electrons and nuclei [3.26].

Hint: Use (3.4.11–16) setting $k = n$, $\quad m = n'$, $\quad M_x = eu$, $\quad M_y = M_z = 0$, $\quad \hbar\omega_{n,n''} = E_{n''} - E_n'$ and [Ref. 3.21, Eq. (2.97)], namely

$$(M_x)_{nn''} = e\left(\frac{\hbar}{2m\omega_0}\right)^{1/2}[(n+1)^{1/2}\delta_{n,n''-1} + n''^{1/2}\delta_{n,n''-1}]. \tag{3.4.19}$$

In order to prove (a) show that

$$(c^0)_{nn} = \frac{e^2/3m}{\omega_0^2 - \omega_L^2} = \alpha_n^{(0)}; \quad (c_{xx}^s) = 2\alpha_n^{(0)}. \tag{3.4.20}$$

$$(c_{yy}^s)_{nn} = (c_{zz}^s)_{nn} = -\alpha_n^{(0)}.$$

To prove (b) show that $(c_{xx}^s)_{n,n\pm2} = 0$.

---

[6] This type of scattering is also known as the ionic Raman effect. It arises from the modulation of the ionic part of the polarizability by the displacements of the ions [Ref. 3.2, p. 347].

4. Brillouin Spectroscopy

In Raman and Brillouin scattering of photons by phonons, the frequency of the scattered radiation differs from that of the incident radiation by the frequency of the phonons. From an empirical point of view, the two types of scattering differ only in that optical phonons are involved in Raman scattering and that acoustical modes are involved in Brillouin scattering. In both phenomena, the intensity of the scattered radiation depends on the change in the electronic polarizability (or susceptibility) of the crystal which is induced by the phonons.

If the wave vector q of an acoustic mode is exactly zero we are dealing with rigid displacements of the entire crystal lattice [Ref. 4.1, Sect. 3.2]. Such displacements lead to no changes in the polarizability and hence do not contribute to Brillouin scattering. To obtain inelastic scattering of light it is necessary to consider the small but nonzero wave vector q which corresponds to sound waves. Since for each wave vector q there are three acoustic modes, the Brillouin spectrum will, in general, consist of three lines at each side of the central Rayleigh peak (Stokes and anti-Stokes lines). Due to the small frequencies of acoustic phonons for small q vectors, these Brillouin lines are separated by small frequency shifts, of the order of less than one $1\,\mathrm{cm}^{-1}$ from the Rayleigh line. For this reason it is not possible to use a grating monochromator as for Raman scattering (Sect. 3.1.1) but rather a Fabry-Pérot interferometer.

4.1 Experimental Techniques

Figure 4.1 shows schematically the apparatus used for Brillouin measurements. The beam of the laser L is focused onto the sample S. The light scattered at 90° is collected by the lens L_1 and the parallel beam passes the Fabry-Pérot interferometer FP. The light leaving the interferometer is collected by the lens L_2 and focused onto the photomultiplier PM. The resulting signal is then processed by a photon-counting equipment.

The *Fabry-Pérot* étalon [4.2, 3] usually consists of two parallel discs of glass or quartz, with the inner faces worked optically so that the gap between them is constant over the whole area to better than 1/50 to 1/100 of a wavelength of green light (Fig. 4.2).

These flat surfaces are coated with some material so that most of the light incident upon them is reflected and only a small fraction is transmitted.

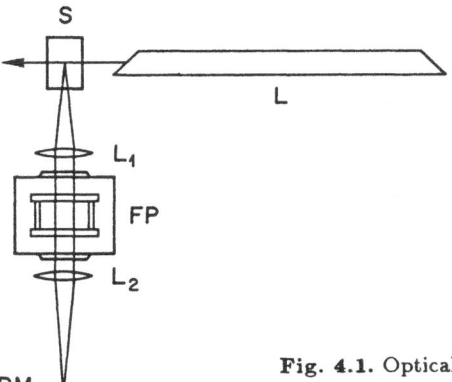

Fig. 4.1. Optical layout of a Brillouin scattering apparatus

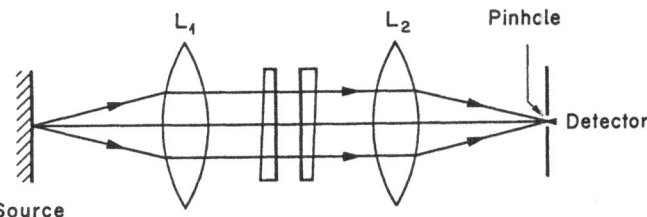

Fig. 4.2. Fabry-Pérot interferometer used as a spectrometer with photoelectric detection

Coating materials are chosen to have as small an absorption as possible. The transmitted intensity I_t for normal incidence is given by the so-called "Airy formula" [4.2, 3]

$$I_t(\delta) = I_i \left(\frac{T}{1-R}\right)^2 \frac{1}{1 + \frac{4R}{(1-R)^2} \sin^2 \frac{\delta}{2}}. \tag{4.1}$$

In this relation the effects of the medium between the plates is neglected. T and R are the fraction of the incident intensity I_i transmitted and reflected at each surface and

$$\delta = 4\pi l/\lambda, \tag{4.2}$$

is the phase change associated with the path $2l$, where l is the distance between the plates and λ the wavelength of the light. According to (4.1, 2) interference maxima occur for orders of interference m given by $\delta/2 = m\pi$ $(m = 1, 2, \ldots)$. Figure 4.3 shows $I_t(\delta)$ if monochromatic light enters the Fabry-Pérot interferometer. If light containing a few wavelengths λ_k is analyzed, as is the case for Brillouin scattering studies with a laser, the condition for a maximum of $I_t(\delta)$ is given by $l_k = m\lambda_k/2$ for each wavelength λ_k. Thus, by mechanical scanning of the plate separation l one obtains the required Brillouin spectrum such as shown in Fig. 4.5.

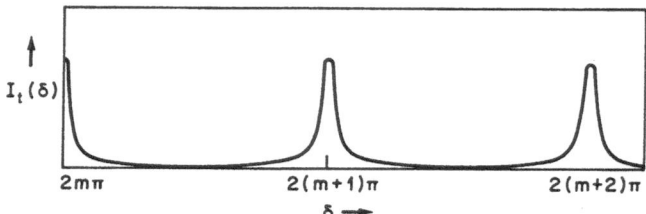

Fig. 4.3. The Airy function

According to (4.1) the sharpness of the peaks (interference fringes) in Fig. 4.3 and hence the resolving power of the instrument depend critically on R and, in particular, on the parameter $\beta = 4R/(1 - R)^2$ occurring in (4.1). The larger the reflectivity R, the larger is β and the sharper are the peaks. For $R = 0.86$ one obtains $\beta \cong 175$.

Since solids give rise to more spurious scattering than liquids the contrast of the Fabry-Pérot interferometer is particularly important in studies of the solid state. Defining the contrast C to be I_{max}/I_{min} and using (4.1) we obtain

$$C = \frac{I_{max}}{I_{min}} = 1 + \beta \cong \beta. \qquad (4.3)$$

With a single Fabry-Pérot interferometer $C \cong 10^2$ to 10^3 which is often not sufficient for Brillouin scattering of solids. A higher contrast may be achieved by using two Fabry-Pérot interferometers in series, or alternatively, by using two or more passes through the same interferometer. *Sandercock* [4.4, 5] has constructed a high-performance stabilized multi-pass interferometer which has a contrast several orders of magnitude greater than that of a single-pass

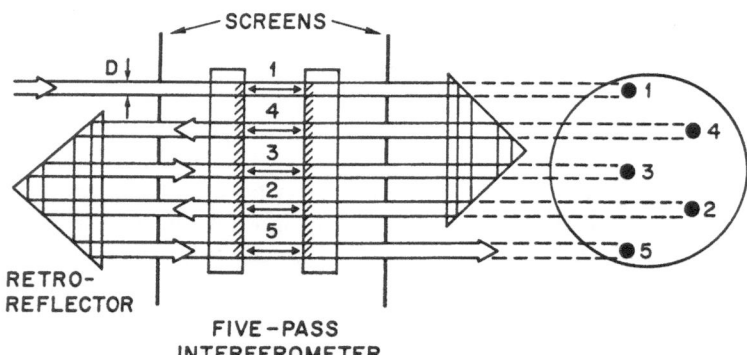

Fig. 4.4. Experimental arrangement for a five pass interferometer showing screens to reduce scattered light. The beam diameter has been reduced for clarity – in practice a greater fraction of the mirror area may be utilized [4.4]

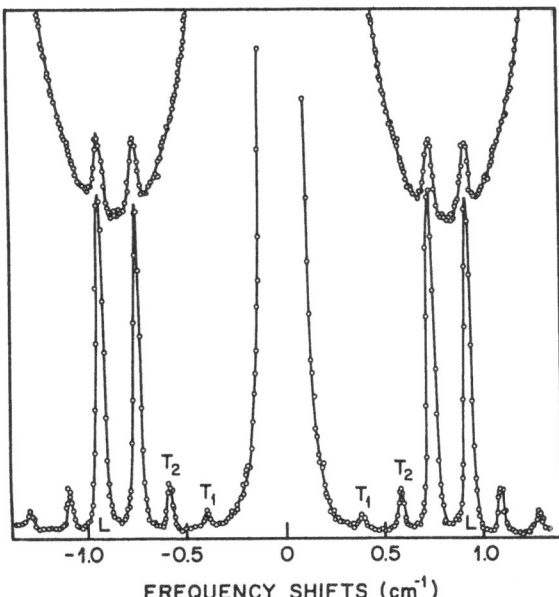

Fig. 4.5. Brillouin back scattering spectrum of SbSI using a double passed interferometer. The longitudinal peaks (L) from neighbouring orders are overlapped. The upper recording (the base line is the same) shows the spectrum obtained using a single pass through the same interferometer. Vertical scale linear [4.4, 5]

system (Fig. 4.4). Figure 4.5 shows the Brillouin back-scattering spectrum of SbSI using the interferometer shown in Fig. 4.4, but with only two passes [4.4]. The increase in contrast as compared with the single-pass spectrum is very impressive.

4.2 Kinematics and Origin of Brillouin Scattering

In terms of the corpuscular theory of light, first-order Brillouin scattering corresponds to an inelastic collision of a photon $(\omega_L, \mathbf{k}_L)$ with an acoustic phonon $(\omega_j, \mathbf{q})$. As for Raman scattering, the photon either loses a quantum of vibrational energy (Stokes line) or acquires such a quantum (anti-Stokes line). The conservation of energy and momentum are, see (3.40, 41),

$$\omega_L = \omega_{\mathrm{sc}} \pm \omega_j(\mathbf{q}), \tag{4.4}$$

$$\mathbf{k}_L = \mathbf{k}_{\mathrm{sc}} \pm \mathbf{q}, \tag{4.5}$$

where ω_{sc} is the frequency of the scattered photon with wave vector $\mathbf{k}_{\mathrm{sc}}$. The Stokes and anti-Stokes processes are illustrated in Fig. 3.21. As discussed in

Sect. 3.3.1, for light in the visible or UV region we have $k_L \cong k_{sc} \ll \pi/a$, where a is the lattice parameter. Therefore, only acoustic modes with small (but non-zero) wave vectors $\boldsymbol{q}$, i.e. sound waves, are observed in first-order Brillouin scattering. The dispersion of these acoustic modes is given by [Ref. 4.1, Sect. 3.7]

$$\omega_s = v_s q, \tag{4.6}$$

where v_s is the phase velocity of the phonon $s = (\boldsymbol{q}, j)$. From (4.4, 5) it follows (Sect. 3.3.5) that q is given by

$$q = 2 \frac{\omega_L n(\omega_L)}{c} \sin \theta. \tag{4.7}$$

In this equation $n(\omega_L)$ is the refractive index at the laser frequency ω_L, and 2θ is the scattering angle, i.e. the angle between $\boldsymbol{k}_L$ and $\boldsymbol{k}_{sc}$ (Fig. 3.27). Substituting (4.7) into (4.6) gives

$$\omega_s = 2 \frac{v_s}{c} \omega_L n(\omega_L) \sin \theta. \tag{4.8}$$

This relation shows that the positions of the Brillouin lines depend on ω_L and θ. Furthermore, due to the factor v_s/c, the frequency ω_s is extremely small compared with ω_L.

It is not difficult to understand why a beam of light should interact with acoustic waves. Consider first a crystal with all atoms at their equilibrium positions. The electric field of the light will produce a polarization P_0 which depends on the positions and polarizabilities of all the atoms in the unstrained crystal. If a long-wavelength acoustic wave propagates through the crystal, the positions of the atoms are slightly displaced from their equilibrium positions, i.e. the crystal is in a strained state, and this leads to a corresponding spatial variation $\delta P(\boldsymbol{r}, t)$ in the polarization: $P(\boldsymbol{r}, t) = P_0 + \delta P(\boldsymbol{r}, t)$. Since the dielectric constant ε is given by $\varepsilon = 1 + 4\pi P/E$, the variation $\delta P(\boldsymbol{r}, t)$ gives rise to a corresponding variation $\delta \varepsilon(\boldsymbol{r}, t)$. Thus, in Brillouin scattering, acoustic modes of long wavelengths produce strain and thereby modulate the dielectric constant (or refractive index) of the medium.

The situation is particularly simple for the case of longitudinal (or compression waves). Such waves create a succession of planes of higher and lower density, moving along with sound velocity v_s. Since the refractive index $n = \varepsilon^{1/2}$ depends on the density ϱ, the spatial variation in ϱ produces in term a spatial variation in n or ε. The planes of different refractive index act as "mirror planes" which reflect waves in a selective way (Fig. 4.6).

The general relation for the strain dependence of ε is discussed in Sect. 4.3. There it will be shown that also transverse waves which do not create a change in density, produce a spatial variation in ε, as shown in

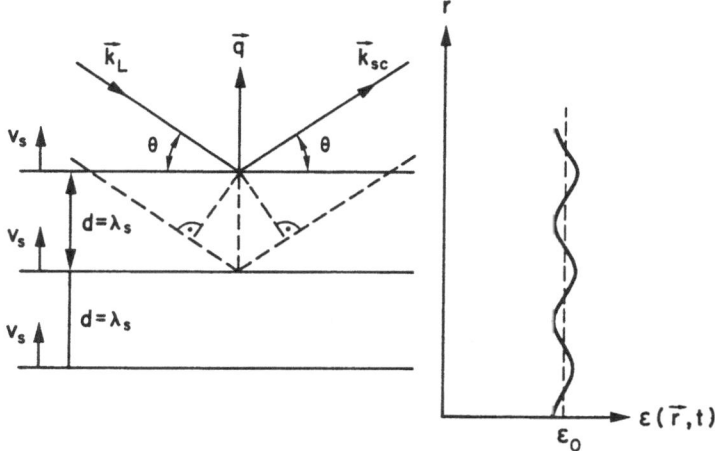

Fig. 4.6. In Brillouin scattering acoustic modes of long wavelengths λ_s produce strain and thereby modulate the dielectric constant ε (or refrative index) of the medium. The planes of different refractive index act as "mirror planes" moving with sound velocity v_s which reflect waves in a selective way. The figure also shows the spatial variation of the dielectric constant $\varepsilon(\mathbf{r}, t)$ associated with the elastic wave

Fig. 4.6. If the pattern of atomic displacements created by the sound waves were at rest, this would be the usual problem of X-ray reflection on crystal planes according to Bragg's law

$$2d \sin \theta = m\lambda_L. \tag{4.9}$$

In fact, Bragg's law with order $m = 1$ is identical with (4.7) as can be verified by putting $d = \lambda_s$, $q = 2\pi/\lambda_s$, $\lambda_L = \lambda_v/n(\omega_L)$ and $\omega_L\lambda_v = 2\pi c$ (λ_v: wavelength in vacuum). Since the sound wave has a frequency ω_s and propagates with velocity $\pm v_s$, we expect that the frequency of the scattered light is $\omega_{sc} = \omega_L \pm \omega_s$; this frequency change may be considered as a Doppler effect which is due to the "moving mirror planes" of velocity $\pm v_s$ (Problem 4.5.1). It should be emphasized that if $\mathbf{k}_L, \omega_L$ and 2θ are fixed by the experiment, only one set of sound waves is involved, namely that which is selected by Bragg's law (4.9) with $m = 1$. Its direction is found by inspection of Fig. 4.6, the magnitude of $\mathbf{q}$ follows from (4.7) and the three frequencies associated with $\mathbf{q}$ are given by (4.6), where $v_s = v(\mathbf{q}, j)$ with polarization or branch indices $j = 1, 2, 3$.

4.3 Strain Dependence of Dielectric Constant

In the following we confine ourselves to cubic crystals. In this case the time average optical dielectric constant is a scalar ε_∞ times the unit tensor, and the refractive index n_∞ is independent of the direction of propagation.

However, at a given instant of time, the thermal fluctuations in the crystal (which change very slowly compared with the electric field of the light) cause off-diagonal components to appear in the dielectric constant, so that we must regard $\delta\varepsilon$ as being a tensor whose elements fluctuate in time. A particular sound wave with frequency ω_s and wave vector $\boldsymbol{q}$ will produce the contribution

$$\delta\varepsilon_{\alpha\beta}^{(s)}(\boldsymbol{r}, t) = \delta\varepsilon_{\alpha\beta}^{(s)} \cos{(\boldsymbol{q}\cdot\boldsymbol{r} - \omega_s t)}, \tag{4.10}$$

to $\delta\varepsilon(\boldsymbol{r}, t)$. Here, α and β denote the coordinate axes parallel to the axes of the cube, i.e. $\alpha, \beta = 1, 2, 3$. In general, for small strains, we may express the change in the dielectric tensor components $\delta\varepsilon_{\alpha\beta}(\boldsymbol{r}, t)$ as a linear function of the elastic strain components $\epsilon_{\gamma\lambda}(\boldsymbol{r}, t)$. For cubic crystals one obtains [4.6–10]

$$-\frac{1}{\varepsilon_{\infty}^2}\delta\varepsilon_{\alpha\beta}(\boldsymbol{r}, t) = \sum_{\gamma\lambda} p_{\alpha\beta,\gamma\lambda}\epsilon_{\gamma\lambda}(\boldsymbol{r}, t). \tag{4.11}$$

The coefficients $p_{\alpha\beta,\gamma\lambda}$ are the Pockels elasto-optical constants [4.7]. As discussed in [Ref. 4.1, Sect. 3.6], the strain components $\epsilon_{\gamma\lambda}(\boldsymbol{r}, t)$ are related to the elastic displacements $\boldsymbol{u}(\boldsymbol{r}, t)$ in the medium by[1]

$$\epsilon_{\gamma\lambda}(\boldsymbol{r}, t) = \frac{1}{2}\left(\frac{\partial u_{\gamma}(\boldsymbol{r}, t)}{\partial x_{\lambda}} + \frac{\partial u_{\lambda}(\boldsymbol{r}, t)}{\partial x_{\gamma}}\right), \tag{4.12}$$

where $\boldsymbol{r} = (x_1, x_2, x_3)$. In a cubic crystal in which each cube axis has a fourfold symmetry, there are only three independent constants in the elasto-optic tensor, namely [4.7]

$$p_{11,11} = p_{22,22} = p_{33,33} = p_{11},$$
$$p_{11,22} = p_{22,11} = p_{22,33} = p_{33,22} = p_{33,11} = p_{11,33} = p_{12},$$
$$p_{12,12} = p_{21,21} = p_{12,21} = p_{21,12} = p_{23,23} = p_{23,32} = p_{32,23}$$
$$= p_{32,32} = p_{31,31} = p_{31,13} = p_{13,31} = p_{13,13} = p_{44}. \tag{4.13}$$

We may therefore write (4.11) as[2]

$$-\frac{1}{\varepsilon_{\infty}^2}\delta\varepsilon_{\alpha\beta} = 2p_{44}\epsilon_{\alpha\beta} + \delta_{\alpha\beta}(p_{11} - p_{12} - 2p_{44})\epsilon_{\alpha\alpha}$$
$$+ \delta_{\alpha\beta}p_{12}\sum_{\gamma}\epsilon_{\gamma\gamma}. \tag{4.14}$$

---

[1] In order to avoid confusion with the dielectric constant ε we use here the symbol ϵ for the strain tensor (ϵ corresponds to ε in [Ref. 4.1, Sect. 3.6])

[2] There is an error in [Ref. 4.9, Eq. (25)]. Equation (4.14) differs from Eq. (25) in [Ref. 4.9] by the factor of 2 multiplying p_{44}; see also [4.10]

The last term in (4.14) is closely related with the relative change in volume V or density ϱ at the point $\boldsymbol{r}$. In fact, we have [4.11]

$$\sum_\gamma \epsilon_{\gamma\gamma} = \delta V / V_0 = -\delta\varrho/\varrho_0. \tag{4.15}$$

On the other hand, from (4.12) we obtain

$$\sum_\gamma \epsilon_{\gamma\gamma} = \nabla\cdot\boldsymbol{u}, \tag{4.16}$$

from which it follows that the local change in density at the point $\boldsymbol{r}$ and time t is given by

$$\delta\varrho(\boldsymbol{r},t) = -\varrho_0\nabla\cdot\boldsymbol{u}(\boldsymbol{r},t), \tag{4.17}$$

where ϱ_0 is the density of the undeformed body.

In order to illustrate (4.14) we first consider the alteration of the optical dielectric tensor produced by an adiabatic hydrostatic compression. In such a compression $\epsilon_{11} = \epsilon_{22} = \epsilon_{33}$, $\epsilon_{\alpha\beta} = 0$ for $\alpha \neq \beta$, and from (4.15) one obtains

$$\epsilon_{\alpha\beta} = -\tfrac{1}{3}\delta_{\alpha\beta}\delta\varrho/\varrho,$$

from which we find

$$\delta\varepsilon_{\alpha\beta} = \delta_{\alpha\beta}\delta\varepsilon_\infty, \quad \text{where}$$

$$\delta\varepsilon_\infty = \tfrac{1}{3}\varepsilon_\infty^2(p_{11} + 2p_{12})\delta\varrho/\varrho. \tag{4.18}$$

From (4.18) if follows that the change of the refractive index produced by a hydrostatic pressure is given by

$$\delta n_\infty \cong \tfrac{1}{6}n_\infty^3(p_{11} + 2p_{12})\delta\varrho/\varrho. \tag{4.19}$$

Here n_∞ is the refractive index of the undeformed body and $n_\infty^2 = \varepsilon_\infty$.

Next consider a longitudinal acoustic wave propagating along (100). Using (4.14) for a strain $\epsilon_{11} \neq 0$ and all other $\epsilon_{\alpha\beta} = 0$, we obtain

$$\delta\varepsilon_{11} = -\varepsilon_\infty^2 p_{11}\epsilon_{11}, \tag{4.19a}$$

$$\delta\varepsilon_{22} = \delta\varepsilon_{33} = -\varepsilon_\infty^2 p_{12}\epsilon_{11}. \tag{4.19b}$$

The corresponding changes in the refractive index are given by

$$\delta n_1 \cong -\tfrac{1}{2}n_\infty^3 p_{11}\epsilon_{11}, \tag{4.19c}$$

$$\delta n_2 = \delta n_3 \cong -\tfrac{1}{2} n_\infty^3 p_{12} \epsilon_{11}. \tag{4.19d}$$

Thus a simple strain ϵ_{11} produces uniaxial birefringence.

We now consider a shear wave, say ϵ_{23}, which propagates along (010) with particle displacements along (001). From (4.14) we obtain

$$\delta \varepsilon_{23} = -2\varepsilon_\infty^2 p_{44} \epsilon_{23}. \tag{4.20a}$$

Such a shear produces biaxial birefringence with the axis of the index ellipsoid in the directions $x_1' = x_1 = (100)$, $x_2' = (011)$, and $x_3' = (01\bar{1})$; the corresponding changes of n are given by

$$\delta n_1' = 0, \tag{4.20b}$$

$$\delta n_2' = -n_\infty^3 p_{44} \epsilon_{23}, \tag{4.20c}$$

$$\delta n_3' = +n_\infty^3 p_{44} \epsilon_{23}. \tag{4.20d}$$

It is also informative to examine the form of the relation (4.14) for isotropic solids such as amorphous compounds. In this case

$$p_{44} = \tfrac{1}{2}(p_{11} - p_{12}), \tag{4.21}$$

and there are only two independent Pockels constants. The relation (4.14) reduces to

$$-\frac{\delta \varepsilon_{\alpha\beta}}{\varepsilon_\infty^2} = (p_{11} - p_{12})\epsilon_{\alpha\beta} + \delta_{\alpha\beta} p_{12} \sum_\gamma \epsilon_{\gamma\gamma}. \tag{4.22}$$

In a liquid $p_{44} = 0$, and hence there is only one elasto-optical coefficient $p_{11} = p_{12}$, and the fluctuations in the dielectric constant tensor are just

$$\delta \varepsilon_{\alpha\beta} = \delta_{\alpha\beta} \delta \varepsilon, \quad \text{where}$$

$$\delta \varepsilon = -\varepsilon_\infty^2 p_{11} \nabla \cdot \boldsymbol{u} = \varepsilon_\infty^2 p_{11} \delta \varrho / \varrho, \tag{4.23}$$

and where we have used (4.16, 17).

From (4.19, 20) it follows that p_{11} and p_{23} are involved in a longitudinal acoustic wave (compression), and that p_{44} is involved in a purely transverse wave (shear). The Pockels constants p_{11} and p_{12} are usually considerably larger than p_{44} and therefore larger elasto-optic effects are expected for longitudinal waves than for transverse waves. This is physically appealing since longitudinal waves are associated with local changes in the density, but not transverse waves.

Many theoretical studies have been performed with the aim to explain the photo-elastic behaviour and in particular to evaluate the photo-elastic constants p_{11}, p_{12} and p_{44} for cubic crystals [4.12–16]. Such calculations require the evaluation of the polarization P or the dielectric constant ε_∞ for the strained crystal. The simplest models are based on the Clausius-Mosotti relation for ε_∞ [Ref. 4.1, Sect 4.2.2]. This relation, together with (4.18) allows an estimate of the quantity $p_0 = p_{11} + 2p_{12}$ (Problem 4.5.2). In more elaborate models the constants p_{ij} are expressed in terms of the parameters of the three-body-force shell model [4.13].

In the next section it will be shown that information about the photo-elastic constants p_{ij} can be obtained from the observed intensities of the Brillouin components.

4.4 Intensities of Brillouin Components

As mentioned at the beginning of this chapter, Raman and Brillouin scattering differ only in that optical phonons are involved in Raman scattering and acoustic modes in Brillouin scattering. The calculation of the intensities of the Brillouin components is therefore similar to that of the Raman lines. In both cases the electric field of the incident light, E, induces a dipole moment $M = \alpha E$, α being the electronic polarizability (or susceptibility). The dipole moment emits radiation whose intensity is proportional to the time average $\langle |\ddot{M}(t)|^2 \rangle_t$ as discussed in Sect. 3.2. In the case of Raman scattering α is modulated by optical phonons and in the case of Brillouin scattering by long wavelength acoustic phonons. In both cases the intensity is proportional to $|\delta\alpha|^2$, $\delta\alpha$ being the change of α produced by the phonons involved. It is, in fact possible to develop a microscopic theory of Brillouin scattering on the basis of the discrete lattice which involves the individual atomic or ionic polarizabilities [4.17]. From a practical point of view, however, it is more convenient to base the calculation of the intensities of the Brillouin components on a macroscopic model which considers the scattering system as an elastic medium (long-wavelength limit). In this approach the intensities of the Brillouin components are proportional to $|\delta\varepsilon|^2$, where $\delta\varepsilon$ is the change of the optical dielectric constant ε_∞ produced by the long-wavelength acoustic phonons (sound waves). For a particular acoustic mode $s = (qj)$ and at high temperatures ($\hbar\omega_s \ll k_B T$), the intensity is given by [4.7, 8]

$$I_s(e, f) \sim \frac{k_B T}{\omega_s^2} |f \delta\varepsilon(s) e|^2, \tag{4.24}$$

where e and f are the unit polarization vectors of the incident and scattered light, and $\delta\varepsilon(s)$ is the modulation of the dielectric constant produced by the mode s. $\delta\varepsilon(s)$ is obtained from (4.11, 12) by transforming the spatially slowly-changing displacement field $u(r, t)$ to normal coordinates Q_s of the acoustic vibrations ($j = 1, 2, 3$). According to (4.11) this leads to expressions for the intensities which involve the Pockels constants $p_{\alpha\beta,\gamma\lambda}$.

The explicit expression of the Brillouin components associated with the acoustic phonon $s = (\boldsymbol{q}, j)$ for unpolarized light can be written in the form [4.7]

$$I_s/I_0 = \frac{A}{\varrho v_s^2} \sum_{ik} \Big(\sum_{\alpha\beta} f_{i\alpha} g_{\alpha\beta}(s) e_{k\beta} \Big)^2. \tag{4.25}$$

In this expression I_0 is the intensity of the incident beam, ϱ is the density of the scattering medium, and v_s is the phase velocity of the acoustic phonon $s = (\boldsymbol{q}, j)$. The vectors $\boldsymbol{e}_k$ ($k = 1, 2$) with components $e_{k\beta}$ are two mutually perpendicular unit vectors both perpendicular to the direction of incidence; similarly, the vectors $\boldsymbol{f}_i$ ($i = 1, 2$) with components $f_{i\alpha}$ are two mutually perpendicular unit vectors both perpendicular to the direction of observation. The constant A is given by

$$A = \frac{V k_{\mathrm{B}} T \omega_L^4}{8\pi^2 c^4}, \tag{4.26}$$

where V is the scattering volume and ω_L the frequency of the incident light. The coefficients $g_{\alpha\beta}(s)$ are defined by

$$g_{\alpha\beta}(s) = \frac{1}{2^{3/2}} \sum_{\gamma\lambda} k_{\alpha\beta,\gamma\lambda} e_\gamma(s) h_\lambda. \tag{4.27}$$

In this equation $e_\gamma(s)$ are the components of the unit polarization vectors for elastic waves belonging to the phonon s and h_λ are the components of the unit vector parallel to the wave vector $\boldsymbol{q}$. The coefficients $k_{\alpha\beta,\gamma\lambda}$ are given by

$$k_{\alpha\beta,\gamma\lambda} = - \sum_{\mu\nu} (\varepsilon_\infty)_{\alpha\mu} p_{\mu\nu,\gamma\lambda} (\varepsilon_\infty)_{\nu\beta}, \tag{4.28}$$

where ε_∞ is the optical dielectric constant and $p_{\mu\nu,\gamma\lambda}$ are the elasto-optical constants discussed in Sect. 4.3.

The intensity relation (4.25) has been derived on the basis of the following approximations:

a) Since $\omega_s \ll \omega_L$, the factor $(\omega_L \pm \omega_s)^4$ has been replaced by ω_L^4 in (4.26).

b) Classical theory has been used. This is a good approximation, since except at the very lowest temperature, $k_{\mathrm{B}} T \gg \hbar\omega_s$ for long-wavelength acoustic phonons. According to classical theory the intensity vanishes as $T \to 0$. A quantum-mechanical calculation shows that the spectrum of the scattered light becomes asymmetrical as $k_{\mathrm{B}} T \lesssim \hbar\omega_s$ [4.9]. As $T \to 0$, the amount of light in the low-frequency (Stokes) side of each doublet becomes independent

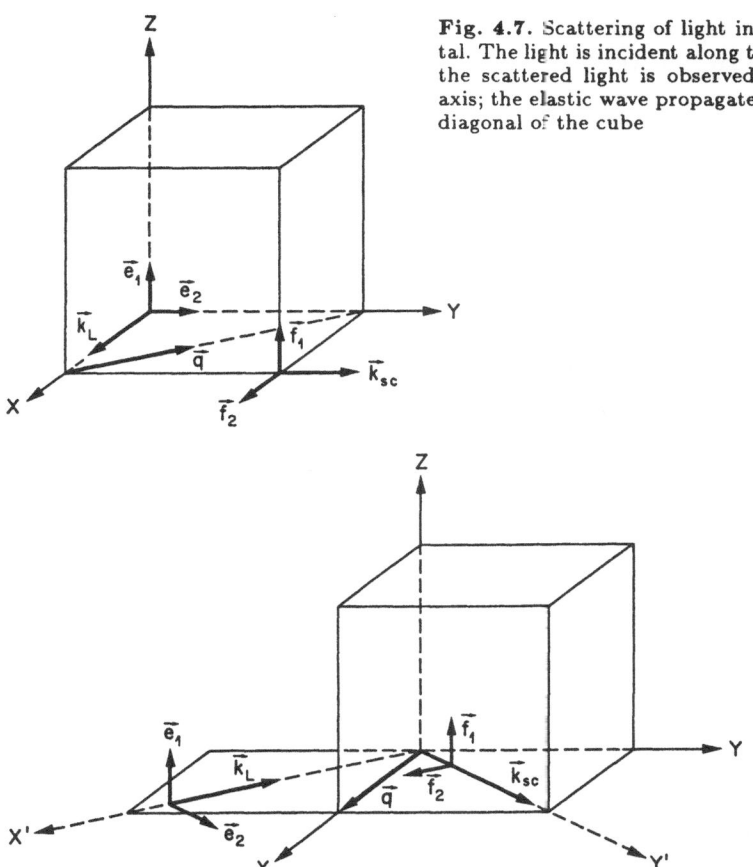

Fig. 4.7. Scattering of light in a cubic crystal. The light is incident along the X axis and the scattered light is observed along the Y axis; the elastic wave propagates along a face diagonal of the cube

Fig. 4.8. Scattering of light in a cubic crystal. The light is incident along a face diagonal and the scattered light is observed at an angle of 90° along another face diagonal of the cube; the elastic wave propagates along the X axis

on temperature, while the high-frequency (anti-Stokes) component steadily gets weaker in intensity like $\exp(-\hbar\omega_s/k_B T)$. Quantum mechanically this results because one can create phonons even at $T = 0$ (Stokes component); however, the annihilation of phonons (anti-Stokes components) is proportional to the number of excited phonons and this number goes to zero as $\exp(-\hbar\omega_s/k_B T)$ as $T \to 0$. A similar temperature dependence has been obtained for Stokes and anti-Stokes Raman lines, see (3.87–89).

c) The relation (4.25) gives an approximate expression for the intensity inside the scattering medium. Since the intensity is measured outside the medium, corrections should be made for transmission and reflection losses of the scattered light at the boundaries of the crystal.

In the following we illustrate the scattering formula (4.25) for the case of cubic crystals for which $(\varepsilon_\infty)_{\alpha\mu} = \delta_{\alpha\mu}\varepsilon_\infty$, and (4.28) reduces to

$$k_{\alpha\beta,\gamma\lambda} = -\varepsilon_\infty^2 p_{\alpha\beta,\gamma\lambda}. \tag{4.29}$$

Substituting (4.29) into (4.27) gives

$$g_{\alpha\beta}(s) = -\frac{\varepsilon_\infty^2}{2^{3/2}} \sum_{\gamma\lambda} p_{\alpha\beta,\gamma\lambda} e_\gamma(s) h_\lambda. \tag{4.30}$$

We consider two cases: (I) the light is incident along the X axis of the cube (axis of symmetry of fourth order) and is obseved along the Y axis of the cube (Fig. 4.7). (II) The light is incident along a face diagonal of the cube and the scattered light is observed at an angle of 90°along another face diagonal of the cube (Fig. 4.8).

Table 4.1a. Eigenfrequencies and eigenvectors for elastic waves propagating along $(-1, 1, 0)$ in a cubic crystal (Fig. 4.7). (T: transverse, L: longitudinal)

j	$\omega_s = v_s q$	$e_1\binom{q}{j}$	$e_2\binom{q}{j}$	$e_3\binom{q}{j}$	Polarization
1	$(C_{44}/\varrho)^{1/2}q$	0	0	1	T_1
2	$[(C_{11} - C_{12})/2\varrho]^{1/2}q$	$2^{-1/2}$	$2^{-1/2}$	0	T_2
3	$[(C_{11} + C_{12} + 2C_{44})/2\varrho]^{1/2}q$	$-2^{-1/2}$	$2^{-1/2}$	0	L

Table 4.1b. Relative intensities for Brillouin scattering in a cubic crystal in the case in which the light is incident along the X axis and observed along the Y axis. In the scattering configuration $X(ZX)Y$, for example, the polarizer is parallel to the Z axis and the analyzer parallel to the X axis, as explained in Sect. 3.1.2

Mode	Configuration $X(ZZ)Y$	$X(ZX)Y$	$X(YZ)Y$	$X(YX)Y$
$j = 1$, T_1	0	$\dfrac{p_{44}^2}{2C_{44}}$	$\dfrac{p_{44}^2}{2C_{44}}$	0
$j = 2$, T_2	0	0	0	0
$j = 3$, L	$\dfrac{2p_{12}^2}{C_{11}+C_{12}+2C_{44}}$	0	0	$\dfrac{2p_{44}^2}{C_{11}+C_{12}+2C_{44}}$

Case I: We have $k_L = (k,0,0)$, $k_{sc} = (0,k,0)$ and according to (4.5) $q = \pm(-k,k,0)$. The vectors e_k and f_i in (4.25) are chosen as follows: $e_1 = (0,0,1)$, $e_2 = (0,1,0)$, $f_1 = (0,0,1)$, $f_2 = (1,0,0)$. Using the results obtained for elastic waves in cubic crystals [Ref. 4.1, Problem 3.8.5]one easily obtains the eigenfrequencies $\omega_s = \omega_j(q)$ and eigenvectors $e\binom{q}{j}$ for the acoustic modes $j = 1, 2, 3$. The results are summarized in Table 4.1a. It is now straightforward to calculate the intensities I_s as given by (4.25, 26, 30). Using (4.13) for the elasto-optical constants of cubic crystals we obtain the results given in Table 4.1b. The absolute intensities are obtained by multiplying the terms in Table 4.1b by the common factor $A\varepsilon_\infty^4/8$, where A is defined by (4.26). Table 4.1b shows that the intensity of one of the components of the fine structure, corresponding to the transverse T_2 mode which is polarized in the plane of incidence, is equal to zero. Therefore, in this case, there will be only two Stokes and two anti-Stokes satellites. From Table 4.1b it follows that for irradiation of the crystal with natural (unpolarized) light and observation without an analyzer, the total intensity of the two Stokes and the anti-Stokes components is given by

$$I/I_0 = \frac{V k_B T \omega_L^4 \varepsilon_\infty^4}{32\pi^2 c^4} \left(\frac{2(p_{12}^2 + p_{44}^2)}{C_{11} + C_{12} + 2C_{44}} + \frac{p_{44}^2}{C_{44}} \right). \tag{4.31}$$

This relation has already been derived by *Leontovich* and *Mandelstam* [4.18].

Case II: We have $k_L = 2^{-1/2}(-k, k, 0)$, $k_{sc} = 2^{-1/2}(k, k, 0)$ and according to (4.5) $q = \pm 2^{1/2}(-k, 0, 0)$. The polarization vectors of the light e_k and f_i are chosen as follows: $e_1 = (0, 0, 1)$, $e_2 = 2^{-1/2}(1, 1, 0)$; $f_1 = (0, 0, 1)$, $f_2 = 2^{-1/2}(-1, 1, 0)$ (Fig. 4.8). The eigenfrequencies and eigenvectors are summarized in Table 4.2a. The results for the relative intensities are summarized in Table 4.2b. The absolute intensities are obtained by multiplying the terms in Table 4.2b by the common factor $A\varepsilon_\infty^4/8$ with A given by (4.26). The transverse mode T_2 polarized in the plane of incidence has

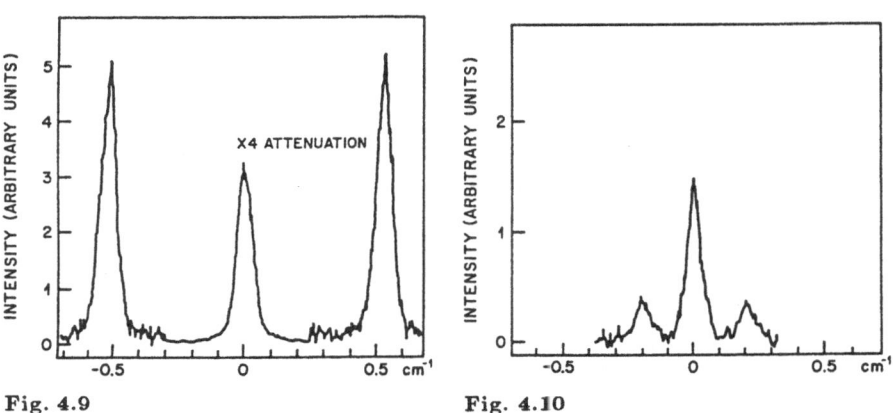

Fig. 4.9   Fig. 4.10

Fig. 4.9. Brillouin scattering from NaCl by a longitudinal elastic wave propagating along a face diagonal of the cube (Fig. 4.7); the scattering geometry is $X(ZZ)Y$ (Table 4.1b) [4.10]

Fig. 4.10. Brillouin scattering from KCl by a transverse elastic wave propagating along the X axis of the cube (Fig. 4.8); the scattering geometry is $X'(Y'Z)Y'$ (Table 4.2b) [4.10]

Table 4.2a. Eigenfrequencies and eigenvectors for elastic waves propagating along the (1,0,0) direction of a cubic crystal. (T: transverse, L: longitudinal)

j	$\omega_s = v_s q$	$e_1\binom{q}{j}$	$e_2\binom{q}{j}$	$e_3\binom{q}{j}$	Polarization
1	$(C_{44}/\varrho)^{1/2}q$	0	0	1	T_1
2	$(C_{44}/\varrho)^{1/2}q$	0	1	0	T_2
3	$(C_{11}/\varrho)^{1/2}q$	1	0	0	L

Table 4.2b. Relative intensities for Brillouin scattering in a cubic crystal in the case in which the light is incident along the face diagonal X' and observed along the face diagonal Y' of the cube (Fig. 4.8). In the scattering configuration $X'(ZX')Y'$, for example, the polarizer is parallel to the Z axis and the analyzer parallel to the X' axis

Mode	Configuration	$X'(ZZ)Y'$	$X'(ZX')Y'$	$X'(Y'Z)Y'$	$X'(Y'X')Y'$
$j = 1$, T_1		0	$\dfrac{p_{44}^2}{2C_{44}}$	$\dfrac{p_{44}^2}{2C_{44}}$	0
$j = 2$, T_2		0	0	0	0
$j = 3$, L		$\dfrac{p_{12}^2}{C_{11}}$	0	0	$\dfrac{(p_{11}-p_{12})^2}{4C_{11}}$

again zero intensity. From Table 4.2b the total intensity for the Stokes and anti-Stokes lines with unpolarized light is in this case given by

$$I/I_0 = \frac{V k_B T \omega_L^4 \varepsilon_\infty^4}{32\pi^2 c^4}\left(\frac{4p_{12}^2 + (p_{11}-p_{12})^2}{4C_{11}} + \frac{p_{44}^2}{C_{44}}\right). \tag{4.32}$$

A comparison of Cases I and II shows that for the Brillouin components the cubic crystal is not isotropic. This is the case for both the frequencies and the intensities and originates from the fact that there are anisotropies of the elastic and elasto-optical properties of cubic crystals.

Figures 4.9 and 10 show the Brillouin spectra of NaCl and KCl for selected scattering geometries [4.10]. The spectrum in Fig. 4.9 is due to light scattering by a longitudinal elastic wave propagating along a face diagonal of the cube (Fig. 4.7), and the scattering geometry is $X(ZZ)Y$ (Table 4.1b). In the case of KCl, light is scattered by a transverse wave propagating along the X axis of the cube (Fig. 4.8) and the scattering geometry is $X'(Y'Z)Y'$ (Table 4.2b).

From the observed frequencies of the Brillouin components and the density it is possible to determine the elastic constants C_{11}, C_{12} and C_{44} (Tables 4.1a, b). The photoelastic constants p_{11}, p_{12} and p_{44} can be determined from the observed intensities (Tables 4.1b, 4.2b). For this purpose

Table **4.3.** Elastic and photo-elastic constants determined from Brillouin scattering. The elastic constants are in units of GPa [4.19]

Crystal	NaCl	KBr	LiF
C_{11}	49.4±0.2	35.1±0.2	113.6±0.9
p_{11}	0.112±0.019	0.22±0.03	0.032±0.005
C_{12}	12.7±0.3	6.1±0.4	47±2
p_{12}	0.158±0.012	0.171±0.013	0.137±0.012
C_{44}	12.7±0.2	5.1±0.1	62.9±0.6
p_{44}	−0.013±0.002	−0.019±0.002	−0.051±0.008

it is necessary to determine the absolute intensities of the Brillouin components by comparing the Brillouin intensities with the intensities scattered from suitable liquids such as toluene or benzene which have well-known photoelastic constants [4.10]. Table 4.3 contains the elastic and photoelastic constants of NaCl, KBr, and LiF as determined from Brillouin scattering [4.19].

4.5 Problems

4.5.1 Doppler Effect and Brillouin Frequency Shifts

Referring to Fig. 4.6 show that as a consequence of the Doppler effect on the "moving mirror planes" of the sound velocity $\pm v_s$, the frequency of the scattered light is $\omega_{sc} = \omega_L \pm \omega_s$, where

$$\omega_s = 2(v_s/c)\omega_L n(\omega_L) \sin \theta. \qquad (4.5.1)$$

Hint: For an observer moving with the sound wave of velocity $+v_s$, the frequency of the incident light is $\omega_1 = (1+v_s'/v_L)\omega_L$, whereas the frequency of the scattered light is $\omega_2 = (1 - v_s'/v_L)\omega_L$, where $v_s' = v_s \sin \theta$ and $v_L = c/n$. The total frequency change is $\omega_s = \omega_1 - \omega_2$. If the sound wave is moving in the opposite direction, the frequency shift has opposite sign.

4.5.2 Strain Dependence of ε_∞ Under the Effect of Hydrostatic Pressure for Alkali Halides

a) Show that the strain derivative of the electronic dielectric constant ε_∞ under the effect of hydrostatic pressure is given by

121

$$\frac{R}{\varepsilon_\infty}\frac{d\varepsilon_\infty}{dR} = -\frac{1}{\varepsilon_\infty}(\varepsilon_\infty - 1)(\varepsilon_\infty + 2)(1 - \lambda), \qquad (4.5.2)$$

where the strain polarizability parameter λ is given by

$$\lambda = \frac{1}{3}\frac{R}{\alpha}\frac{d\alpha}{dR}. \qquad (4.5.3)$$

Here, R is the interionic separation and α is the electronic polarizability per unit cell.

b) From (4.18) show that

$$\frac{R}{\varepsilon_\infty}\frac{d\varepsilon_\infty}{dR} = -\varepsilon_\infty(p_{11} + 2p_{12}), \quad \text{and hence} \qquad (4.5.4)$$

$$p_0 = p_{11} + 2p_{12} = \frac{1}{\varepsilon_\infty^2}(\varepsilon_\infty - 1)(\varepsilon_\infty + 2)(1 - \lambda). \qquad (4.5.5)$$

Hint: Use the Clausius-Mosotti relation

$$\frac{\varepsilon_\infty - 1}{\varepsilon_\infty + 2} = \frac{4\pi\alpha}{3v_a}, \qquad (4.5.6)$$

derived in [Ref. 4.1, Sect. 4.2.2] and $v_a = 2R^3$. Note that the paramter λ is a measure of the ionic overlap and the nature of the bonding; its evaluation is not simple but theoretical estimates based on more elaborated models have been worked out [4.20]. Good agreement is obtained for experimental and theoretical values of p_0 for many alkali halides.

4.5.3 Brillouin Intensities for Isotropic Solids and Liquids

a) Based on the discussion given in Sect. 4.3 show that for isotropic solids such as glasses and amorphous compounds, for which $C_{44} = (C_{11} - C_{12})/2$, the total Brillouin intensity of Stokes and anti-Stokes lines is given by

$$I/I_0 = K\left(\frac{4p_{12}^2 + (p_{11} - p_{12})^2}{4C_{11}} + \frac{(p_{11} - p_{12})^2}{2(C_{11} - C_{12})}\right), \qquad (4.5.7)$$

where

$$K = V k_B T \omega_L^4 \varepsilon_\infty^4 / 32\pi^2 c^4. \qquad (4.5.8)$$

b) Show that the corresponding intensity for liquids is given by

$$I/I_0 = K p_{11}^2 / C_{11}. \qquad (4.5.9)$$

5. Interaction of X-Rays with Phonons

We have mentioned in Chap. 1 that the scattering of X-rays by phonons is not a very suitable technique to study phonon dispersion curves. If we wish to obtain detailed information about lattice vibrations for which the value of the wave vector q is anywhere in the Brillouin zone we must study their interaction with radiation of wavelength comparable with interatomic distances and having an energy quantum $\hbar\omega$ comparable with phonon energies. X-rays satisfy the first of these conditions, but not the second since $\hbar\omega = 10^4$ eV for X-rays while $\hbar\omega_j(q) \cong 0.02$ eV for phonons (Fig. 1.1). However, a beam of thermal neutrons having a velocity $v_n = 2 \cdot 10^3$ m/s has a wave vector of magnitude $k = m_n v_n / \hbar$ or wavelength $\lambda = 2\pi/k = 2.0$ Å, and the energy quantum is $m_n v_n^2 / 2 = 0.02$ eV. The scattering of a beam of mono-energetic neutrons of about this wavelength provides the most powerful method of studying lattice vibrations and measuring phonon energies. Such experiments are experiments in neutron spectroscopy which will be discussed in Chap. 6.

The reader may then ask why a whole chapter is devoted to the scattering of X-rays by phonons. There are essentially three reasons:

1) As is well known, the scattering of X-rays by a perfectly ordered crystal gives rise to Bragg scattering; if the intensities of the Bragg reflections are known it is possible to determine the crystal structure. Now consider a disordered crystal. The disorder may be "static" such as in a truely disordered solid and/or "dynamic" due to thermal motion of the atoms. Dynamic disorder will, of course, always be present, independent of whether the mean positions of the atoms form an ordered lattice or a disordered arrangement. Any departure from strict regularity in the arrangement of the scattering atoms will tend to weaken the intensities of the Bragg reflections; but there can be no loss of scattered energy on the whole, and there must be an increase in the scattered intensity in directions not allowed by Bragg's law. This extra scattering which is due to static and/or dynamic disorder is called *diffuse scattering*. Diffuse scattering of X-rays is a widely used tool for the investigation of short-range order in disordered solids [5.1–4], superstructures due to lattice instabilities [5.5, 6]and of ion-ion correlations in superionic conductors [5.7–15]. In such experiments the presence of thermal diffuse scattering remains a constant potential complication as one is generally unable to account quantitatively for the phonon induced scattering (dynamic disorder). Phonon scattering may mask the diffuse scattering due

to static short-range order in disordered solids, so that it may in many cases be impossible to deduce any information about the latter from diffuse X-ray scattering. Static and dynamic disorder can, in principle, be distinguished by measuring the temperature dependence of the total diffuse scattering. The intensity of the diffuse scattering produced from static disorder is essentially independent on temperatur while thermal diffuse scattering decreases with decreasing temperature. These considerations show that thermal diffuse scattering of X-rays is not only a possible method for studying phonons, but is also of considerable importance in connection with the determination of the structure of ordered and disordered solids [5.16, 17].

2) The future use of *synchrotron X-radiation* together with multiple Bragg reflection monochromators might offer a very fascinating tool to study phonon dispersion curves. In this technique the scattered X-ray beam is directly energy analyzed as in the case of neutron scattering; it was already demonstrated that an energy resolution of 10 meV may be obtained. If the resolution can be improved, the method allows the study of low frequency excitations from extremely small single crystals [5.18].

3) The theory of thermal diffuse X-ray scattering developed in this chapter serves as an introduction to the scattering of neutrons by phonons. It will be instructive to discuss the similarities and differences of the expressions for the cross sections for inelastic X-ray and neutron scattering (Chap. 6).

5.1 The Static Approximation

It should be noted that there is an impotant difference between X-rays scattering and neutron scattering: X-ray photons propagate with the speed of light and the time required to pass through a unit cell is of the order of 10^{-18}s; for thermal neutrons with an energy of 5 meV the velocity is about 10^5cm/s and the corresponding time is of the order of 10^{-13}s. X-rays propagate with such a high speed that each photon samples an instantaneous static configuration of the atoms so that for X-rays the thermal motion of the atoms can be replaced by a sequence of static configurations because during the short time intervall of 10^{-18}s the atoms do not move. On the other hand, the atoms have changed considerably their positions during the time of $\sim 10^{-13}$s required for a neutron to pass a unit cell. In addition, the energy of thermal neutrons is comparable with the energies of phonons which can give rise to a large energy exchange between neutrons and phonons. The energy change of X-ray photons is extremely small and can usually be neglected.

When X-rays of frequency ω_0 and wave vector k_0 are scattered by phonons, the frequency ω and wave vector k of the scattered radiation satisfy the laws of energy and momentum conservation

$$\hbar(\omega - \omega_0) = \pm\hbar\omega_j(\boldsymbol{q}), \tag{5.1}$$

$$\hbar(\boldsymbol{k} - \boldsymbol{k}_0) = \hbar(\boldsymbol{\tau}\pm\boldsymbol{q}). \tag{5.2}$$

Here, $\omega_j(\boldsymbol{q})$ is a phonon frequency and $\boldsymbol{\tau}$ is a reciprocal lattice vector. For future work with synchrotron radiation [5.18], Eqs. (5.1, 2) will form the basis for the determination of phonon dispersion curves. If, however, conventional X-ray sources are used the energy resolution is not sufficient to observe the small phonon frequencies $\omega_j(\boldsymbol{q})$. In this case information about phonon dispersion can be obtained only by measuring the *intensity* of scattered X-rays. In order to calculate the intensities from first principles, the theory should be based on time-dependent perturbation theory. According to this theory the intensities are directly related to the transition probabilities $|\langle n|\boldsymbol{M}|m\rangle|^2$ between the states $|n\rangle$ and $|m\rangle$ caused by the perturbation. As for Raman scattering, $\boldsymbol{M} = \alpha\boldsymbol{E}$ is the dipole moment induced by the electric field $\boldsymbol{E}$ of the X-rays and α is the electronic polarizability of the system. On this basis *Born* has calculated the intensity of thermal diffuse scattering [5.19]. We shall not base the theory on *Born*'s treatment but we shall rather use the fact that since $\omega_j(\boldsymbol{q})\ll\omega_0$ we can omit the time dependence of processes and treat the system as static although we cannot then expect to recover the condition of (5.1) but instead $\omega = \omega_0$. The theory of X-ray scattering is then much simplified [5.16, 20, 21]. Since $\omega_0 = ck_0$ and $\omega = ck$ it follows that $k = k_0$. Thus in the "static approximation" we have

$$\omega = \omega_0, \tag{5.3}$$

$$k = k_0. \tag{5.4}$$

In the following we introduce the *scattering vector* $\boldsymbol{K}$ defined by

$$\boldsymbol{K} = \boldsymbol{k} - \boldsymbol{k}_0. \tag{5.5}$$

Using $k = k_0 = 2\pi/\lambda_0$ where λ_0 is the wavelength of the incident X-rays we obtain

$$K = 2k_0 \sin \theta = \frac{4\pi}{\lambda_0} \sin \theta, \tag{5.6}$$

where 2θ is the scattering angle (Fig. 5.1). The scattering of X-rays can be illustrated by using Ewald's construction (Fig. 5.1): The vector $\boldsymbol{k}_0$ is drawn in the direction of the incident X-ray beam and it terminates at any reciprocal lattice point. A sphere of radius $k_0 = 2\pi/\lambda_0$ is drawn about the origin of $\boldsymbol{k}_0$: this is the scattering surface of X-rays or *Ewald's sphere*. It will be shown in Sect. 5.4 that the only scattering from a perfect static crystal (which is of course unattainable) is in the direction defined by

$$\boldsymbol{K} = \boldsymbol{\tau}, \tag{5.7}$$

where $\boldsymbol{\tau}$ is any reciprocal lattice vector. This is simply a statement of Bragg's

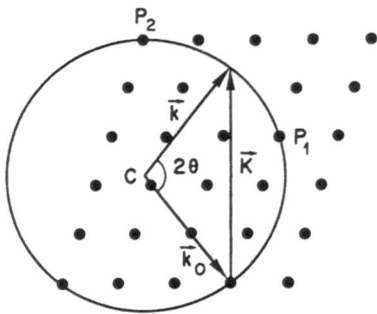

Fig. 5.1. The scattering surface for X-rays or Ewald's sphere. The points represent reciprocal lattice points of the crystal

law [5.22]. To satisfy the condition $\boldsymbol{K} = \boldsymbol{\tau}$, a point of the reciprocal lattice must fall on the surface of the sphere. Thus in general, when a beam of monochromatic X-rays falls on a perfect static crystal there will be no scattered beam. When, however, the crystal is turned around, its reciprocal lattice turns with it, and a reciprocal lattice point will eventually touch the sphere and a scattered beam will appear in the direction of the vector $\boldsymbol{k}$ (the points P_1 and P_2 in Fig. 5.1).

5.2 Experimental Technique

Figure 5.2 shows the experimental set-up used for the observation of diffuse X-ray scattering. The X-rays generated by a suitable source (usually MoK_α radiation) are monochromatized by Bragg reflection using a singly bent or doubly bent LiF crystal. The monochromatic X-ray beam passes through a slit and is scattered by the sample. The scattered radiation is either detected by a NaI(Tl) scintillation counter followed by an amplifier and a single-channel pulse height analyzer, or by a plane film. Excessive blackening from the direct beam is prevented by the presence of a beam stop placed before the film. If a plane film is used, the region of reciprocal space

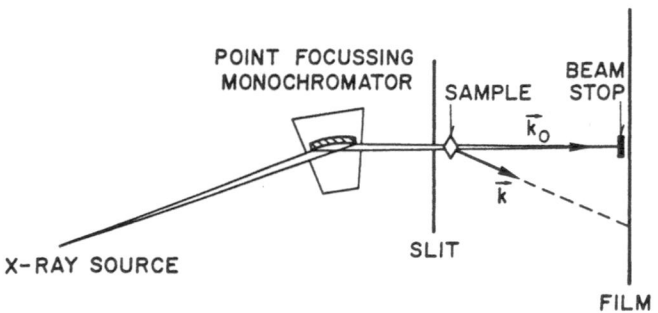

Fig. 5.2. Monochromatic Laue set-up for diffuse X-ray scattering

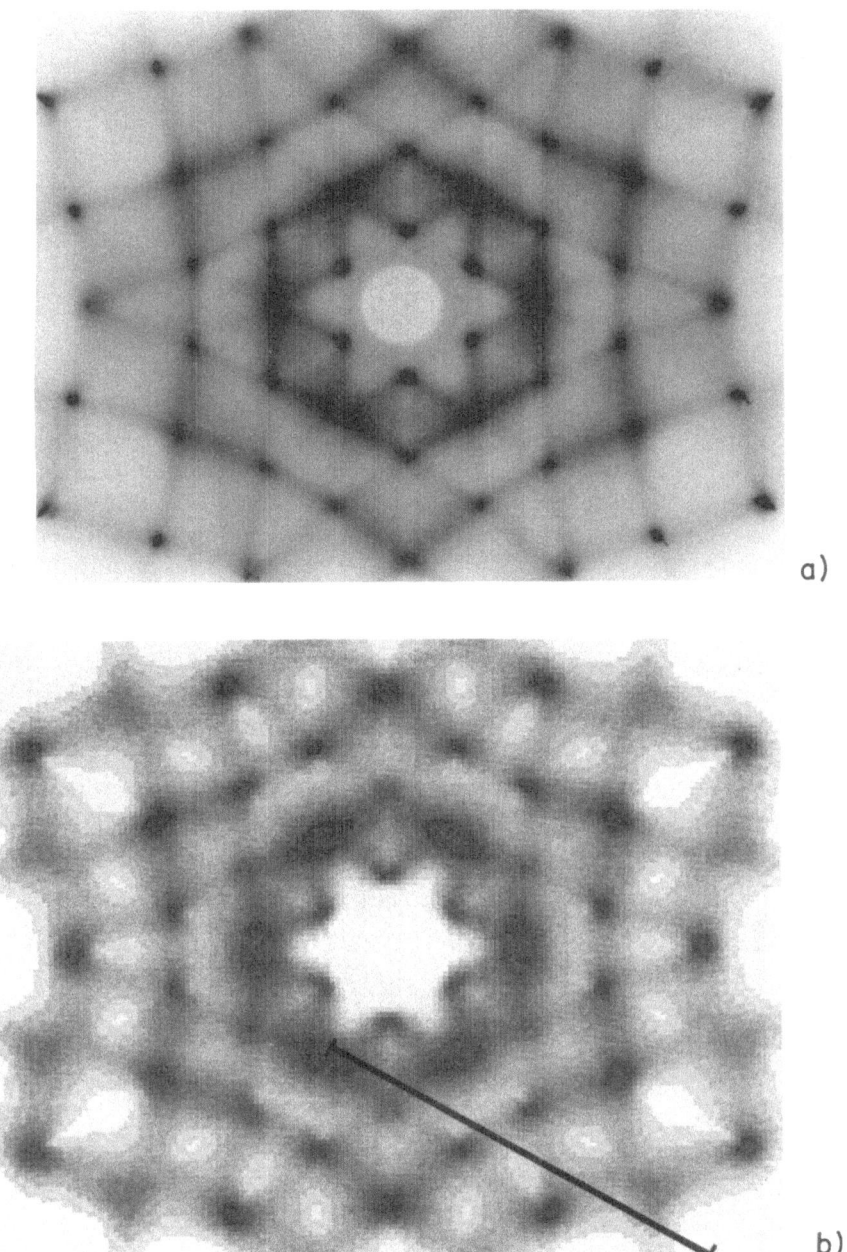

Fig. 5.3. (a) Experimental diffuse X-ray scattering in β-AgI. (b) Computed thermal diffuse scattering intensity in the geometry of Fig. 5.3a. The trace refers to a direction in q-space along which thermal scattering has been studied in more detail [5.14]

which is projected onto the film is the surface of the Ewald sphere (Fig. 5.1) in reciprocal space with radius equal to $k_0 = 2\pi/\lambda_0$ where $\lambda_0 = 0.7107\text{Å}$ for $\text{Mo}K_\alpha$ X-rays. Figure 5.3a shows the experimental diffuse scattering of β-AgI which crystallizes in the ordered hexagonal Wurtzite structure ([Ref. 5.23], Fig. 3.7, [5.14]). The picture has been taken in a steady-crystal-steady-film geometry (monochromatic Laue set-up, Fig. 5.2). The scattering intensity thus corresponds to that on a sphere tangential to the reciprocal basal plane [(001) Ewald sphere]. The striking feature of the photograph are the strong diffuse "streamers" between the Bragg spots. These diffuse streamers originate from scattering by very low-frequency dispersionless lattice modes, as illustrated in [Ref. 5.23, Fig. 3.12]. Such low-frequency lattice modes have large amplitudes of vibration and therefore modify considerably the direction, amplitude and phase of the reflected X-rays. The effect of the thermal motion on a X-ray beam traversing the crystal has been compared with the effect of the agitated surface of the sea on the image of the setting sun. There is no sharp reflection, but a diffuse ribbon of light stretching towards the observer. This diffusion is obviously produced by the innumerable waves of various length and direction. Figure 5.3b shows the computed thermal diffuse scattering in the geometry of Fig. 5.3a [5.14]. This calculation is based on the relation (5.56) for one-phonon scattering discussed in Sect. 5.6 and on a lattice dynamical model ([5.24], Ref. 5.23, Figs. 3.12, 13). In comparing computed and experimental data one has to keep in mind that the latter data contain incoherent Compton scattering and scattering in air. In addition, only one-phonon scattering has been included in the calculation, second- and higher-order scattering are neglected (Sect. 5.5). Nevertheless, the visual comparison represents an adequate proof of the thermal origin of the observed diffuse scattering.

5.3 Interaction Mechanism

In order to illustrate the interaction mechanism of X-rays with the system we consider the scattering by a single electron bound to its nucleus (Fig. 5.4). The electric field of the incident X-rays at the position $\boldsymbol{r}$ of the electron is

$$\boldsymbol{E}(\boldsymbol{r}, t) = \boldsymbol{E}_0 \exp[i(\omega_0 t - \boldsymbol{k}_0 \cdot \boldsymbol{r})], \tag{5.8}$$

and the classical equation of motion is

$$\ddot{\boldsymbol{s}} + \gamma \dot{\boldsymbol{s}} + \omega_e^2 \boldsymbol{s} = \frac{e}{m} \boldsymbol{E}(\boldsymbol{r}, t). \tag{5.9}$$

Here e and m is the charge and the mass of the electron, respectively, γ is a damping constant and ω_e is the resonance frequency of the bound electron. Under the influence of the electric field the electron executes a forced oscillation and the solution to (5.9) is

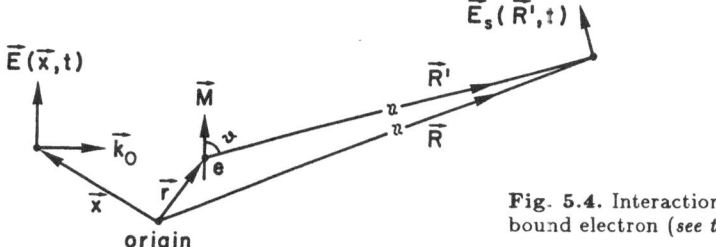

Fig. 5.4. Interaction of X-rays with a bound electron (*see text*)

$$s = \frac{e}{m} \frac{E_0 \exp[i(\omega_0 t - \mathbf{k}_0 \cdot \mathbf{r})]}{\omega_e^2 - \omega_0^2 + i\gamma\omega_0}. \tag{5.10}$$

The dipole associated with the oscillating electron is

$$\mathbf{M} = e\mathbf{s} = \frac{e^2}{m} \frac{E_0 \exp[i(\omega_0 t - \mathbf{k}_0 \cdot \mathbf{r})]}{\omega_e^2 - \omega_0^2 + i\gamma\omega_0} = \alpha(\omega_0)\mathbf{E}, \tag{5.11a}$$

where

$$\alpha(\omega_0) = \frac{e^2/m}{\omega_e^2 - \omega_0^2 + i\gamma\omega_0}, \tag{5.11b}$$

is the *polarizability* of the system. This oscillating dipole emits radiation and according to classical theory the amplitude of the electric field of this radiation at the large distance $\mathbf{R}$ from the electron (Fig. 5.4) is given by [5.25, 26]

$$E_s(R', t) = \frac{\omega_0^2}{c^2} \sin \vartheta M_0 \frac{e^{-ik_0 R'}}{R'} \exp(i\omega_0 t)$$

$$= \frac{e^2 \sin \vartheta}{mc^2} \frac{\omega_0^2}{\omega_e^2 - \omega_0^2 + i\gamma\omega_0} \frac{e^{-ik_0 R'}}{R'} E(\mathbf{r}, t). \tag{5.12}$$

This equation shows that the oscillating electron becomes the centre of a spherical scattered wave. The result for scattering by a free electron is obtained by putting $\omega_e = 0$, $\gamma = 0$ and gives

$$E_s(R', t) = -\frac{e^2}{mc^2} \sin \vartheta \frac{e^{-ik_0 R'}}{R'} E(\mathbf{r}, t). \tag{5.13}$$

Writing $E_s = f_b E$ and $E_s = f_f E$ for the bound and free electrons, respectively, we define $f = f_b/f_f$ as the scattering factor of a bound damped electron. From (5.12, 13) we obtain

$$f = \frac{\omega_0^2}{\omega_0^2 - \omega_e^2 - i\gamma\omega_0}. \tag{5.14}$$

For normally used X-ray wavelengths only the most tightly bound electrons in an atom have resonance frequencies ω_e approaching or exceeding ω_0, so that for the atomic scattering factor only the contributions of the inner most electrons differ appreciably from their normal values calculated on the assumption of free electrons. The scattering factor of an atom with several electrons may therefore be written in the form

$$f = f_0 + \Delta f' + \Delta f'', \tag{5.15}$$

where f_0 is the free-electron contribution, and $\Delta f'$, $\Delta f''$ are corrections which are often small enough to be ignored. Only in those cases in which the frequency ω_0 of the primary radiation is just above that corresponding to an absorption edge of the scattering atom can the dispersion terms $\Delta f'$ and $\Delta f''$ lead to important effects. Values of $\Delta f'$ and $\Delta f''$ for various elements and wavelengths are given in the literature [5.27]. In the following we shall neglect $\Delta f'$ and $\Delta f''$ and use the free-electron relation (5.13).

Since the intensity of the incident radiation is $I_0 = |E|^2/2$ and that of the scattered radiation is $I_0 = |E_s|^2/2$ we find from (5.13)

$$I_s(R') = \left(\frac{e^2}{mc^2} \right)^2 \frac{\sin^2 \vartheta}{R'^2} I_0. \tag{5.16}$$

Since $I_s \sim 1/m^2$, only the electrons but not the protons contribute to the scattered intensity.

In the unmodified scattering considered in this chapter, the momentum exchange takes place between the incident photon and the atom as a whole; the wavelength change is then zero for selective reflection by a crystal (strictly coherent scattering); it is extremely small for scattering outside the Bragg reflections due to thermal vibrations of the atoms and in the "static approximation" we put $k = k_0$ according to (5.4). There exists an other type of scattering, called *Compton scattering*, incoherent or modified scattering. The scattering of X-rays by the *Compton effect* is due to the collision, that is the exchange of energy and momentum, between an incident photon of wavelength λ_0 and relatively loosely bound electrons. The electron is expelled from the atom and the photon has, after the collision, a slightly smaller energy, that is a wavelength λ slightly larger [5.28]. For a free electron the change $\Delta\lambda$ of the wavelength can easily be calculated based on energy and momentum conservation. If 2θ is the scattering angle one obtains (Problem 5.8.1)

$$\Delta\lambda = \lambda - \lambda_0 = \frac{2h}{mc} \sin^2 \theta, \tag{5.17}$$

with $h/mc = 0.02426$ Å. The change $\Delta\lambda$ is therefore independent of the wavelength of the incident photon and of the atomic nature of the scatterer. For bound electrons the situation is more complex. There exists tables which give the Compton-scattering intensities for the atoms in different electronic states. These calculated intensities are based on Hartree-Fock wave functions [5.28]. In order to determine experimentally the non-Bragg (diffuse) scattering due to thermal vibrations of the atoms, order-disorder phenomena, etc., it is necessary to subtract from the values measured for the total diffuse scattering the theoretical values of the continuous incoherent scattering due to the Compton effect.

5.4 Scattering by a Perfectly Ordered Crystal

We consider a small parallellepiped of the crystal through which there sweeps a train of plane waves (Fig. 5.5). As discussed in Sect. 5.3 each lattice point becomes the centre of a spherical scattered wavelet described by (5.13), and our problem is to find the combined effect of these wavelets at a point R outside the crystal, and at a distance from it large in comparison with its linear dimensions. In the elementary discussion of the problem certain important simplifying assumptions are introduced:

1) We suppose that the incident X-ray beam travels through the crystal with the velocity of light in vacuum, that is we, in fact, take the refractive index of the crystal for X-rays to be unity. In so far as the positions of the diffraction maxima are concerned the errors introduced by doing so are very small and can be neglected.

2) We shall assume that each scattered wavelet travels through the crystal without being rescattered by other lattice points. Such rescattering must occur, and its effect may become important. For small crystals, however, we make little errors by neglecting it, and the conditions under which it is permissable to do so are discussed in the literature [5.16, 17].

3) We shall suppose that no absorption either of the incident or the scattered radiation takes place in the crystal considered, an assumption that again limits the validity of the calculation to small crystals.

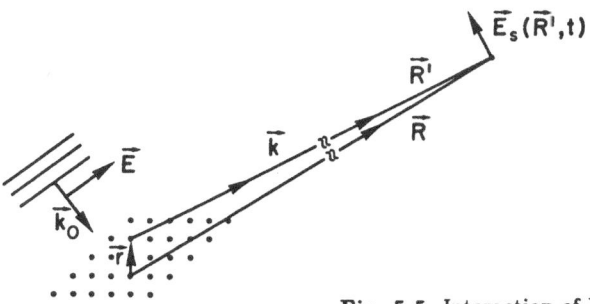

Fig. 5.5. Interaction of X-rays with a crystal (*see text*)

If the electric field of the incident wave at the lattice point r (Fig. 5.5) is $E(r,t) = E_0 \exp[i(\omega_0 t - k_0 \cdot r)]$, Eq. (5.13) suggests that the field scattered by the lattice point r at the point R' is given by

$$E_s(R',t) = CE_0 e^{-ik_0 \cdot r} \frac{e^{-ik_0 R'}}{R'} e^{i\omega_0 t}. \tag{5.18}$$

The constant C depends on the nature of the scattering center and contains an angular factor. If the scattering center is a free electron then $C = -(e^2/mc^2) \sin \vartheta$ according to (5.13). Since $R \cong R' \gg r$ we have

$$R' \cong R - r \cos (R, r). \tag{5.19}$$

Substituting (5.19) into (5.18) the phase factor becomes

$$\exp[-i(k_0 \cdot r + k_0 R')] = \exp(-ik_0 R) \exp\{-i[k_0 \cdot r - k_0 r \cos (R, r)]\}.$$

Since $k_0 \cong k$ by (5.4) and since the direction of R is nearly parallel to the direction of k we can write

$$k_0 r \cos (R, r) \cong kr \cos (k, r) = k \cdot r,$$

hence

$$E_s(R,t) \cong CE_0 \frac{e^{-ikR}}{R} e^{i\omega_0 t} e^{iK \cdot r}, \tag{5.20}$$

where $K = k - k_0$ is the scattering vector introduced in (5.5). In the denominator of (5.20) we have replaced R' by R without introducing an appreciable error.

The total wave scattered in a given direction by all the lattice points of a Bravais crystal is the sum of all the individual waves scattered by the lattice points $r(l)$. According to (5.20) the interesting quantity is the sum of the phase factors

$$S(K) = \sum_l e^{iK \cdot r(l)}. \tag{5.21}$$

From the discussion given above we have seen that the electrons do the scattering. In the general case of an arbitrary electronic charge distribution it is reasonable to assume that the scattering originating from the volume element $d^3 r$ of the crystal is proportional to the local electron charge density $\varrho(r)$; $\varrho(r) d^3 r$ is the probability that an electron is found in the volume element $d^3 r$ at the position r. From these considerations and from (5.20) we see that the amplitude of the radiation scattered from the crystal and received at the point R (Fig. 5.5) will be proportional to

$$\frac{e^{ikR}}{R} \int\limits_{\text{crystal}} \varrho(\mathbf{r}) e^{i\mathbf{K}\cdot\mathbf{r}} d^3 r.$$

In the following we omit the factor $(1/R)\exp(ikR)$ because it is constant over the volume and write for the amplitude of the scattered wave

$$A(\mathbf{K}) = \int\limits_{\text{crystal}} \varrho(\mathbf{r}) e^{i\mathbf{K}\cdot\mathbf{r}} d^3 r. \qquad (5.22)$$

For the following we write the electron density as the superposition of electron densities $\varrho_{l\kappa}$ associated with each atom κ in the unit cell l. If $\mathbf{r}\binom{l}{\kappa}$ is the vector of the center of atom $\binom{l}{\kappa}$, then the function $\varrho_{l\kappa} = \varrho_\kappa(\mathbf{r} - \mathbf{r}\binom{l}{\kappa})$ defines the contribution of that atom to the electron density at $\mathbf{r}$ (Fig. 5.6). The total density at $\mathbf{r}$ due to all the atoms is the sum

$$\varrho(\mathbf{r}) = \sum_{l\kappa} \varrho_\kappa(\mathbf{r} - \mathbf{r}\binom{l}{\kappa}). \qquad (5.23)$$

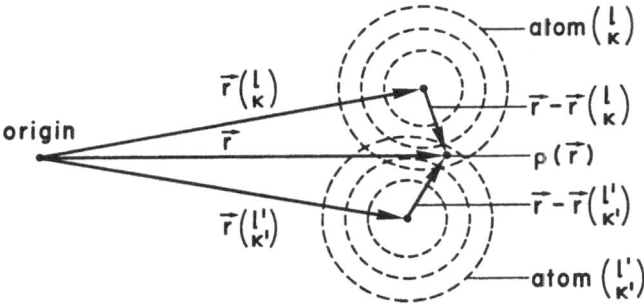

Fig. 5.6. For two atoms $\binom{l}{\kappa}$ and $\binom{l'}{\kappa'}$ the total electron density at $\mathbf{r}$ is given by $\varrho(\mathbf{r}) \cong \varrho_\kappa[\mathbf{r} - \mathbf{r}\binom{l}{\kappa}] + \varrho_{\kappa'}[\mathbf{r} - \mathbf{r}\binom{l'}{\kappa'}]$

Figure 5.6 shows the situation for two atoms $\binom{l}{\kappa}$ and $\binom{l'}{\kappa'}$. The decomposition of $\varrho(\mathbf{r})$ is not unique, for we cannot always say how much charge is associated with each atom. For the present purpose, however, this difficulty is of no importance. Substituting (5.23) into (5.22) and using $\boldsymbol{x} = \mathbf{r} - \mathbf{r}\binom{l}{\kappa} = \mathbf{r} - \mathbf{r}(l) - \mathbf{r}(\kappa)$, $d^3 r = d^3 x$, we obtain

$$A(\mathbf{K}) = F(\mathbf{K}) \sum_l e^{i\mathbf{K}\cdot\mathbf{r}(l)} = NF(\mathbf{K})\Delta(\mathbf{K} - \boldsymbol{\tau}), \qquad (5.24)$$

where

$$F(\mathbf{K}) = \sum_\kappa f_\kappa(\mathbf{K}) \cdot e^{i\mathbf{K}\cdot\mathbf{r}(\kappa)}, \qquad (5.25)$$

133

is the *structure factor* and

$$f_\kappa(\boldsymbol{K}) = \int \varrho_\kappa(\boldsymbol{x})\mathrm{e}^{i\boldsymbol{K}\cdot\boldsymbol{x}}d^3x, \tag{5.26}$$

is the *atomic scattering factor* or *atomic form factor*. For an atom with spherical charge distribution, $f_\kappa(\boldsymbol{K})$ can easily be calculated (Problem 5.8.2).

In deriving (5.24) we have made use of the relation (Appendix G)

$$S(\boldsymbol{K}) = \sum_l \mathrm{e}^{i\boldsymbol{K}\cdot\boldsymbol{r}(l)} = N\varDelta(\boldsymbol{K} - \boldsymbol{\tau}) = \begin{cases} N & \text{for } \boldsymbol{K}=\boldsymbol{\tau} \\ 0 & \text{for } \boldsymbol{K}\neq\boldsymbol{\tau} \end{cases}, \tag{5.27}$$

where $\boldsymbol{\tau}$ is a vector of the reciprocal lattice and N is the number of unit cells in the crystal. The intensity of the scattered wave is proportional to $|A(\boldsymbol{K})|^2$. For the static crystal in which each atom is located at its equilibrium position $\boldsymbol{r}(\begin{smallmatrix}l\\\kappa\end{smallmatrix})$ we find for the intensity (Appendix G)

$$I(\boldsymbol{K})_{\mathrm{static}} \sim |A(\boldsymbol{K})|^2 = |F(\boldsymbol{K})|^2\,|S(\boldsymbol{K})|^2. \tag{5.28}$$

In this expression $S(\boldsymbol{K})$ is given by (5.21) and for the *Laue interference function* $|S(\boldsymbol{K})|^2$ we obtain (Appendix G)

$$|S(\boldsymbol{K})|^2 = \prod_{i=1}^{3} \frac{\sin^2\left[N_i(\boldsymbol{a}_i\cdot\boldsymbol{K})/2\right]}{\sin^2\left[(\boldsymbol{a}_i\cdot\boldsymbol{K})/2\right]}. \tag{5.29}$$

This function is strongly concentrated at the points $\boldsymbol{\tau}$ of the reciprocal lattice. When $N = N_1N_2N_3$ is large, each peak acquires the characteristics of the $\varDelta$ function defined by (5.27). It has absolute maxima equal in value to N^2 when $\boldsymbol{K} = \boldsymbol{\tau}$, a vector of the reciprocal lattice. The width of each maximum is proportional to $1/N_i$ and can be extremely narrow for an ideal crystal of macroscopic dimensions (Appendix G). We may therefore express (5.28) formally by writing

$$I(\boldsymbol{K})_{\mathrm{static}} \sim N^2|F(\boldsymbol{K})|^2\varDelta(\boldsymbol{K} - \boldsymbol{\tau}). \tag{5.30}$$

The presence of the factor $\varDelta(\boldsymbol{K} - \boldsymbol{\tau})$ in this relation shows that the only scattering from a perfect static crystal is in the directions defined by $\boldsymbol{K} = \boldsymbol{\tau}$. According to the definition (5.25) the structure factor $F(\boldsymbol{K})$ depends on the positions of the atoms in the unit cell. The relation (5.30) is therefore the basis for the determination of crystal structures. The condition

$$\boldsymbol{K} = \boldsymbol{k} - \boldsymbol{k}_0 = \boldsymbol{\tau}, \tag{5.31}$$

is another statement of the Bragg law: From (5.6) we have $K = (4\pi/\lambda)\sin\theta$

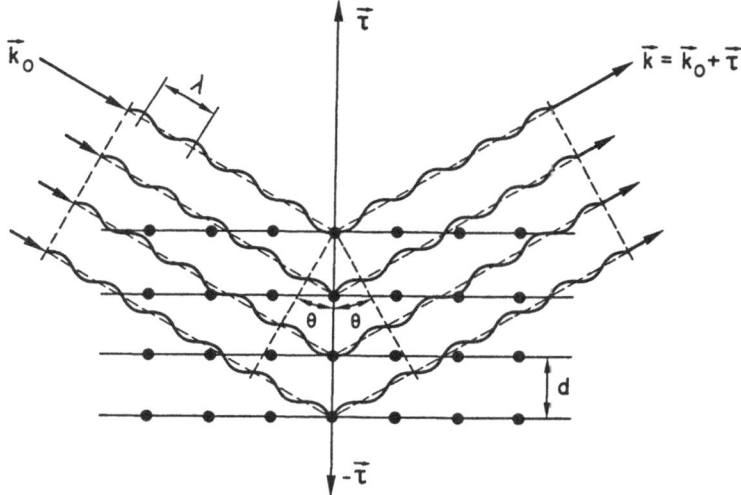

Fig. 5.7. Bragg reflection from parallel planes of atoms in a crystal. In the collision process described by $\boldsymbol{k} = \boldsymbol{k}_0 + \boldsymbol{\tau}$ the crystal recoils with the momentum $-\hbar\boldsymbol{\tau}$

and it can be shown that $\tau = 2\pi n/d$ [5.22], where n is an integer and d is the distance between neighbouring lattice planes perpendicular to $\boldsymbol{\tau}$. Thus (5.31) gives the Bragg law

$$n\lambda = 2d \sin \theta. \tag{5.32}$$

In the explanation of the diffracted beam presented by Bragg the incident waves are reflected specularly from parallel planes of atoms in the crystal, with each plane reflecting only a very small fraction of the radiation, like a lightly silvered mirror. The diffracted beams are found when the reflections from parallel planes of atoms interfere constructively, as in Fig. 5.7. The relation (5.31) can be interpreted as the conservation of momentum : Before the collision of the photon with the crystal, the momentum of the photon is $\hbar\boldsymbol{k}_0$ and that of the crystal is zero; after the collision, the momentum of the scattered photon is $\hbar\boldsymbol{k}$ and the crystal momentum is $-\hbar\boldsymbol{\tau}$. The conservation of momentum requires that

$$\hbar\boldsymbol{k}_0 = \hbar\boldsymbol{k} - \hbar\boldsymbol{\tau}, \tag{5.33}$$

which is identical with (5.31). Thus in the collision process described by $\boldsymbol{k} = \boldsymbol{k}_0 + \boldsymbol{\tau}$ the crystal recoils with the momentum $-\hbar\boldsymbol{\tau}$ (Fig. 5.7). The speed of recoil of a crystal with a mass of $M = 1\,\mathrm{gr}$ and $\tau = 1 \cdot 10^8 \mathrm{cm}^{-1}$ is $v = \hbar\tau/M \cong 10^{-19}\mathrm{cm/s}$ which is too small to be detected.

5.5 Thermal Diffuse Scattering for a Bravais Crystal

We consider now the effects of thermal motion on the scattering of X-rays. In order to simplify the treatment we consider a crystal with only one atom in the unit cell. The calculation of the scattered intensity involves some algebra given in Appendix H. We shall concentrate here on the physical significance of the results. For a Bravais crystal we can choose $r(\kappa) = 0$ in (5.25) and the structure factor reduces to the atomic scattering factor $f(K)$ defined by (5.26). If $u(l)$ is the small thermal displacement of atom l, the actual position of this atom is $r(l) + u(l)$, and according to (5.24) the amplitude of the scattered wave becomes

$$A(K) = f(K) \sum_l \exp\{iK\cdot[r(l) + u(l)]\}. \tag{5.34}$$

Here we have assumed rigid atoms, that is the charge distribution of an atom and hence the atomic scattering factor $f(K)$ is independent on the configuration of the surrounding atoms. The theory for a lattice of deformable atoms is more complicated [5.29]. The intensity of the scattered light or the scattering cross section is proportional to $\langle |A(K)|^2 \rangle$, where triangular brackets denote the thermal or ensemble average over a canonical ensemble defined by the harmonic Hamiltonian [Ref. 5.23, Eq. (3.71)]. It is also the time average over our crystal of N unit cells. Thus using (5.34) we have for the dynamical case

$$I(K)_{\text{dyn}} \sim \langle |A(K)|^2 \rangle$$

$$= |f(K)|^2 \sum_{ll'} \exp\{iK[r(l) - r(l')]\} \langle \exp\{iK[u(l) - u(l')]\} \rangle. \tag{5.35}$$

For the evaluation of (5.35), $u(l)$ is expressed in terms of creation and annihilation operators. For a Bravais crystal we obtain [Ref. 5.23, Eq. (3.72)]

$$u(l) = \left(\frac{\hbar}{2Nm} \right)^{1/2} \sum_{qj} \frac{e\binom{q}{j}}{[\omega_j(q)]^{1/2}} e^{iq\cdot r(l)} [a^+\binom{-q}{j} + a\binom{q}{j}]. \tag{5.36}$$

In (5.36) N is the number of unit cells or atoms with mass m, $\omega_j(q)$ and $e\binom{q}{j}$ are the frequency and polarization vector of the phonon $\binom{q}{j}$, and $a^+\binom{-q}{j}$ and $a\binom{q}{j}$ are the creation and annihilation operators, respectively. It is shown in Appendix H that the intensity $I(K)_{\text{dyn}}$ can be written as a sum of zero-order, first-order, second-order, etc. intensities:

$$I(K)_{\text{dyn}} = I_0(K) + I_1(K) + I_2(K) + \dots . \tag{5.37}$$

$I_0(K)$ is the intensity for Bragg scattering or elastic scattering and is given by

$$I_0(K) = N^2 |f(K)|^2 \Delta(K - \tau) e^{-2W(K)}$$

$$= I(K)_{\text{static}} e^{-2W(K)}, \qquad (5.38)$$

where we have used (5.30). The exponential factor is the *Debye-Waller factor* and $W(K)$ is given by

$$2W(K) = \frac{1}{Nm} \sum_{qj} |K \cdot e(\tfrac{q}{j})|^2 \frac{\overline{E}(\tfrac{q}{j})}{\omega_j^2(q)} , \qquad \text{where} \qquad (5.39a)$$

$$\overline{E}(\tfrac{q}{j}) = \hbar \omega_j(q) \left[\overline{n}_j(q) + \frac{1}{2} \right], \qquad (5.39b)$$

is the mean energy of the phonon $(\tfrac{q}{j})$ with $\overline{n}_j(q)$ given by [Ref. 5.23, Eq. (2.126)]. From (5.37) we see that the intensity $I(K)_{\text{static}}$ for the rigid lattice is reduced by the Debye-Waller factor. Reflections of low $|K|$ are affected less than reflections of high K according to (5.38). On the other hand, the presence of the Laue function in (5.38) shows that the sharpness of the interference lines do not suffer from thermal agitation: the width of the Bragg reflections is the same as for the static crystal described by (5.28).

The intensity $I_0(K)$ is that of the coherent diffraction or the elastic scattering in the well-defined Bragg directions given by $K = \tau$. *The intensity lost from these directions appears as a diffuse background* such as shown in Fig. 5.3. Thus the intensity lost reappears in the inelastic scattering or incoherent scattering in which the X-ray photon cause the excitation or de-excitation of phonons, and the photon changes direction and energy. In other words, the lost intensity from the Bragg reflections appears in the terms $I_1(K)$, etc., of (5.37).

The term $I_1(K)$ describes first-order or *one-phonon scattering*. Since we have not considered the time dependence but rather based our treatment on the "static approximation" expressed by (5.3, 4), the final result does not make explicit the fact that the cross section is the sum of that for two processes, one for which the X-ray energy is increased by absorption (annihilation) of a phonon and one in which it is decreased by emission (creation) of a phonon. Since these two processes have the same value of K for X-rays (recall that the scattering surface, the surface on which k and K end, is the Ewald sphere for X-rays (Fig. 5.1)) their intensities add. The result is

$$I_1(K) = N^2 |f(K)|^2 e^{-2W(K)} \sum_{qj} G_j(K, q) \Delta(K + q - \tau), \qquad (5.40)$$

where

$$G_j(\boldsymbol{K}, \boldsymbol{q}) = \frac{1}{Nm} |\boldsymbol{K} \cdot \boldsymbol{e}(\begin{smallmatrix} \boldsymbol{q} \\ j \end{smallmatrix})|^2 \frac{\overline{E}(\begin{smallmatrix} \boldsymbol{q} \\ j \end{smallmatrix})}{\omega_j^2(\boldsymbol{q})}. \tag{5.41}$$

The sum over $\boldsymbol{q}$ extends on the wave vectors in the *first Brillouin zone*. Thus for a certain scattering vector $\boldsymbol{K} = \boldsymbol{k} - \boldsymbol{k}_0$ given by the experiment, $\boldsymbol{q}$ is the vector which connects the end point of $\boldsymbol{K}$ with the *nearest* reciprocal lattice point (or vice versa). (Vectors $\boldsymbol{q}'$ leading from the end point of $\boldsymbol{K}$ to other reciprocal lattice points can be represented as $\boldsymbol{q}' = \boldsymbol{q} + \boldsymbol{\tau}$ where $\boldsymbol{\tau}$ is a reciprocal lattice vector; thus $\boldsymbol{q}'$ is equivalent to $\boldsymbol{q}$.) (Fig. 5.8).

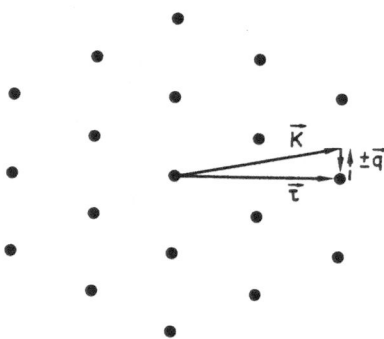

Fig. 5.8. In (5.42) $\boldsymbol{q}$ is the vector which connects the end point of $\boldsymbol{K}$ with the *nearest* reciprocal lattice point (or vice versa)

The intensity at the point $\boldsymbol{K}$ is determined by all the phonons $(\begin{smallmatrix} \boldsymbol{q} \\ j \end{smallmatrix})$ for which $\boldsymbol{q} = \boldsymbol{\tau} - \boldsymbol{K}$ is the wave vector which connects the end point of $\boldsymbol{K}$ with the nearest reciprocal lattice point (or vice versa), and the index j labels the different branches. Summing (5.40) over $\boldsymbol{q}$ there remains a single term, the term with $\boldsymbol{q} = \boldsymbol{\tau} - \boldsymbol{K}$, and using (5.41) we obtain

$$I_1(\boldsymbol{K}) = \frac{N}{m} |f(\boldsymbol{K})|^2 e^{-2W(\boldsymbol{K})} \sum_j |\boldsymbol{K} \cdot \boldsymbol{e}(\begin{smallmatrix} \boldsymbol{q} \\ j \end{smallmatrix})|^2 \frac{\overline{E}(\begin{smallmatrix} \boldsymbol{q} \\ j \end{smallmatrix})}{\omega_j^2(\boldsymbol{q})} \ , \qquad \text{where} \tag{5.42}$$

$$\boldsymbol{q} = \boldsymbol{\tau} - \boldsymbol{K}. \tag{5.43}$$

Equation (5.43) expresses the conservation of momentum, but while $\hbar \boldsymbol{K}$ is certainly the change of photon momentum and $\hbar(\boldsymbol{\tau} - \boldsymbol{q})$ that of the crystal, it is not correct to make a further division and identify $\hbar \boldsymbol{q}$ as literally the phonon momentum since a phonon is a quasi-particle which does not really carry momentum [Ref. 5.23, Sect. 2.2.4]. For the experiments, however, a phonon behaves as if its momentum were $\hbar \boldsymbol{q}$; for this reason $\hbar \boldsymbol{q}$ is also called the quasi-momentum or pseudo-momentum. According to (5.42) the intensity of scattering of a phonon depends on its polarization through the occurrence of the factor $\boldsymbol{K} \cdot \boldsymbol{e}(\begin{smallmatrix} \boldsymbol{q} \\ j \end{smallmatrix})$. This is physically understandable: only that

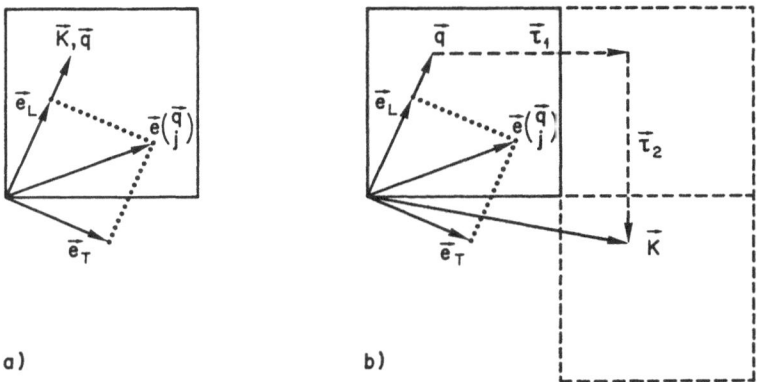

Fig. 5.9. (a) In a normal process $(\tau = 0)$ K is a vector of the first Brillouin zone and only longitudinal phonons can be observed. (b) In an Umklapp process $(\tau \neq 0)$ K is not a vector of the first Brillouin zone and both, longitudinal and transverse phonons can be observed

component of the atomic motion which is in the direction of the change in momentum of the radiation is effective. The same scalar product $K \cdot e\binom{q}{j}$ occurs also for scattering of neutrons and electrons by phonons.

Processes with $\tau = 0$ are called *normal processes* while processes with $\tau \neq 0$ are called *Umklapp processes*. For a normal process $K = \pm q$ is a vector in the first Brillouin zone (Fig. 5.9a). In this case only longitudinal phonons can be observed since $K \cdot e_L\binom{q}{j} \neq 0$ but $K \cdot e_T\binom{q}{j} = 0$ (Fig. 5.9a). In an Umklapp process, however, K is not a vector in the first Brillouin zone and in the general case both longitudinal and transverse phonons can be observed since $K \cdot e_L\binom{q}{j} \neq 0$ and $K \cdot e_T\binom{q}{j} \neq 0$ (Fig. 5.9b). When transverse modes are being measured, K is arranged to be roughly perpendicular to q; in this case the scattering from longitudinal modes is small since $K \cdot e_L\binom{q}{j}$ is small. Using the method outlined in Appendix I the intensity due to *two-phonon processes* involving phonons (q_1, j_1) and (q_2, j_2) can be written in the form [5.22]

$$I_2(K) = \frac{1}{2}|f(K)|^2 e^{-2W(K)} \sum_{q_1 j_1} \sum_{q_2 j_2} |K \cdot e\binom{q}{j_1}|^2$$

$$\times |K \cdot e\binom{q}{j_2}|^2 \frac{\overline{E}\binom{q}{j_1}\overline{E}\binom{q}{j_2}}{m^2 \omega_{j_1}^2(q_1)\omega_{j_2}^2(q_2)}, \qquad \text{where} \qquad (5.44)$$

$$K + q_1 + q_2 = \tau. \qquad (5.45)$$

In order to calculate the intensity at a point $k = \tau - q$ all the phonons for which $q = q_1 + q_2$ must be considered; the first sum over q_1 extends over the first Brillouin zone while the second sum disappears since $q_2 = q - q_1 = \tau - K - q_1$.

To get an idea of the order of magnitude of the different terms in (5.37) and their temperature dependence let us consider the high-temperature limit $\overline{E}\binom{q}{j} = k_B T$ and the Einstein model of a crystal with an Einstein frequency ω_E independent of q. It will be shown in Sect. 5.7 that in this case the quantity $2W(K)$ defined by (5.38) is given by

$$2W = \frac{k_B T}{m\omega_E^2} K^2. \tag{5.46}$$

From (5.42, 44) we then obtain

$$I_1(K) = 2N(fe^{-W})^2 W, \tag{5.47a}$$

$$I_2(K) = 2N(fe^{-W})^2 W^2, \quad \text{and} \tag{5.47b}$$

$$\frac{I_2}{I_1} = W = \frac{1}{2}K^2 \frac{k_B T}{m\omega_E^2}. \tag{5.48}$$

Thus at moderate temperatures the major contribution to thermal scattering is the first-order scattering given by (5.42).

According to (5.42) the intensity of thermal scattering is related with the phonon frequencies $\omega_j(q)$. For simple crystals it is possible to determine $\omega_j(q)$ from the observed intensities. For Bravais crystals the sum over j involves only three acoustic modes, and each polarization vector $e\binom{q}{j}$ is a unit vector making an angle α_j with K. Equation (5.42) may be written as

$$I_1(K) = N|f(K)|^2 K^2 e^{-2W(K)} \sum_{j=1}^{3} \frac{\cos^2 \alpha_j \, \overline{E}\binom{q}{j}}{m\omega_j^2(q)}. \tag{5.49}$$

Thus for a general value of K and therefore of q, the contributions of the three modes cannot be separated. However suppose that K lies in the direction (100) of a cubic crystal. Modes for which q lies along (100) have their polarization determined by the symmetry of the crystal, they are purley longitudinal ($\alpha_L = 0$ or purely transverse ($\alpha_T = \pi/2$) [Ref. 5.23, Fig. 3.17]). Therefore in this situation only the longitudinal modes can contribute to the first-order intensity. By chosing other special directions for K it is possible to find situations where only a transverse mode contributes appreciably (Fig. 5.9b), and in fact the frequencies of all longitudinal and transverse modes can be measured for q parallel to (100), (110) and (111). Correction must be made for second-order scattering which according to (5.44) involves the frequencies and polarization vectors of all modes. An approximate model has to be used therefore to calculate this correction, and a

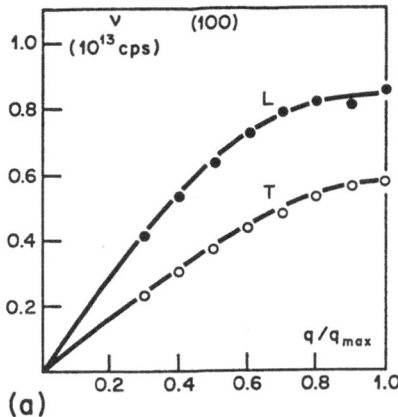 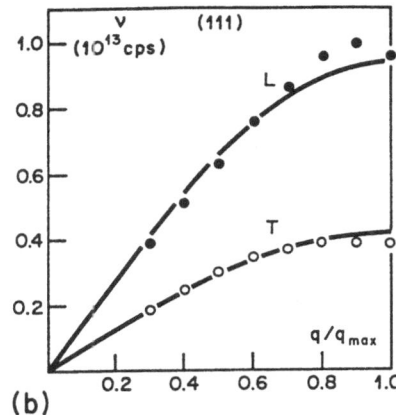

Fig. 5.10a,b. Phonon dispersion curves observed for aluminum by means of thermal diffuse scattering of X-rays. The measured data for the longitudinal and transverse waves are shown, respectively, by the solid and open circles. The smooth curves represent the fitted solutions of the eigenvalue problem [Ref. 5.23, Sect. 3.7]. (a) Direction of propagation along the [100] axis; (b) Direction of propagation along the [111] axis [5.30]

correction must also be made for Compton scattering. Figure 5.10 shows the dispersion curves for phonons propagating along the (100) axis in aluminum by means of thermal diffuse scattering of X-rays [5.30].

5.6 Thermal Scattering for a Crystal with Basis

We consider now a crystal with n atoms in the primitive unit cell. Replacing $r\binom{l}{\kappa}$ by $r\binom{l}{\kappa} + u\binom{l}{\kappa}$ in (5.24), the amplitude of the scattered X-rays are given by

$$A(K) = \sum_{l\kappa} f_\kappa(K) \exp\{iK \cdot [r(l) + r(\kappa) + u\binom{l}{\kappa})]\}. \tag{5.50}$$

Here $r(\kappa)$ $(\kappa = 1 \dots n)$ define the position of the atoms in the primitive unit cell and $u\binom{l}{\kappa}$ is the displacement of atom κ in the unit cell l. Generalizing (5.36) we express $u\binom{l}{\kappa}$ in terms of creation and annihilation operators [Ref. 5.23, Eq. (3.74)] and write

$$u\binom{l}{\kappa} = \left(\frac{\hbar}{2Nm_\kappa}\right)^{1/2} \sum_{qj} \frac{e(\kappa|_j^q)}{[\omega_j(q)]^{1/2}}$$

$$\times \exp[iq \cdot r\binom{l}{\kappa})][a^+\binom{-q}{j} + a\binom{q}{j}]. \tag{5.51}$$

Substituting (5.51) into (5.50) and following a derivation paralleling that given in Appendix H one obtains [5.19, 21b, 31]

$$I(\mathbf{K})_{\text{dyn}} = I_0(\mathbf{K}) + I_1(\mathbf{K}) + I_2(\mathbf{K}) + \dots . \qquad (5.52)$$

For the zero-order or Bragg scattering we obtain

$$I_0(\mathbf{K}) = N^2 |F(\mathbf{K})|^2 \Delta(\mathbf{K} - \boldsymbol{\tau}), \quad \text{where} \qquad (5.53)$$

$$F(\mathbf{K}) = \sum_{\kappa=1}^{n} f_\kappa(\mathbf{K}) \exp[-2W_\kappa(\mathbf{K})] \exp[i\mathbf{K}\cdot\mathbf{r}(\kappa)], \qquad (5.54)$$

is the structure factor which should be compared with (5.25) for the case of the static crystal. The exponent $2W_\kappa(\mathbf{K})$ of the Debye-Waller factor of atom κ is given by

$$2W_\kappa(\mathbf{K}) = \frac{1}{N m_\kappa} \sum_{qj} |\mathbf{K}\cdot\boldsymbol{e}(\kappa|_j^q)|^2 \frac{\overline{E}(_j^q)}{\omega_j^2(q)}. \qquad (5.55)$$

This is the generalization of (5.38). The intensity for one-phonon scattering is given by

$$I_1(\mathbf{K}) = N \sum_j \left| \sum_\kappa f_\kappa(\mathbf{K}) \exp[-W_\kappa(\mathbf{K})] \right.$$
$$\left. \times \frac{\mathbf{K}\cdot\boldsymbol{e}(\kappa|_j^q)}{m_\kappa^{1/2}} \exp[i\boldsymbol{\tau}\cdot\mathbf{r}(\kappa)] \right|^2 \frac{\overline{E}(_j^q)}{\omega_j^2(q)}, \qquad (5.56)$$

where

$$q = \boldsymbol{\tau} - \mathbf{K}. \qquad (5.57)$$

The relation (5.56) has been applied to determine phonon dispersion curves of alkali halides such as for example NaF [5.32]. The measured intensities must again be corrected for the multi-phonon contributions to the scattering, as well as for Compton scattering. Furthermore the method is not readily applicable to crystals with several atoms in the unit cell because the intensities cannot always be separated unambiguously into the contributions from the different modes.

The relation (5.56) is also the basis for the calculation of thermal diffuse scattering observed in β-AgI (Fig. 5.3b) [5.14].

5.7 The Debye-Waller Factor

Let us consider the Debye-Waller factor $\exp[-2W(\mathbf{K}, T)]$ in some more detail. For a Bravais crystal we have according to (5.39a)

$$2W(\boldsymbol{K}) = \frac{1}{Nm} \sum_{\boldsymbol{q}j} |\boldsymbol{K}\cdot\boldsymbol{e}(\genfrac{}{}{0pt}{}{\boldsymbol{q}}{j})|^2 \frac{\overline{E}(\genfrac{}{}{0pt}{}{\boldsymbol{q}}{j})}{\omega_j^2(\boldsymbol{q})}. \tag{5.59a}$$

This expression can also be written in the physically more appealing form (Appendix H)

$$2W(\boldsymbol{K}) = \langle|\boldsymbol{K}\cdot\boldsymbol{u}(l)|^2\rangle = \sum_\alpha K_\alpha^2 \langle u_\alpha^2\rangle, \tag{5.59b}$$

where $\langle u_\alpha^2\rangle$ are the components of the mean square amplitudes of vibration of an atom referred to principal axes and $\langle\ldots\rangle$ denotes the thermal average.

For a cubic Bravais crystal (5.59) can also be written in the form (Problem 5.8.3)

$$2W(K) = \frac{K^2}{3Nm} \int \overline{E}(\omega)\omega^{-2}g(\omega)d\omega, \tag{5.60}$$

where $g(\omega)$ is the density of states [Ref. 5.23, Sect. 3.5.1] and $\overline{E}(\omega)$ is the mean energy given by (5.39b)

$$\overline{E}(\omega) = \hbar\omega\left[\overline{n}(\omega, T) + \frac{1}{2}\right], \quad \text{and} \tag{5.61}$$

$$\overline{n}(\omega, T) = [\exp(\hbar\omega/k_{\mathrm{B}}T) - 1]^{-1}, \tag{5.62}$$

is the mean occupation number. The evaluation of (5.60) requires the knowledge of the density of states. We shall evaluate $2W(K)$ for the Einstein and the Debye model.

In the *Einstein model* we have [Ref. 5.23, Eq. (3.89)]

$$g_{\mathrm{E}}(\omega) = 3N\delta(\omega - \omega_{\mathrm{E}}), \tag{5.63}$$

and we obtain immediately

$$2W_{\mathrm{E}}(K, T) = \frac{K^2}{m\omega_{\mathrm{E}}^2}\overline{E}(\omega_{\mathrm{E}}, T). \tag{5.64}$$

At high temperatures $\overline{E} = k_{\mathrm{B}}T$ and

$$2W_{\mathrm{E}}(K, T) = \frac{K^2}{m\omega_{\mathrm{E}}^2}k_{\mathrm{B}}T. \tag{5.65a}$$

At low temperatures $\overline{n} = \exp(-\hbar\omega_{\mathrm{E}}/k_{\mathrm{B}}T)$ and

$$2W_{\mathrm{E}}(K, T) = \frac{\hbar K^2}{m\omega_{\mathrm{E}}}[\exp(-\hbar\omega_{\mathrm{E}}/k_{\mathrm{B}}T) + \tfrac{1}{2}], \tag{5.65b}$$

and at $T = 0$

$$2W_E(K) = \frac{\hbar K^2}{2m\omega_E} .$$ (5.65c)

The intensities of the Bragg reflections are therefore reduced even at $T = 0$ due to the zero-point motion of the atoms.

In the Debye-model the density of states is given by [Ref. 5.23, Eq. (3.94)]

$$g_D(\omega) = \begin{cases} 9N\omega^2/\omega_D^3 & \text{for} \quad \omega \leq \omega_D \\ 0 & \text{for} \quad \omega > \omega_D, \end{cases}$$ (5.66)

where ω_D is the Debye frequency defined by

$$\hbar\omega_D = k_B\theta_D, \quad \omega_D^3 = \frac{6\pi^2}{v_a}v^3.$$ (5.67)

Here θ_D is the Debye temperature, v is the mean sound velocity and v_a the volume of the unit cell. Substitution of (5.61, 62, 66) into (5.60) gives

$$2W_D(K) = \frac{3\hbar K^2}{m\omega_D^3} \int_0^{\omega_D} \omega \left(\frac{1}{\exp(\hbar\omega/k_BT) - 1} + \frac{1}{2} \right) d\omega.$$ (5.68)

At high temperatures one obtains (Problem 5.8.3)

$$2W_D(K, T) = 3\frac{\hbar^2 K^2}{mk_B\theta_D^2}T,$$ (5.69a)

while at $T = 0$ the result is

$$2W_D(K, T = 0) = \frac{3}{4}\frac{\hbar^2 K^2}{mk_B\theta_D} .$$ (5.69b)

We note that the effect due to the zero-point motion is not negligible: From (5.69) it follows that $W_D(K, T = 0) \cong W_D(K, \theta_D)/4$.

According to (5.38) we have

$$I_0(\tau, T) = I(\tau)_{\text{static}}e^{-2W(\tau,T)},$$ (5.70)

for the intensity of the Bragg reflections, $I(\tau)_{\text{static}}$ being the intensity from the rigid lattice at the reciprocal lattice point $K = \tau$. From (5.65, 69, 70) we expect that for a given τ the logarithm of $I_0(\tau, T)$ decreases linearly with T at high temperatures. Furthermore, at a given temperature the intensity is expected to decrease with an increasing magnitude of the reciprocal lattice vector τ associated with the reflection: the larger $|\tau|$ (or K) is, the weaker the reflection. An experimental verification of these predicitons is given in Fig. 5.11 which shows the temperature dependence of the ($h00$) X-ray

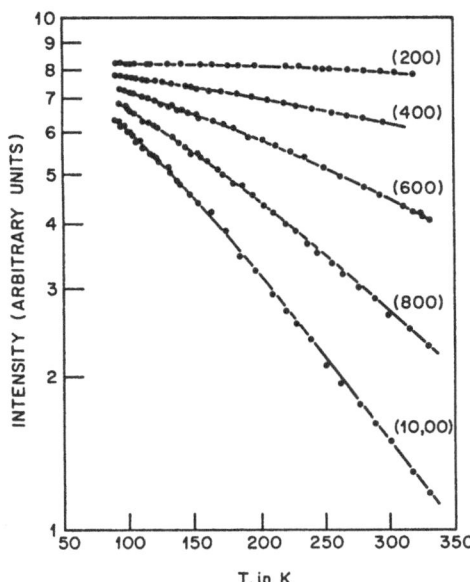

Fig. 5.11. The dependence of intensity on temperature for the $(h00)$ X-ray reflections of aluminum [5.33]

reflections of aluminum [5.33]. When there is more than one atom per unit cell, the Debye-Waller factors for the different atoms $\kappa = 1 \ldots n$ in the unit cell cannot be found from the frequency distribution $g(\omega)$ alone, as was the case for a simple cubic Bravais crystal according to (5.60). In fact, from (5.55) we have

$$2W_\kappa(\boldsymbol{K}) = \frac{1}{Nm_\kappa} \sum_{qj} |\boldsymbol{K} \cdot \boldsymbol{e}(\kappa|_j^q)|^2 \frac{\overline{E(_j^q)}}{\omega_j^2(\boldsymbol{q})}, \qquad (5.71a)$$

and this expression can only be evaluated when both the eigenfrequencies and eigenvectors are known. Generalizing (5.59b) we can also write (5.71a) in the form

$$2W_\kappa(\boldsymbol{K}) = \sum_\alpha K_\alpha^2 \langle u_\alpha^2(_\kappa^l) \rangle, \qquad (5.71b)$$

where $\langle u_\alpha^2(_\kappa^l) \rangle$ are the components of the mean square amplitudes of vibration of the atom $(_\kappa^l)$ referred to its principal axes which define the ellipsoid of vibration of that atom. Referred to any system of axes the expression becomes

$$2W_\kappa(\boldsymbol{K}) = \sum_{\alpha\beta} K_\alpha K_\beta \langle u_\alpha(_\kappa^l) u_\beta(_\kappa^l) \rangle. \qquad (5.71c)$$

The equivalence of (5.71c and a) is demonstrated by (H.25). The thermal averages occuring in (5.71b, c) are, of course, independent on the unit cell index l, as can be verified explicitly, see (H.25).

145

5.8 Problems

5.8.1 Compton Scattering

Let k_0 and k be the wave vectors of the incident and scattered radiation, 2θ the scattering angle, $\hbar\omega_0$ and $\hbar\omega$ the energies of the incident and scattered photons, m the mass of the electron and v the recoil velocity of the electron after the collision with the photon. The electron is assumed to be at rest before the collision. If relativistic effects are neglected, the energy and momentum conservations are

$$\hbar\omega_0 = \hbar\omega + \tfrac{1}{2}mv^2, \tag{5.8.1}$$

$$\hbar k_0 = \hbar k + mv. \tag{5.8.2}$$

Show that the change of wavelength, $\Delta\lambda = \lambda - \lambda_0$, is given by (5.17), namely $\Delta\lambda = (2h/mc)\sin^2\theta$.

5.8.2 Atomic Scattering Factor

Show that for a spherical symmetrical charge distribution $\varrho_\kappa(x) = \varrho_\kappa(x)$, the atomic scattering factor (5.26) becomes

$$f_\kappa(K) = 4\pi \int x^2 \varrho_\kappa(x) \frac{\sin Kx}{Kx} dx. \tag{5.8.3}$$

Note that $f_\kappa(K = 0) = z_\kappa$, the number of electrons in the atom.

Hint: Write $\boldsymbol{K}\cdot\boldsymbol{x} = Kx\cos\alpha$ and $d^3x = 2\pi x^2 \sin\alpha\,dx\,d\alpha$.

5.8.3 Debye-Waller Factor for a Cubic Bravais Crystal

a) Show that for a cubic Bravais crystal the Debye-Waller factor can be written in the form (5.60), namely

$$2W(K) = \frac{K^2}{3Nm} \int \overline{E}(\omega)\omega^{-2}g(\omega)d\omega. \tag{5.8.4}$$

Hint: From (5.59b) we have $2W(\boldsymbol{K}) = \sum_\alpha K_\alpha^2 \langle u_\alpha^2 \rangle$. Choosing $\boldsymbol{K} = (0,0,K)$ gives $2W(K) = K^2 \langle u_z^2 \rangle$. In a cubic Bravais crystal we have $\langle u_x^2 \rangle = \langle u_y^2 \rangle = \langle u_z^2 \rangle = \frac{1}{3}\langle u^2 \rangle$. Use (H.26) to obtain $\langle u^2 \rangle$ and replace the sum over $\boldsymbol{q}$ and j by an integral over ω according to [Ref. 5.23, Eq. (3.79)]

b) From the result derived in (a) show that in the Debye-model one finds for high temperatures

$$2W_D(K,T) = 3\frac{\hbar^2 K^2}{mk_B\theta_D^2}T, \tag{5.8.5}$$

and for $T \to 0$

$$2W_D(K, T = 0) = \frac{3}{4} \frac{\hbar^2 K^2}{mk_B\theta_D}.$$ (5.8.6)

5.8.4 Correlation Function

In Appendix H we have derived the following expression for the correlation function (H.26)

$$g(ll') = \langle u(l) \cdot u(l') \rangle = \frac{1}{Nm} \sum_{qj} \exp(i q \cdot r_{ll'}) \frac{\overline{E}(_j^q)}{\omega_j^2(q)},$$ (5.8.7)

where $r_{ll'} = r(l) - r(l')$ is the distance between two atoms l and l'. At high temperatures $\overline{E}(_j^q) = k_BT$, and in the Debye-model we write $\omega_j(q) = vq$, $r_{ll'} = r$ and obtain after summing over the three degenerate acoustic branches

$$g_D(r) = \langle u(0) u(r) \rangle = \frac{3k_BT}{Nmv^2} \sum_q q^{-2} e^{i q \cdot r}.$$ (5.8.8)

Show that for the three-dimensional Debye crystal the correlation function is given by

$$g_D(x) = \langle u^2(0) \rangle \frac{Si(x)}{x}$$ (5.8.9)

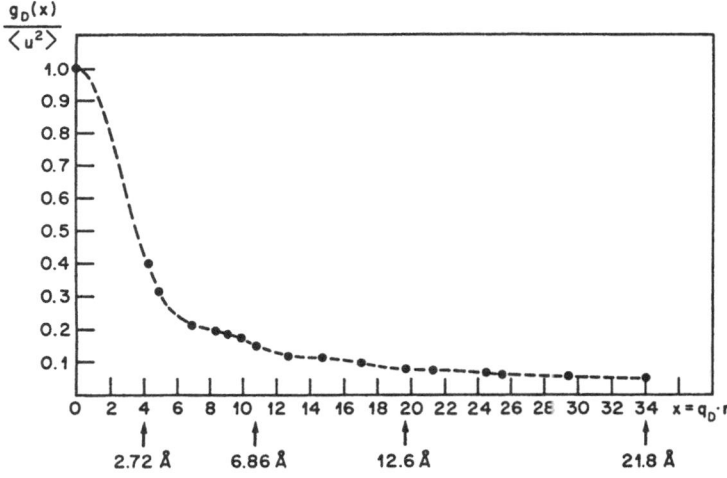

Fig. 5.12. Normalized correlation function $g_D(x)/\langle u^2 \rangle$ calculated on the basis of the Debye model and at high temperatures (- - -). $x = q_D r$, where q_D is the Debye cut-off wavenumber and r is the interatomic distance. The arrows indicate some interatomic distances for Mo with $q_D = 1.560\,\text{Å}^{-1}$, and the points are the results obtained from the Born-von Karman model [5.34]

where

$$\langle u^2(0)\rangle = 9\frac{k_B T}{m\omega_D^2} \quad \text{and}$$

$$Si(x) = \int\limits_0^x \frac{\sin t}{t}\,dt.$$

Here $x = q_D r$ where q_D is the magnitude of the Debye wave vector and ω_D is the Debye frequency [Ref. 5.23, Sect. 3.5.1].

Hint: Transform the sum over $\boldsymbol{q}$ into an integral [Ref. 5.23, Eq. (3.77)]. Write $\boldsymbol{q}\cdot\boldsymbol{r} = qr\cos\vartheta$ and use $d^3q = 2\pi q^2 \sin\vartheta\,d\vartheta\,dq$ for the volume element of the Debye sphere with radius q_D. Figure 5.12 shows a plot of $g_D(x)/\langle u^2\rangle$. Since $q_D = (6\pi^2 n/v_a)^{1/3}$ [Ref. 5.23, Eq. (3.95)], we obtain for Mo (bcc) with $v_a = a^3/2$, $n = 1$ and $a = 3.14\,\text{Å}$ a value of $q_D = 1.56\,\text{Å}^{-1}$. The points marked in Fig. 5.12 indicate some interatomic lattice distances for Mo. The agreement of $g_D(x)$ with $g(x)$ derived on the basis of a 7th neighbour Born-von Karman force model for Mo [5.34] is remarkably good.

6. Inelastic Neutron Scattering

By W. Bührer

Experiments with thermal neutrons are most useful in the study of lattice vibrations. The advantage stems from the basic nature of the neutron: The De Broglie wavelength of a thermal neutron is of the same order of magnitude as the interatomic distances in crystals, and the energy of a thermal neutron is of the same order as the energy of a lattice excitation. Table 6.1 gives some characteristical values and physical constants of thermal neutrons. In a scattering process, the neutron matches the lattice vibrations with respect to energy and wave vector (Fig. 1.1), and there exist pronounced interference phenomena. Coherent neutron scattering allows direct measurement of the dispersion relation, i.e., of the frequencies for distinctive polarizations and wave vectors in the Brillouin-zone, and hence gives the best proof for the existence of phonons.

The neutron interacts with the target via nuclear forces, and there are no restrictions with respect to the type of interatomic binding forces: metallic, ionic, covalent, molecular and hydrogen-bounded crystals can be investigated. The neutron probes directly the motion of the nucleus, whereas in optical measurements the nuclear motions are observed through the coupling of the photon with the electrons or ions. The neutron is uncharged, it penetrates deeply into the target and allows the study of bulk properties; there are no limitations due to surface effects.

One disadvantage of neutron scattering is the relatively poor energy and wave vector resolution, in the order of 1 % for a normal experiment, due to the low intensity of available neutron beams. The necessity of having

Table 6.1. Physical constants and characteristical data of the neutron

mass	m_n	$1.675 \cdot 10^{-27}$ kg
charge		0
spin		1/2

$$E = (1/2)m_n v^2 = \hbar^2/2m_n\lambda^2 = \hbar^2 k^2/2m_n$$

Standard values for thermal neutrons (convention)

$v = 2200 \, m/s$
$E = 25.3 \, meV$
$T = 293 \, K$
$\lambda = 1.798 \, Å$
$k = 3.49 \, Å^{-1}$

149

a large experimental facility, a nuclear reactor or a spallation source, makes the method somewhat exclusive.

Nevertheless a large number of dispersion curves of all types of crystals have been measured and greatly stimulated theoretical research. Very useful bibliographies of lattice-dynamical studies with neutrons have been compiled by the Japanese atomic energy research institute [6.1], and for insulating crystals are given in the "phonon atlas" [6.2].

This chapter describes those aspects of the basic principles and experimental techniques of inelastic neutron scattering, which are mostly used for the investigation of lattice dynamics. Hence we restrict ourselves to nuclear scattering; magnetic neutron scattering is not discussed at all. The relevant cross sections are given, the derivation is similar to the one for diffuse X-ray scattering (Chap. 5); for a thorough discussion the interested reader is referred to the books of *Squires* [6.3] or *Marshall* and *Lovesey* [6.4]. In neutron scattering, the notations and definitions of variables are somewhat different from those used in X-ray scattering, and in this chapter we use the "neutron standard", for a better comparison with the literature. In the experimental section the presently most versatile instrument, the triple-axis spectrometer, is presented. More details of experimental techniques and additional references have been given by *Dolling* [6.5]. Finally some illustrative examples of coherent and incoherent scattering experiments are given.

6.1 Basic Properties

6.1.1 Cross Section

Suppose that a beam of monoenergetic thermal neutrons is incident on a cristalline sample, as shown in Fig. 6.1. The neutrons may either be scattered, be absorbed, or pass without interaction.

The incident and the scattered neutrons are characterized by the momentum vectors $\hbar k_0$ and $\hbar k_1$, respectively, where $k = 2\pi/\lambda$, and λ is the neutron wavelength. Every nucleus in the sample is a center of scattering, and the scattered neutrons will interfere and combine to a total field. If in

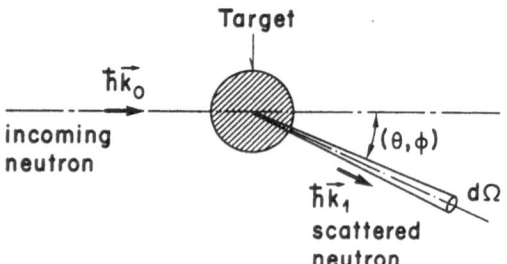

Fig. 6.1. Neutron scattering process

this field neutrons are found with their momentum changed by

$$\hbar Q = \hbar(k_0 - k_1),\tag{6.1}$$

and their energy changed by

$$\hbar\omega = \frac{\hbar^2}{2m_n}(k_0^2 - k_1^2) = E_0 - E_1,\tag{6.2}$$

then the cirstalline sample must be able to change its dynamical properties by the same quantities. In Eqs. (6.1, 2) m_n is the neutron mass, and Q is the scattering vector [a definition different from X-ray scattering, see (5.5)].

The process of scattering is characterized by the cross section. It tells us how likely a neutron is scattered as a function of energy or momentum transfer. The cross sections have dimensions of an area and are usually given in barns (1 barn $= 10^{-24}\,\mathrm{cm}^2$). Depending on the details of information in the scattering process, different scattering cross sections are relevant. The most general quantity, the total cross section, defined by

$$\sigma_{tot} = \frac{\text{number of neutrons scattered per second}}{\text{incident flux}}\tag{6.3}$$

gives the number of neutrons scattered in all directions. As an example the total cross section of oxygen is shown in Fig. 6.2a [6.6]; it is independent of the neutron energy over a large range down to approximately 10 meV.

The angular distribution of the scattered neutrons is given by the differential cross section, defined by

$$\frac{d\sigma}{d\Omega} = \frac{\text{number of neutrons scattered per second in }d\Omega(\theta,\phi)}{\text{incident flux}}.\tag{6.4}$$

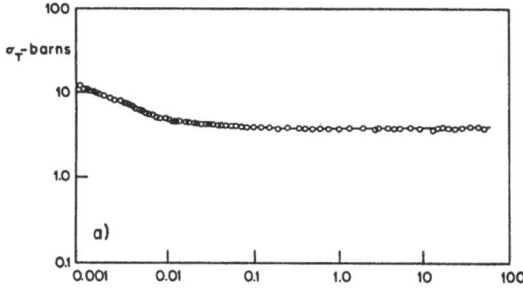

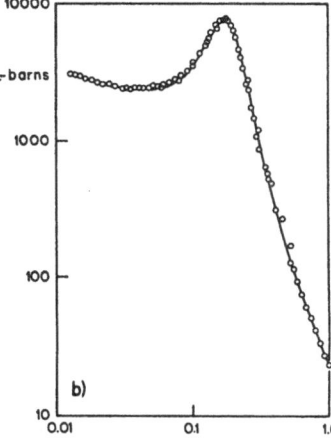

Fig. 6.2a,b. Energy dependence of neutron cross sections [6.6] (a) $_8$O, almost constant for energies $10\,\mathrm{meV} < E < 100\,\mathrm{eV}$ (b) $_{48}$Cd, resonance absorption near 175 meV

151

In the process of scattering the neutron has changed its momentum; if the scattering was not elastic, the energy of the neutron is changed, too. The scattered intensity is then described by the partial differential cross section

$$\frac{d^2\sigma}{d\Omega\, dE_1} = \frac{\begin{array}{c}\text{number of neutrons scattered per second into } d\Omega(\theta,\phi)\\ \text{with energy between } E_1 \text{ and } E_1{+}dE_1\end{array}}{\text{incident flux}}, \tag{6.5}$$

which gives the most detailed information of the scattering process. In a neutron scattering experiment we essentially determine the differential cross section, and the basic problem is to derive a theoretical expression for this quantity.

We consider now the scattering of a neutron by a single nucleus bound at the origin. The interaction potential must take account of the interaction via nuclear forces. (The interaction of the neutron through its magnetic moment with electrons in atoms with a magnetic moment is of the same order of magnitude; it is important for the determination of magnetic structures, magnetic excitations and magnetoelastic interactions, but not for pure lattice dynamics). Nuclear forces have a very short range nature as compared to the neutron wavelength, therefore the scattering is isotropic, and the asymptotic form of the wave function at large distances is

$$\psi_0 + \psi_1 = \exp(ik_0 z) - \frac{b}{r}\exp(ik_1 r) \tag{6.6}$$

where ψ_0 is the incident neutron wave function (z axis along $\mathbf{k}_0$), and ψ_1 is the scattered neutron wave function. By treating the scattering in a first-order Born approximation, the true interaction potential between the neutron and the nucleus at position $\mathbf{R}$ is replaced by the Fermi pseudopotential

$$V(\mathbf{r}) = \frac{2\pi\hbar^2 b}{m_n}\delta(\mathbf{r} - \mathbf{R}) \tag{6.7}$$

to give the desired isotropic scattering. For the scattering cross section (6.3) we then obtain

$$\sigma_{\text{tot}} = \frac{4\pi r^2 v \left|\frac{b}{r}\exp(ik_1 r)\right|^2}{v\left|\exp(ik_0 z)\right|^2} = 4\pi b^2, \tag{6.8}$$

where v is the neutron velocity, equal before and after scattering at a fixed nucleus.

The strength of a neutron-nucleus interaction is described by the scattering length b, which in general is a complex quantity

$$b = b' - ib'', \tag{6.9}$$

the imaginary part giving the neutron absorption.

Table 6.2. Some representative scattering lengths b_l [6.7]

isotope	$b_l[10^{-12} \text{ cm}]$
H	−0.3741
D	0.6674
^{16}O	0.5805
^{51}V	−0.0414
^{58}Ni	1.44
^{60}Ni	0.28
^{62}Ni	−0.87
^{64}Ni	−0.38
^{197}Au	0.736
^{238}U	0.855

In the absorption process, the neutron is captured by the nucleus and a compound nucleus is formed which then decays by the emission of radiation. Fortunately absorption is only restrictive for a few isotopes such as ^{3}He, ^{10}B, ^{113}Cd, which have a resonance absorption in the thermal neutron region. Figure 6.2b displays the total cross section of Cd, it is strongly energy dependent with a resonance energy near 175 meV. For most of the other nuclei the absorption is small, and for the following discussion the scattering length is assumed to be a real number independent of energy. The scattering length depends on the isotope and, if the nuclear spin is not zero, will depend on whether the spin of the neutron and nucleus are parallel or antiparallel. Some characteristic values of scattering lengths are given in Table 6.2 [6.7]. The true neutron-nucleus interaction potential is actually not known; replacing it by the Fermi pseudopotential gives the required result, but the "scattering length" is an empirical quantity in magnitude and sign. In X-ray scattering, the analogous quantity is the atomic scattering factor (5.26) which, however, unlike the scattering length, depends on the scattering vector (the dimension of the electron density distribution is comparable to the X-ray wavelength), and increases monotonically with the number of electrons in the atom. The neutron scattering length varies randomly in the periodic system and has the same order of magnitude for light and heavy elements.

Quantum mechanically the cross section is evaluated in the Born approximation [6.8]. The incoming neutrons are represented by plane wave functions $|k_0\rangle$, and they are incident on the sample in the state $|i\rangle$. In the scattering process the neutron changes to state $|k_1\rangle$ and leaves the sample in state $|f\rangle$. The transition probability is given by Fermi's golden rule

$$W(k_0, i \rightarrow k_1, f) \cong |\langle k_1, f|V|k_0, i\rangle|^2 \, \delta(E_i - E_f + \hbar\omega), \tag{6.10}$$

where the δ-function represents the conservation of energy in the neutron-target system during the scattering process, and V is the interaction potential. The cross section is defined by

$$d\sigma = \frac{W(\boldsymbol{k}_0, i \rightarrow \boldsymbol{k}_1, f)}{\text{incident flux}};$$

(6.11)

together with (6.10), and by considering the density of final scattering states we get

$$\left(\frac{d^2\sigma}{d\Omega\,dE_1}\right)_{(\boldsymbol{k}_0, i \rightarrow \boldsymbol{k}_1, f)} = \frac{k_1}{k_0}\left(\frac{m_\mathrm{n}}{2\pi\hbar^2}\right)^2$$

$$\times |\langle \boldsymbol{k}_1, f|V|\boldsymbol{k}_{0,i}\rangle|^2 \times \delta(E_i - E_f + \hbar\omega).$$

(6.12)

In the experiment we actually measure the cross section (6.5), and to obtain this quantity for all possible scattering processes, we have to sum over all final states $|f\rangle$ and to average over all initial states $|i\rangle$ occuring with probabilities given by Boltzmann's statistics.

By substituting the Fermi pseudopotential (6.7) into (6.12) the cross section can be evaluated and the result is [6.3]

$$\left(\frac{d^2\sigma}{d\Omega\,dE_1}\right) = \frac{k_1}{k_0}\frac{1}{2\pi\hbar}\sum_{ll'}b_l b_{l'}$$

$$\times \int\limits_{-\infty}^{\infty} \langle \exp[-i\boldsymbol{Q}\cdot\boldsymbol{R}_{l'}(0)]\exp[i\boldsymbol{Q}\cdot\boldsymbol{R}_l(t)]\rangle \exp(-i\omega t)dt.$$

(6.13)

This basic equation includes products of two functions, one depending only on the nuclear interaction (scattering lengths), while the other is determined by the structural and dynamical properties of the atoms in the undisturbed sample, clearly showing the correlation between spatial and time behavior. The properties of the scattering system are described by the thermal averages of the Heisenberg operators $\boldsymbol{R}_l(t)$.

6.1.2 Coherent and Incoherent Scattering

As already mentioned, the scattering length b depends on the isotope and the combined neutron-nucleus spin, and generally it is not the same for all nuclei even in a single elemental scattering system. Nuclei having different scattering lengths will statistically be distributed in the crystal, and therefore the neutron sees a crystal in which the scattering length varies from one point to the other. The cross section (6.13) contains products of pairs of scattering lengths, which can be written as the sum of a correlated part and an uncorrelated part

$$\overline{b_l b_{l'}} = |\bar{b}|^2 + \delta_{ll'}(\overline{b^2} - |\bar{b}|^2).$$

(6.14)

Only the system with average scattering length can give interference effects; this coherent scattering is proportional to the average value $\bar{b}$. Incoherent scattering arises from the random distribution of the deviations from $\bar{b}$ and shows no interference effects.

The cross section (6.13) can be split in a coherent part, given by the average system

$$\left(\frac{d^2\sigma}{d\Omega\, dE_1}\right)_{\text{coh}} \cong \bar{b}^2 \frac{k_1}{k_0} \sum_{ll'} \int_{-\infty}^{\infty} \langle \exp[-i\boldsymbol{Q}\cdot\boldsymbol{R}_{l'}(0)]$$

$$\times \exp[\boldsymbol{Q}\cdot\boldsymbol{R}_l(t)]\rangle \exp(-i\omega t)\, dt, \qquad (6.15)$$

and an incoherent part

$$\left(\frac{d^2\sigma}{d\Omega\, dE_1}\right)_{\text{inc}} \cong (\overline{b^2} - \bar{b}^2) \frac{k_1}{k_0} \sum_{l} \int_{-\infty}^{\infty} \langle \exp[-i\boldsymbol{Q}\cdot\boldsymbol{R}_l(0)]$$

$$\times \exp[i\boldsymbol{Q}\cdot\boldsymbol{R}_l(t)]\rangle \exp(-i\omega t)\, dt. \qquad (6.16)$$

These equations show that the coherent scattering (6.15) is determined by the correlation between the positions of the same atom at different times and by the correlation between the positions of different atoms at different times. Coherent scattering from crystals arises only if the geometrical conditions (tanslational symmetry) are satisfied. The incoherent or single-atom scattering (6.16) depends only on the correlation between the positions of the same nucleus at different times.

6.2 Phonon Dispersion-Relation Measurements

6.2.1 Coherent Scattering Cross Section

We consider a crystal as a target for the thermal neutron beam and evaluate the coherent scattering cross section (6.15). The instantaneous position of a nucleus in a Bravais crystal is given by

$$\boldsymbol{R}_l = \boldsymbol{r}_l + \boldsymbol{u}_l(t), \qquad (6.17)$$

where l is the cell index, $\boldsymbol{r}_l$ is the equilibrium position of the atom, and the displacements $\boldsymbol{u}_l$ are due to the thermal fluctuations [Ref. 6.9, Sect. 3.3]. By inserting (6.17) into (6.15) and considering that in a Bravais crystal the correlation between atoms at $\boldsymbol{r}_l$ and $\boldsymbol{r}_{l'}$ depends only on $\boldsymbol{r}_l - \boldsymbol{r}_{l'}$, we obtain

$$\left(\frac{d^2\sigma}{d\Omega\, dE_1}\right)_{\text{coh}} \cong \bar{b}^2 \frac{k_1}{k_0} \sum_l \exp[i\boldsymbol{Q}\cdot\boldsymbol{r}_l]$$

$$\times \int_{-\infty}^{\infty} \langle \exp[-i\boldsymbol{Q}\cdot\boldsymbol{u}_0(0)]\exp[i\boldsymbol{Q}\cdot\boldsymbol{u}_l(t)]\rangle \exp(-i\omega t)\, dt. \tag{6.18}$$

This expression is very similar to (5.35) for the diffuse X-ray intensity; however in (6.18) the time dependence of the displacements has to be considered because for thermal neutrons $\hbar\omega \cong \hbar\omega(\boldsymbol{q})$ (see discussion in Sect. 5.1). A further evaluation of (6.18) parallels that for X-ray scattering, as given in Chap 5. In the harmonic approximation the displacements can be expressed by the set of normal coordinates [Ref. 6.9, Eq. (3.74)]

$$\boldsymbol{u}_l = \sqrt{\frac{\hbar}{2Nm}} \sum_{q,j} \frac{1}{[\omega_j(\boldsymbol{q})]^{1/2}} \boldsymbol{e}\binom{\boldsymbol{q}}{j}\exp(i\boldsymbol{q}\cdot\boldsymbol{r}_l)[a^+\binom{-\boldsymbol{q}}{j} + a\binom{\boldsymbol{q}}{j}]. \tag{6.19}$$

Here $\omega_j(\boldsymbol{q})$ and $\boldsymbol{e}\binom{\boldsymbol{q}}{j}$ are the frequencies and orthonormalized eigenvectors of the normal mode with wave vector $\boldsymbol{q}$, respectively, obtained by solving the dynamical matrix equation [Ref. 6.9, Eq. (3.28)]

$$D(\boldsymbol{q})\boldsymbol{e}\binom{\boldsymbol{q}}{j} = \omega_j^2(\boldsymbol{q})\boldsymbol{e}\binom{\boldsymbol{q}}{j}, \tag{6.20}$$

where m is the mass of the atom at position $\boldsymbol{R}_{l'}$ and $a^+\binom{-\boldsymbol{q}}{j}$ and $a\binom{\boldsymbol{q}}{j}$ are the phonon creation and annihilation operators [Ref. 6.9, Sect. 3.4].

The evaluation of the average in the angular brackets is outlined in Appendix H, and for neutrons (with time dependence of displacements taken into account) we get

$$\langle \exp[-i\boldsymbol{Q}\cdot\boldsymbol{u}_0(0)]\exp[i\boldsymbol{Q}\cdot\boldsymbol{u}_l(t)]\rangle$$

$$= \exp[\langle Q^2 u_0^2(0)\rangle]\exp[\langle \boldsymbol{Q}\cdot\boldsymbol{u}_0(0)\boldsymbol{Q}\cdot\boldsymbol{u}_l(t)\rangle], \tag{6.21}$$

where the first factor $\exp\langle Q^2 u_0^2(0)\rangle$ corresponds to the Debye-Waller factor $\exp(-2W)$ (Appendix H.3). By inserting (6.21) into (6.18) the partial cross section is then given by

$$\left(\frac{d^2\sigma}{d\Omega\, dE_1}\right)_{\text{coh}} \cong \bar{b}^2 \frac{k_1}{k_0} e^{-2W} \sum_l \exp[i\boldsymbol{Q}\cdot\boldsymbol{r}_l]$$

$$\times \int_{-\infty}^{\infty} \exp[\langle \boldsymbol{Q}\cdot\boldsymbol{u}_0(0)\boldsymbol{Q}\cdot\boldsymbol{u}_l(t)\rangle]\exp(-i\omega t)\, dt. \tag{6.22}$$

The initial state $|i\rangle$ of the sample crystal is given by the quantum numbers of the $3N$ oscillators corresponding to the normal modes [Ref. 6.9,

Sect. 2.2.4]. In a general scattering process the crystal changes to the state $|f\rangle$ characterized by another set of quantum numbers. The second factor on the right-hand-side of (6.21) can be expanded in a power series

$$e^{\langle x\rangle} = 1 + \langle x\rangle + \frac{1}{2!}\langle x\rangle^2 + \dots \, , \tag{6.23}$$

the different terms describing zero, first, second-order processes (Appendix H.2):

$$\left(\frac{d^2\sigma}{d\Omega\, dE_1}\right)_{\text{coh}} = 0\text{-ph} +1\text{-ph} +2\text{-ph} + \dots \, . \tag{6.24}$$

In this multi-phonon expansion the zero-th term corresponds to elastic scattering (no change of quantum numbers), the first-order term to the one-phonon process (change of only one state $n_j \rightarrow n_j + 1$), the second-order term to the two-phonon process and so on.

Elastic scattering is obtained with the first term of the expansion (6.23):

$$\left(\frac{d^2\sigma}{d\Omega\, dE_1}\right)_{\text{coh}}^0 \cong \bar{b}^2 e^{-2W} \sum_l \exp(i\boldsymbol{Q}\cdot\boldsymbol{r}_l) \int_{-\infty}^{\infty} \exp(-i\omega t)dt. \tag{6.25}$$

The integration with respect to t gives the condition for energy conservation $\delta(\hbar\omega)$, and by integrating with respect to E_1 we obtain the differential cross section

$$\left(\frac{d\sigma}{d\Omega}\right)_{\text{coh}}^0 \cong \bar{b}^2 e^{-2W} \sum_l \exp(i\boldsymbol{Q}\cdot\boldsymbol{r}_l). \tag{6.26}$$

Using (5.27) the lattice sum can be rewritten and we get finally

$$\left(\frac{d\sigma}{d\Omega}\right)_{\text{coh}}^{\text{el}} \cong \bar{b}^2 e^{-2W} \sum_\tau \Delta(\boldsymbol{Q} - \tau), \tag{6.27}$$

in analogy to the X-ray expression, see (H.19), the coherent neutron scattering length replacing the X-ray form-factor.

The one-phonon scattering is given by the second term $\langle x\rangle$ in the expansion (6.23). The average of the displacements can be expressed as function of the creation and annihilation operators as outlined in Appendix H:

$$\langle \boldsymbol{Q}\cdot\boldsymbol{u}_0(0)\boldsymbol{Q}\cdot\boldsymbol{u}_l(t)\rangle = \frac{\hbar}{2mN} \sum_{\boldsymbol{q},j} \frac{|\boldsymbol{Q}\cdot\boldsymbol{e}({}^{\boldsymbol{q}}_j)|^2}{\omega_j}$$
$$\times (\exp[-i(\boldsymbol{q}\cdot\boldsymbol{r}_l - \omega_j t)]\langle n_j + 1\rangle + \exp[i(\boldsymbol{q}\cdot\boldsymbol{r}_l - \omega_j t)]\langle n_j\rangle), \tag{6.28}$$

and by inserting into (6.22) we obtain for the partial differential cross section

157

$$\left(\frac{d^2\sigma}{d\Omega\,dE_1}\right)^1_{\text{coh}} \cong \bar{b}^2\frac{k_1}{k_0}\frac{1}{m}e^{-2W}\sum_l \exp(i\boldsymbol{Q}\cdot\boldsymbol{r}_l)$$

$$\times\sum_{q,j}\frac{|\boldsymbol{Q}\cdot\boldsymbol{e}(^q_j)|^2}{\omega_j}\int_{-\infty}^{\infty}(\exp[-i(\boldsymbol{q}\cdot\boldsymbol{r}_l - \omega_j t)]$$

$$\times\langle n_j + 1\rangle + \exp[i(\boldsymbol{q}\cdot\boldsymbol{r}_l - \omega_j t)]\langle n_j\rangle)\exp(-i\omega t)dt\,. \quad (6.29)$$

Again the integration with respect to the time t and the summation over l can be performed, and we finally get

$$\left(\frac{d^2\sigma}{d\Omega\,dE_1}\right)^1_{\text{coh}} \cong \bar{b}^2\frac{k_1}{k_0}e^{-2W}\sum_{q,j}\sum_\tau\frac{|\boldsymbol{Q}\cdot\boldsymbol{e}(^q_j)|^2}{\omega_j}$$

$$\times\langle n_j + \tfrac{1}{2}\pm\tfrac{1}{2}\rangle\delta(\omega\mp\omega_j(\boldsymbol{q}))\Delta(\boldsymbol{Q}\mp\boldsymbol{q} - \tau), \quad (6.30)$$

where $\langle n_j\rangle$ is the Bose-population factor

$$\langle n_j\rangle = \frac{1}{\exp[h\omega_j(\boldsymbol{q})/k_\text{B}T] - 1}. \quad (6.31)$$

The upper and lower sign in (6.30) refer to phonon creation and annihilation, respectively. At low temperature the cross section for phonon annihiliation tends to zero, but that for creation tends to a finite value. Equation (6.30) contains two δ-functions for energy and crystal momentum conservation, and these two conditions allow the determination of the phonon dispersion curves $\omega_j(\boldsymbol{q})$.

Coherent one-phonon scattering can, in a more pictorial way, be considered as elastic scattering on a moving "special target crystal". Let us in a first stage consider the effect of a static displacement of the atoms in the crystal, given by a single wave vector $\boldsymbol{q}$ with polarization $\boldsymbol{e}(^q_j)$. The elastically scattered neutron intensity of such a system will, besides the Bragg points, show additional intensity due to interference effects produced by the sinusoidal variation of the displacements. The positions of the additional lines are given by

$$\boldsymbol{Q} = \tau\pm\boldsymbol{q}\,. \quad (6.32)$$

The optical analogue of this effect are the "ghost-lines" obtained by incorrect ruling of a diffration grating. In the second stage we let now move the displacement pattern with the wave-velocity $\omega_j(\boldsymbol{q})/q$ in the direction of $\boldsymbol{q}$. Transforming the velocities of the incident and scattered neutrons to their values in the laboratory frame is given by

$$\frac{\hbar^2}{2m_\text{n}}(k_0^2 - k_1^2) = \pm h\omega_j(\boldsymbol{q}). \quad (6.33)$$

Equations (6.32, 33) are the conditions for constructive interference for elastically scattered neutrons on our "special crystal", and they are contained in the two δ-functions in (6.30).

Two-phonon and higher-order processes give a broad distribution in momentum and energy space, and can be considered as background for sufficiently low temperatures ($T \cong \theta_D$). The given scattering cross section is correct within the harmonic approximation of the lattice vibrations and within the perturbation theory used for the derivation of the transition probabilities.

The extension of (6.30) to a non-Bravais crystal gives for the partial differential cross section

$$\left(\frac{d^2\sigma}{d\Omega\, dE_1}\right)^1_{\text{coh}} \cong \frac{k_1}{k_0} \sum_{q,j} \sum_{\tau} \left| \sum_{\kappa} \frac{\bar{b}_\kappa}{\sqrt{m_\kappa}} e^{-W_\kappa} \exp(i\boldsymbol{Q}\cdot\boldsymbol{r}_\kappa) \boldsymbol{Q}\cdot\boldsymbol{e}(\kappa|^q_j) \right|^2$$

$$\times \langle n_j + \tfrac{1}{2} \pm \tfrac{1}{2}\rangle \delta(\omega \mp \omega_j(\boldsymbol{q})) \Delta(\boldsymbol{Q} \mp \boldsymbol{q} - \boldsymbol{\tau}) \tag{6.34}$$

where κ is the atom index within the unit cell.

6.2.2 Selection Rules

Infrared and Raman experiments are governed by seletion rules which on one hand help identifying modes but, on the other hand, limit the experimental information. For neutron scattering no such restrictive selection rules exist; the relative intensity of a particular phonon mode $\omega_j(\boldsymbol{q})$, measured at the general point $\boldsymbol{Q}$ in reciprocal space, is given by (6.34):

$$|F_j|^2 = \left| \sum_{\kappa} \frac{\bar{b}_\kappa \exp(-W_\kappa)}{\sqrt{m_\kappa}} \boldsymbol{Q}\cdot\boldsymbol{e}(\kappa|^q_j)\exp(i\boldsymbol{Q}\cdot\boldsymbol{r}_\kappa) \right|^2 \tag{6.35}$$

where the summation extends over all atoms in the unit cell. F_j is periodic in reciprocal space (for atomic positions $\boldsymbol{r}_\kappa$ commensurate with the lattice constants) [Ref. 6.9, Appendix A] and need only be evaluated in the structure zone defined by vectors $\boldsymbol{\tau}'$ such that

$$\exp(2i\pi\boldsymbol{\tau}'\cdot\boldsymbol{r}_\kappa) = 1 \tag{6.36}$$

for all atoms in the unit cell [6.10]. If the structure zone is f times larger than the Brillouin zone, then a phonon may be observed at f inequivalent points.

Phonons with wave vector $\boldsymbol{q}$ belong to an irreducible representation of a subgroup of the crystal that leaves $\boldsymbol{q}$ invariant. The symmetry types and the corresponding representations can be determined by standard procedures [6.11, 12]. Multi-dimensional respresentations show symmetry-related degeneracies of phonon frequencies, and from the characters of the representations, compatibility relations can be deduced. If a representation occurs only once at q, then the polarization vectors are completely determined by symmetry and the structure factor (6.35) of this mode can easily be calculated. If a representation occurs many times, then the polarization vectors are determined by the interatomic forces in the crystal and only the sum of structure factors for modes of the same symmetry can group theoretically be calculated. Within the structure zone, points and directions of high symmetry can usually be found where the structure factor is high for a given representation, but small for all the others. Together with the scalar product $(Q \cdot e(\kappa|_j^q))$, which allows to distinguish between longitudinal and transversal polarisation, it is then generally possible to assign an experimentally observed peak to a specific mode j, and phonons in a complicated system with many branches are experimentally clearly separated despite the relative poor energy resoltuion in the scattering experiment (examples in Sects. 6.2.4b, c).

Besides measuring phonon frequencies $\omega_j(q)$, also the polarisation vectors (= atomic displacements) of a mode can be determined. By measuring the intensity of the one-phonon peak at different points in the structure zone, the components of $e(\kappa|_j^q)$ can be evaluated by a least-squares routine, where the theoretical values (6.35) are fitted to the experimental observed intensities. This, however, is a rather time consuming matter and has only been performed for a few selected, basically important phonon modes (example in Sect. 6.2.4c).

6.2.3 Experimental: Triple Axis Spectrometer

The inelastic neutron experiment is a measurment of neutron intensities as function of energy and momentum transfer. The principle is very simple: an incoming neutron beam of a well defined energy E_0 and momentum $\hbar k_0$ has to be produced, and after the scattering at the target, the detector has to count the number of neutrons with energy E_1, scattered into the solid angle $d\Omega$ in a given direction. Furthermore, the intensity has to be normalized with respect to the incoming neutron flux.

The source of neutrons is the most important part in the experimental setup. The scattering cross sections are rather weak and therefor a high incoming neutron flux is demanded. Nuclear reactors and spallation sources can give the required flux of 10^{14} to 10^{15} n/cm^2/s. The fast fission or spallation neutrons are slowed down in a moderator. In the equilibrium they have a Maxwellian distribution, and the energy of the flux maximum is given by the moderator temperature. Modeated neutrons are classified according to the respective temperature in "cold" ($E < 5$ meV), "thermal" or "hot" ($E > 200$ meV), and lattice-dynamical experiments can be performed with an energy which matches best to the excitation energies being studied.

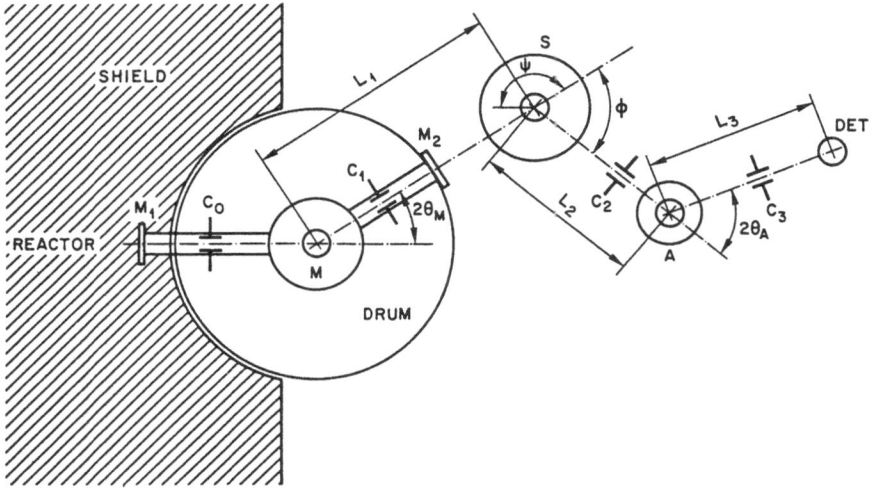

Fig. 6.3. Schematic layout of a triple-axis-spectrometer; (M: Monochromator system with curved crystals, S: Sample table, A: Analyzer system with curved crystals, L_1, L_2, L_3: Spectrometer arms with variable lenghts, C_0, C_1, C_2, C_3: Collimators or diagrams defining the angular divergencies of the neutron beam, and M_1, M_2: monitors for beam control)

Basically there are two ways of defining the energy of thermal neutrons in the order of 1 %, namely (i) by Bragg scattering from a single crystal, or (ii) by chopping of the incident neutron beam in pulses and time-of-flight analysis of the scattered neutrons. We only discuss the first method which is best suited for steady-state operation of the source and which mostly has been used in lattice-dynamical studies. Time-of-flight techniques will have an increasing importance when spallation sources are operational.

The most versatile crystal instument is the "triple-axis spectrometer" originally due to *Brockhouse* [6.13]. Figure 6.3 shows the schematic layout. Neutrons can pass along a beam tube through the biological shielding of the source. A Bragg reflexion (angles θ_M for the crystal and $2\theta_M$ for the arm) at the monochromator system selects a well defined energy E_0 out of the white incoming neutron spectrum. The monochromator crystal (graphite, beryllium, copper) is surrounded by shielding material of about 1 m radius in order to suppress fast neutron and gamma radiation emerging from the source. The flux of the monochromatic neutron beam is usually monitored with a fission chamber of low efficiency. The monochromatic beam hits the sample crystal which can be oriented at any desired angle ϕ with respect to the incoming neutrons $\boldsymbol{k}_0$. The scattered neutrons, deflected by the angle ψ, arrive at the analyzer crystal, those with the selected energy E_1 are reflected in this second Bragg system (angles θ_A and $2\theta_A$) and finally are counted with a ^{3}He detector, again shielded to suppress unwanted background.

Present-day spectrometers are equipped with curved monochromator and analyzer crystal systems in order to focus the neutrons and to increase the flux at the sample position; collimators or slits can be inserted in the neutron beam to define the angular divergencies. In addition, the length of the arms between monochromator, sample table, analyzer and counter can be varied. This gives the possibility to adapt the (Q,ω) resolution of the spectrometer to the requirements of the experiment. The spectrometer is computer controlled and during a measurement, the energy (wave vectors) of the incident and scattered neutrons, the scattering angle and the sample orientation can be varied.

The equation for conservation of energy and crystal momentum in the 1-phonon process are

$$\pm\hbar\omega = E_0 - E_1 \quad \text{and} \tag{6.37}$$

$$\boldsymbol{k}_0 - \boldsymbol{k}_1 = \boldsymbol{Q} = \tau \pm \boldsymbol{q}; \tag{6.38}$$

the energy/wave vector relation of the unknown excitation is

$$\omega = \omega_j(\boldsymbol{q}), \tag{6.39}$$

and if these three conditions are simultanously fulfilled, a maximum in the scattered neutron intensity is expected. This, in principle, infinitely sharp peak is broadened by experimental resolution effects and by the life time of the phonon and what is observed is a "neutron group". The experimental scan consists of a sequence of intensity measurments at points in (Q,ω) space.

On a triple axis spectrometer the number of instrumental parameters (k_0, k_1, ψ, ϕ) is larger than the number of parameters characterizing the dynamical process $(Q_x, Q_y, \hbar\omega)$. this gives the possibility to specify additional parameters in the experiment, and the most common type of scan is the "constant-Q" mode of operation with fixed analyzer energy. The scattering diagram for a set of consecutive points is shown in Fig. 6.4a. For a selected wave vector q with momentum transfer $\hbar Q$, the scattered neutron intensity is mapped as a function of energy transfer, and the observed peak is assigned to the energy of the excitation. The fact that the measurement can be performed at pre-selected q values is very advantageous because symmetry relations of the excitation can be considered and because lattice dynamical theories very often predict a peculiar behaviour or a temperature dependence of a phonon mode for special q-vectors. Furthermore the "const-Q"-mode is best suited for excitations with weak dispersion, whereas for a strong dispersion the "const-ω" techniques (Fig. 6.4b), gives best results.

The experimentally measured intensity does not directly give the cross section. Several corrections have to be applied such as resolution or background effects, multiple scattering or multi-phonon scattering. The resolution function of a triple axis spectrometer has been given in detail by

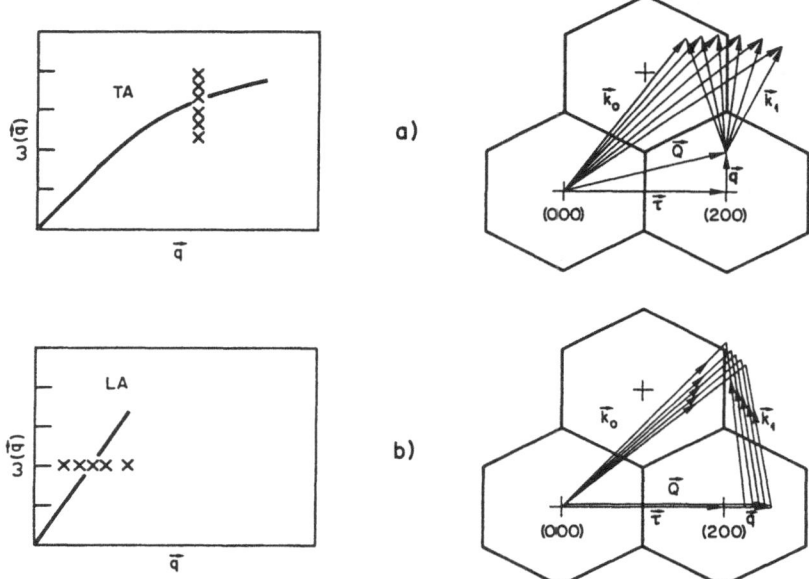

Fig. 6.4a,b. Representations of triple-axis-scans in $(\boldsymbol{Q}, \omega)$ space and corresponding scattering diagrams, taking advantage of the cross section term $\left|\boldsymbol{Q}{\cdot}\boldsymbol{e}\binom{q}{j}\right|^{2}$, see (6.30), for mode determination; (a) const.-$\boldsymbol{Q}$ measurement of a transversal polarized phonon with fixed analyzer energy; (b) const.-ω measurement of a longitudinal polarized phonon

Cooper and *Nathans* [6.14], and is represented by a four-dimensional function in $(\boldsymbol{Q}, \omega)$ space. The extensions are determined by the energies of the neutrons (E_0, E_1), by the mosaic spread of the monochromator and analyzer crystals, and by the vertical and horizontal angular divergencies of the neutron beam along its path from the reactor to the counter. Corrections of the experimental data are essential if the resolution function embraces a region of strongly varying cross section. The observed intensities have then to be deconvoluted in order to get the cross section. Necessary corrections can always be reduced by improving the resolution, which however is at the expense of intensity, and consequently in the experiment a compromise between accuracy and overall measuring time has to be achieved.

6.2.4 Examples

Measured dispersion curves of simple structured substances have been presented in [Ref. 6.9: Argon (FCC) p. 106, lead (FCC) p. 143, NaI (FCC, NaCl-structure) p. 118, Ge and GaAs (FCC, diamond and Zns-structure) p. 130]. In this subsection we first give a "historic" example, copper, with a comparison of results of diffuse X-ray and neutron scattering techniques.

The more complicated system β-silver-iodide has been discussed in detail ([Ref. 6.9, p. 83] and [Ref. 6.15, Chap. 7] and finally results on soft modes in ferroelectric perovskites have been presented in [Ref. 6.15, Chap. 3]).

a) Copper

Copper crystallizes in a face-centered (FCC) cubic lattice with one atom in the primitive unit cell. The vibrational spectrum consists of three acoustical branches, and along high symmetry directions $[x00]$, $[xx0]$, and $[xxx]$ the modes have either purely longitudinal or transversal polarization. The phonon dispersion relation has been determined by diffuse X-ray [6.16a] and inelastic neutron scattering [6.16b] techniques. The dispersion relation along $[x00]$ is shown in Fig. 6.5. The discrepancies near the zone boundary can be attributed to incorrect intensity correction of compton scattering in the X-ray experiment.

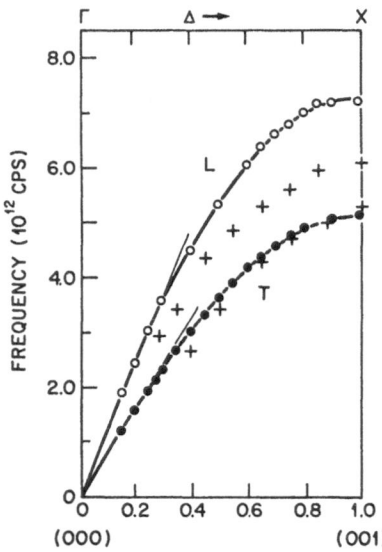

Fig. 6.5. Phonon dispersion relation in copper along $[00x]$ at 296 K: (+) Jacobsen [6.16a], diffuse X-ray; (o–o) *Svensson* et al. [6.16b], inelastic neutrons

The interpretation of diffuse X-ray intensities in terms of phonon dispersion curves is only possible with a lattice dynamical model because (5.42) (first-order process) and (5.44) (second-order processes) contain summations over phonon modes in the zone. Hence one needs an a priori knowledge of the dispersion and an iterative procedure to solve the problem.

Inelastic neutron scattering enables the direct measurement of a phonon frequency for a distinctive wave vector. An example is shown in Fig. 6.6; the scattering diagram for the measurment of the longitudinal acoustic mode for $q = 0.5\,q_{\mathrm{max}}$ along $[00x]$ is drawn in Fig. 6.6a; the experimentally observed "neutron group" of the 1-phonon peak is shown in Fig. 6.6b, where the

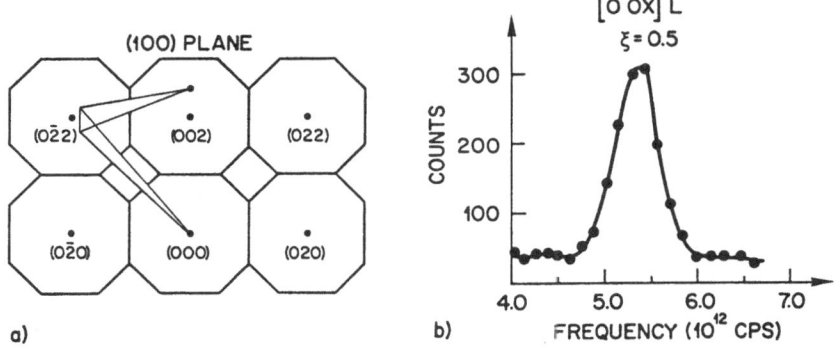

Fig. 6.6a,b. Phonon dispersion determination in copper; (a) Scattering diagram in (100)-plane for const.-Q measurement of $[00x]$ LA mode, $q = 0.5q_{max}$; (b) Neutron group (intensity versus energy transfer) for LA mode, $q = 0.5q_{max}$

intensity is plotted versus the energy transfer. This individual measurement can be interpreted without knowledge of the complete lattice dynamics: the peak position determines the phonon frequency, and the line-width and the integrated intensity of the peak contain information on the life time and the eigenvectors of the observed phonon mode, respectively. With a series of scans of this type the dispersion curves can experimentally be determined along any desired crystallographic direction.

b) β-Silver-Iodide

The structural and dynamical properties of the silver halides are governed by the competition of ionic bonding versus covalent bonding, and especially silver iodide shows a number of crystallographic phases as function of temperature and pressure. At atmospheric conditions silver iodide crystallizes in a hexagonal wurtzite lattice (β-modification). The structure consists of stacked iodine tetrahedrons with a silver ion in the center (or vice versa) and is displayed in Fig. 6.7a. There are two formulae units in the primitive cell giving rise to 12 phonon branches. The Brillouin zone of the hexagonal lattice is shown in Fig. 6.7b. The group theoretical analysis of the phonon modes has been performed with the computer program of *Worlton* and *Warren* [6.12]. The decompositions at some selected high symmetry q vectors are given in Table 6.3. This symmetry decomposition is very helpful and simplifies the experimental work, it tells us, for instance, that the transversal modes along the hexagonal direction Δ are doubly degenerate and that at the point A the transversal and longitudinal modes are 4 times and doubly degenerate, respectively, hence at the point A there are only 4 different frequencies to be measured instead of 12 for a general q vector. Along the directions T and Σ, group theory is less helpful because there

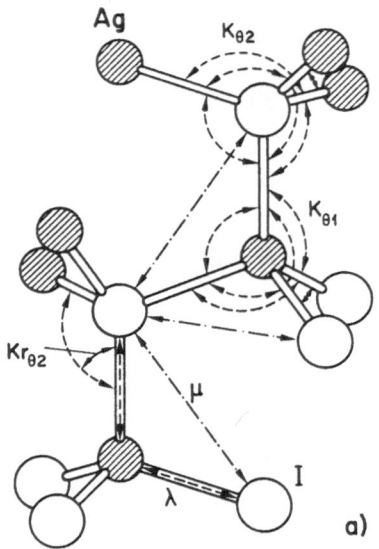

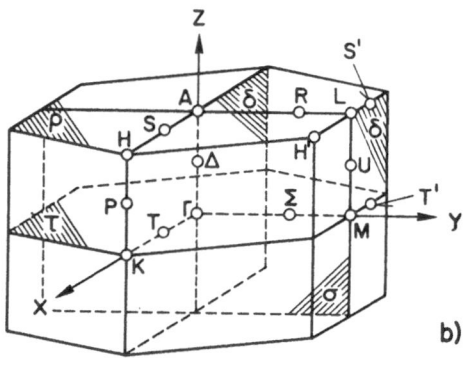

Fig. 6.7a,b. Phonon dispersion determination in β-AgI: **(a)** crystal structure of β-AgI: hexagonal, wurtzite, space group C_{6v}^4; **(b)** Brillouin zone of the hexagonal lattice

Table 6.3. Irreducible representation for β-AgI, the dimensions are given in parentheses; decomposition and notation according to *Worlton* and *Warren* [6.12]

$q = 0$ (Point Γ)	
$2\Gamma_1$	ir, R, ac
$2\Gamma^{(2)}$	ir, R, ac
$2\Gamma_4^{(2)}$	R
$2\Gamma_5$	

ir: infrared active, R: Raman activ, ac: acoustic

$q = \frac{1}{2}c^*$ (Point A)		$q = \mu \cdot \frac{1}{2}a^*$ (line Σ)	
2	$A_1^{(2)}$	8	Σ_1
2	$A_3^{(4)}$	4	Σ_2

$q = \mu \cdot \frac{1}{2}c^*$ (line Δ)		$q = \mu \cdot \frac{1}{2}(a^* + b^*)$ (line T)	
2	Δ_1	6	T_1
2	$\Delta_3^{(2)}$	6	T_2
2	$\Delta_4^{(2)}$		
2	Δ_6		

exist only two different types of representations and both are one dimensional. The structure zone is 24 times larger than the Brillouin zone, and this allows a selection of reciprocal lattice vectors in order to have a good separation of branches and to measure only one specific mode in a scan.

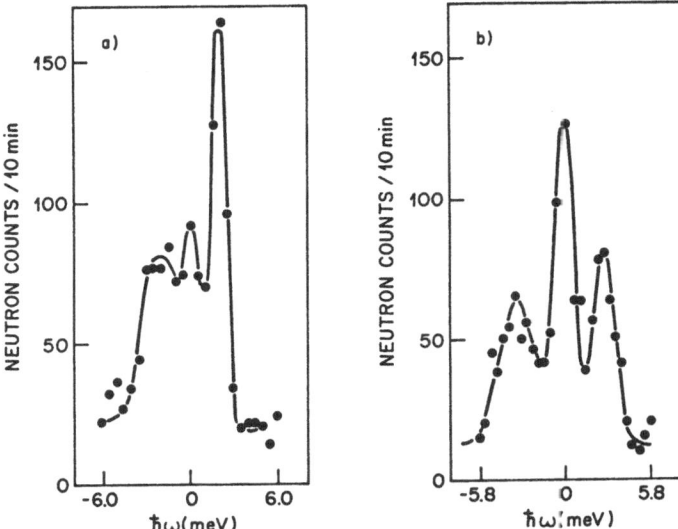

Fig. 6.8a,b. Neutron groups for const.-Q scans in β-AgI at 295 K; $q = 0.3 q_{max}$ along Δ, 3-peak structure: phonon annihilation-elastic incoherent-phonon creation; phonon peak intensities (integrated) are given by the structure factor (6.35), the peak widths are determined by resolution effects [6.14]; (a) transversal acoustic mode, (300) reciprocal zone; (b) lowest energetic transversal optic mode, (301) reciprocal zone

Examples of measured intensities, as obtained with a triple-axis spectrometer in const-Q scans are shown in Fig. 6.8. The intensity distribution of the transversal polarized acoustic mode, measured at the point (300) of the reciprocal lattice, is displayed in Fig. 6.8a, the data of the transversal optic mode, measured at the point (301), is shown in Fig. 6.8b. Due to the very different structur factors of these two modes in the different zones, there is only one mode which is observable in a given spectrometer setting. The scalar product $(Q \cdot e(\kappa|^{q}_{j}))$ excludes any contribution from a longitudinally polarized mode. The diagrams show a three-peak structure: at zero-energy transfer the elastic incoherent contribution (mainly from silver) and at the energies $\pm \hbar \omega$ the one-phonon peaks for phonon creation and annihilation, respectively. The peak shapes are influenced by resolution effects [6.14], and especially the acoustic mode with strong dispersion shows a defocused and a focused contribution on the gain and loss side, respectivellly. The exact peak frequency can be obtained by deconvoluting the measured intensity with the resolution function. The experimental dispersion curves [6.17] along high-symmetry directions are shown in Fig. 6.9, circles and triangles mark transversal and longitudinal modes, and open and full symbols represent optic and acoustic modes. The solid lines through the origin are the velocities of sound, as determined from ultrasonic measurements; the zone-center

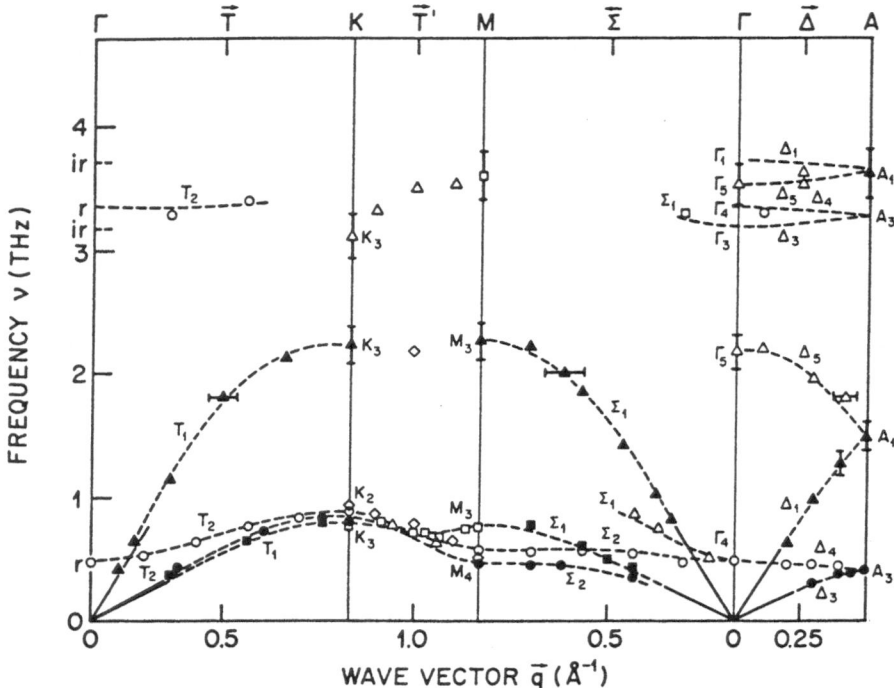

Fig. 6.9. Phonon dispersion curves in β-AgI at 160 K [6.17]; polarization vectors (experimental)

▲ LA   ● TA⊥c ■ TA∥c
△ LO   ○ TO⊥c □ TO∥c
◇ unknown   ir/r   infrared/Raman
(———) velocities of sound (- - -) guide to the eye.
Group theoretical notation according to *Worlton* and *Warren* [Ref. 6.12, Table 3]

frequencies measured by optic methods are marked with "ir" (infrared active) and "r" (Raman active), respectively. The dashed lines are a guide to the eye, they connect points belonging to the same phonon branch as obtained experimentally by considering the selection rules.

c) Soft Modes in Ferroelectrics

Ferroelectricity and ferroelectric phase transitions are treated in detail in [Ref. 6.15, Chap. 3]. The theory predicts a close relationship of the temperature dependence between the dielectric constant and the lowest transversal optic mode [Ref. 6.15, Eqs. (3.37, 38)]:

$$\varepsilon_0 = \frac{4\pi C}{T - T_0} + \varepsilon_\infty, \quad \text{and} \tag{6.40}$$

$$\omega_{TO}^2(\boldsymbol{q} = 0) = \gamma'(T - T_0), \tag{6.41}$$

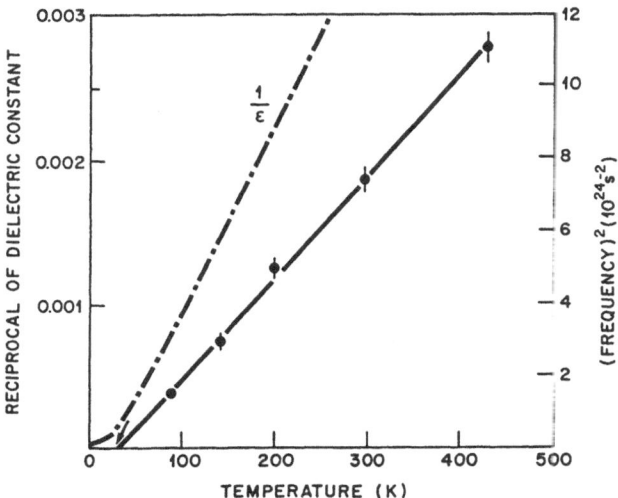

Fig. 6.10. Temperature dependence of frequency squared for the $q = 0$ lowest transversal optic mode in $SrTiO_3$ [6.18]; the dashed-dotted line shows the reciprocal of the static dielectric constant

where C is the Curie constant, T_0 is the transition temperatur and γ' is a constant. *Cowley* [6.18]measured the temperature dependence of the lowest-transverse optic-mode zone-center frequency in $SrTiO_3$. The result, as shown in Fig. 6.10, gives clear evidence for (6.40, 41), and the straight-line extrapolation gives a Curie temperature which is in excellent agreement with the static dielectric constant measurements. For a further discussion of $SrTiO_3$ which is a "near-ferroelectric" or incipient ferroelectric, the reader is referred to [Ref. 6.15, Sects. 3.3, 7].

In the ferroelectric perovskites the interatomic forces of short-range and long-range order are in a delicate balance, and small changes of one component as function of temperature produce softening of the lowest TO mode. From a lattice-dynamical viewpoint it is of great interest to see what kind of atomic vibrational displacements occur in this mode.

Experiments have been performed by *Harada* et al. [6.19]on $SrTiO_3$, $KTaO_3$, and $RbMnF_3$ with a triple-axis spectrometer. Group theory cannot predict the actual displacements because all transverse modes are characterized by the same representation [6.18]. Zone-center ($q = 0$) phonons have been measured at about 15 different reciprocal lattice points of the type (hhl) in the structure zone. The measured intensities have been corrected for instrumental-resolution effects and then analyzed according to (6.35). The components of possible, symmetry restricted eigenvectors have been determined by a least-squares fitting procedure similar to those used in solving structural problems. It turned out that for $SrTiO_3$ and $KTaO_3$,

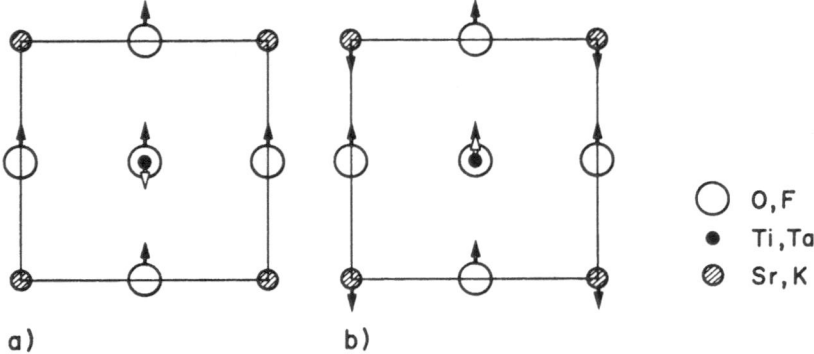

Fig. 6.11a,b. Symmetry coordinates for atomic displacements in zone center soft modes in perovskites; (a) "Slater"-mode [6.20], (b) "Last"-mode [6.21]

the soft mode is mainly of the "Slater"-type of displacements (Fig. 6.11a) [6.20], whereas for the fluorite $RbMnF_3$, the "Last"-type of displacement (Fig. 6.11b) [6.21], is dominant. For the oxides, the center atom (Ta or Ti) moves against the rigid O_6 octahedron, but for $RbMnF_3$, the Rb moves out of phase against the regid MnF_3 cage. The researchers explain this different behavior by the larger polarizabilities and stronger long-range forces in the oxides as compared to the fluorides.

6.3 Phonon Density of States Measurements

6.3.1 Incoherent Scattering Cross Section

The evaluation of the incoherent part of the scattering cross section is similar to the procedure used for the coherent scattering. The basic equation is (6.16). We again consider a Bravais crystal; and for the partial cross section (6.22) we obtain

$$\left(\frac{d^2\sigma}{d\Omega\, dE_1}\right)_{\text{inc}} \cong \sigma_{\text{inc}} \frac{k_1}{k_0} e^{-2W}$$

$$\times \int_{-\infty}^{\infty} \exp[\langle \boldsymbol{Q}\cdot\boldsymbol{u}_0(0)\,\boldsymbol{Q}\cdot\boldsymbol{u}_0(t)\rangle]\exp(-i\omega t)dt, \qquad (6.42)$$

with the "incoherent cross section"

$$\sigma_{\text{inc}} = 4\pi(\overline{b^2} - \bar{b}^2). \qquad (6.43)$$

As befor the exponent can be expanded in a power series, the nth term corresponding to the n-phonon process.

For the elastic scattering we get

$$\left(\frac{d\sigma}{d\Omega}\right)_{\text{inc}}^{\text{el}} \cong \sigma_{\text{inc}}\, e^{-2W}, \tag{6.44}$$

with a dependence of the scattered intensity on temperature and momentum transfer through the Debye-Waller factor $\exp[-2W(Q,T)]$.

The one-phonon term is given by

$$\left(\frac{d^2\sigma}{d\Omega\, dE_1}\right)_{\text{inc}}^{1} \cong \sigma_{\text{inc}}\frac{k_1}{k_0}\, e^{-2W} \sum_{q,j} \frac{|Q\cdot e(_j^q)|^2}{\omega_j}\,(n_j + \tfrac{1}{2}\pm\tfrac{1}{2})\delta(\omega\mp\omega_j), \tag{6.45}$$

where the upper and lower sign refer to phonon creation and annihilation, respectively. As compared to the coherent cross section (6.30), the δ-function for crystal momentum conservation no longer appears, and there is only the δ-function for energy conservation. Hence, a measurement of the incoherent scattering gives only information on the number of phonon modes as function of energy, but not on the dispersioin relations.

The cross section can further be simplified for a cubic Bravais crystal. The summation over all modes (q,j) can be replaced by an integral over all frequencies by introducing the phonon density of states $g(\omega)$. The polarization vector contribution can be averaged and we obtain

$$\left(\frac{d^2\sigma}{d\Omega\, dE_1}\right)_{\text{inc}}^{1} \cong \sigma_{\text{inc}}\frac{k_1}{k_0}\, e^{-2W} \frac{g(\omega)}{\omega}\,(n + \tfrac{1}{2}\pm\tfrac{1}{2}). \tag{6.46}$$

The phonon density of states is proportional to the one-phonon cross section and can in principle directly be measured. The method is best suited for predominantly incoherent scatterers, but the condition "cubic Bravais crystal" is very restrictive for a practical application. Values of σ_{inc} and σ_{coh} for some elements are given in Table 6.4; hydrogen is by far the best incoherent scatterer, almost an order of magnitude stronger than deuterium and the other elements.

6.3.2 Examples

a) Vanadium

One of the few materials where the phonon density of states $g(\omega)$ can be determined directly by experiments is vanadium. It has the desird simple structure (cubic, Bravais lattice) and it is a strong incoherent scatterer. Many experiments have been performed with this substance, the result of *Glaeser* et al. [6.22] is displayed in Fig. 6.12. The intensity curve shows two peaks mainly due to the transversal and longitudinally polarized mode frequencies near the Brillouin-zone boundary. Through (6.46) the correlation of the density of states and the incoherent scattering cross section is very direct, however the measured intensities have to be corrected for multi-

171

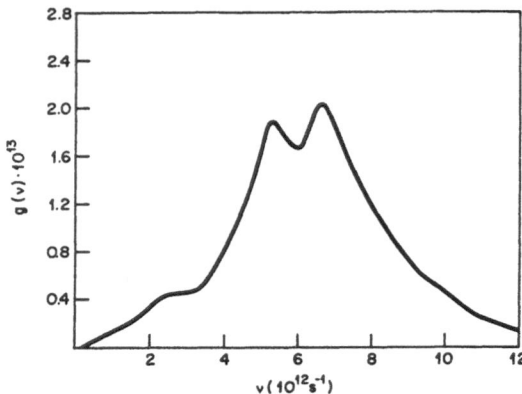

Fig. 6.12. Phonon density of states $g(\omega)$ for vanadium determined by incoherent inelastic neutron scattering [6.22]

phonon and multiple scattering, and there exist some discrepancies between the derived phonon density of states of different researchers, especially in the lowest- and highest-frequency regions [6.22].

b) Molecular Solids

Incoherent scattering is extremely useful for compounds containing hydrogen. Hydrogen is a very strong incoherent scatterer, and a natural application is the spectroscopy of molecular solids. Interesting systems are however not of the cubic Bravais type, and the cross section (6.45) can not directly be evaluated.

As outlined in [Ref. 6.9, Sect. 4.6], the dynamics of molecular crystals can be split into internal and external vibrations. The internal- and high-frequency external motions, in general, show only a weak or even no dispersion at all. As a result the phonon density of states shows a singular behavior at critical frequencies [Ref. 6.9, Eq. (3.86)], and in many cases these peaks can be observed and separated in the incoherently scattered intensity.

The experimental data are then analyzed in terms of modified frequency distributions [6.23], and the incoherent scattering is represented by

$$\left(\frac{d^2\sigma}{d\Omega\,dE_1}\right)^1_{\text{inc}} \cong \frac{k_1}{k_0}\sum_{\kappa}\sigma_{\text{inc},\kappa}G_{\kappa}(\omega)\mathrm{e}^{-2W_{\kappa}}\langle n+\tfrac{1}{2}\pm\tfrac{1}{2}\rangle, \qquad (6.47)$$

where the sum extends over all atoms in the unit cell and G_{κ} is the eigenvector weighted density of states of the atom at $\mathbf{r}_{\kappa}$. Mainly coherent scatterers like oxygen or carbon can be neglected (Table 6.4), and the summation actually reduces to the hydrogens in the unit cell. The experiment probes the hydrogen motion, but if now a part of the hydrogen is replaced by the coherent scatterer deuterium, the measured intensities will drastically change, and peaks disappear. Hence by selective deuteration in a compound, frequencies of different molecular groups can be measured and attributed to individual motions.

Table 6.4. Some incoherent and coherent scattering cross sections [barns] [6.7]

	$\sigma_{\text{inc}} = 4\pi(\overline{b^2} - \overline{b}^2)$	$\sigma_{\text{coh}} = 4\overline{b}^2$
H	80.2	1.8
D	2.0	5.6
C	0.0	5.6
O	0.0	4.2
V	5.0	0.02
Ni	5.0	13.4

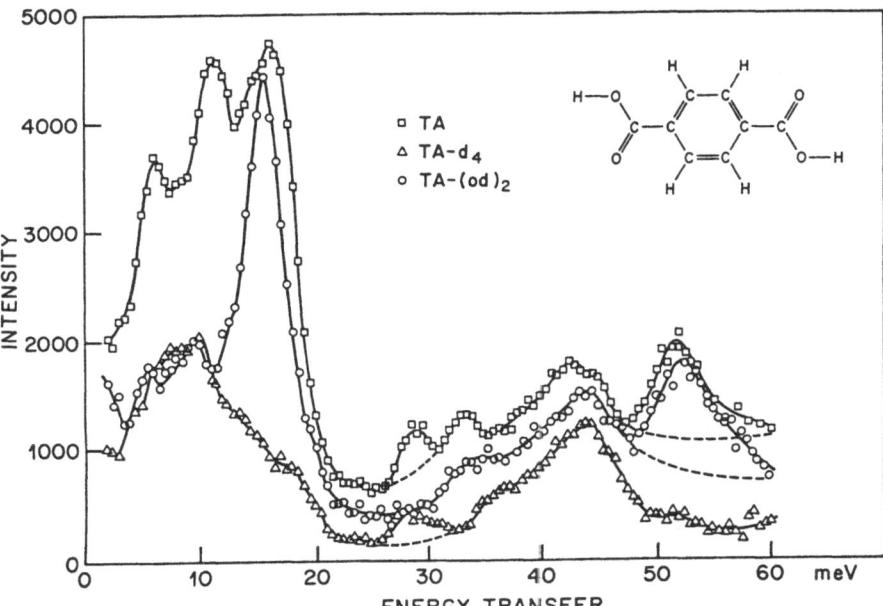

Fig. 6.13. Measured neutron intensities as function of energy transfer in terephthalic acid at 10 K [6.24]; effect by deuteration on different molecular groups; TA: undeuterated; TA-d₄: deuterated on the ring; TA-(od)₂: deuterated on the carboxilic group

Experiments on terephthalic acid $C_8H_6O_4$ are shown in Fig. 6.13 [6.14]. Ta-d₄ is deuterated on the ring, and Ta-(od)₂ is deuterated on the carboxyl group. From the differencies in the observed intensities, the peaks near 16 and 52 meV can directly be assigned to ring motion frequencies. A detailed analysis must be based on a lattice-dynamical model calculation, and frequencies and eigenvectors must be computed in the whole Brillouin zone and the summation over all modes has to be done numerically. The final eigenvector-weighted density of states, convoluted with the spectrometer resolution function, has then to be compared with the experimental data in order to refine the model parameters in an iterative procedure.

7. Other Techniques

In this chapter we consider some other techniques which are used to study phonons, namely ultrasonic methods, inelastic electron tunneling spectroscopy, point-content-spectroscopy, and the spectroscopy of surface phonons, thin films, and adsorbates.

7.1 Ultrasonic Methods

The study of the propagation behaviour of high-frequency elastic waves in solids is a well established method for studying various physical properties of materials. In the most common method a pulse of ultrasonic waves is "injected" into a block of the material from a transducer, and the propagation behaviour of the stress waves is determined by the measurement of the velocity and the attenuation (energy loss) of the stress waves as a function of whatever variables of interest. Such measurements permit one to study the influence, on the velocity and attenuation, of any property of a solid which is sufficiently well coupled to the lattice. Altogether, there are a dozen or more known "interactions" between stress waves and various properties of a solid. Stress waves can interact, for example, with various types of lattice defects, with conduction electrons in metals, with the nuclear spin system, with electron spins of paramagnetic centers, with charge carriers in piezo-electric materials, and with thermal lattice vibrations. It is this latter type of interaction which is of interest in this section, the interaction between the externally produced stress waves with lattice vibrations which can be described in terms of phonon-phonon interactions.

If the velocities of stress waves propagating in different directions and having different polarization properties (LA and TA for example) are known, it is possible to determine the elastic constants of the crystal [Ref. 7.1, Sect. 3.6]. These velocities are the phase velocities of long-wavelength acoustic waves with dispersion $\omega = vq$ [Ref. 7.1, Problem 3.8.5]. On the other hand, from the study of the attenuation of sound waves in a well ordered crystal as a function of temperature one obtains information about phonon-phonon interactions and hence about anharmonic properties of the crystals [Ref. 7.1, Chap. 5].

7.1.1 Experimental Techniques

Ultrasonic waves are introduced into the sample by using a piezoelectric transducer (usually a quartz transducer), as shown in Fig. 7.1. The transducer is bonded to the sample on one of the two parallel faces. A signal from a pulsed transmitter operating at the fundamental frequency of the transducer or at one of its odd harmonics, is applied across the faces of the transducer, and, as a consequence of the piezoelectric effect, a stress wave is produced which propagates into the sample. In Table 7.1 some common transducer materials together with the frequency and wavelength ranges are listed. The single transducer is both the source of the initial pulse and the receiver for all the successive echos that result from the single pulse. Alternatively, a second transducer on the opposite face of the sample may be used as the receiver.

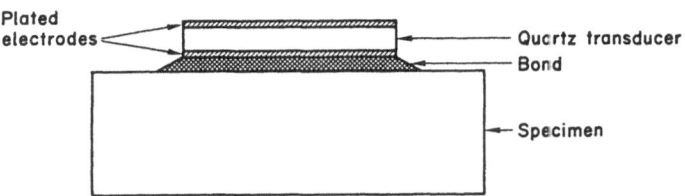

Fig. 7.1. Schematic representation of specimen, bond, and transducer with plated electrodes

Table 7.1. Some common transducer materials with frequency and wavelength ranges

Transducer	Quartz slab	Evaporated sputtered CdS or ZnO film	Tunnel junctions
Frequency	300 kHz–300 Mz	3 GHz	THz region
Wavelength	1 cm–1 μm	~1 μm	~300 Å

The initial pulse sent into the sample is almost perfectly reflected at the opposite air-sample interface, and it returns to the transducer where all but a small fraction of the energy is again returned to the sample. By the time the first echo has arrived back at the transducer, the transmitter has been turned off. The transducer converts a small amount of the energy of the returned pulse back into electrical energy. This electrical signal is then amplified by an appropriate receiver, and the result is displayed on the oscilloscope. In the meantime, the stress wave has been reflected at the transducer-sample interface and has begun another round trip in the sample, so that the process just described is repeated a number of times.

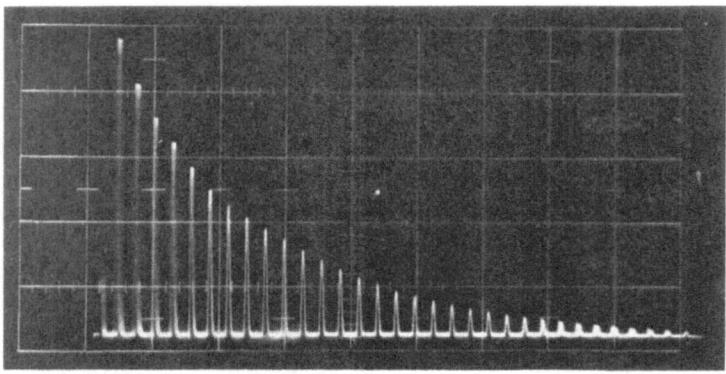

Fig. 7.2. Pulse echo pattern [7.2a](*see text*)

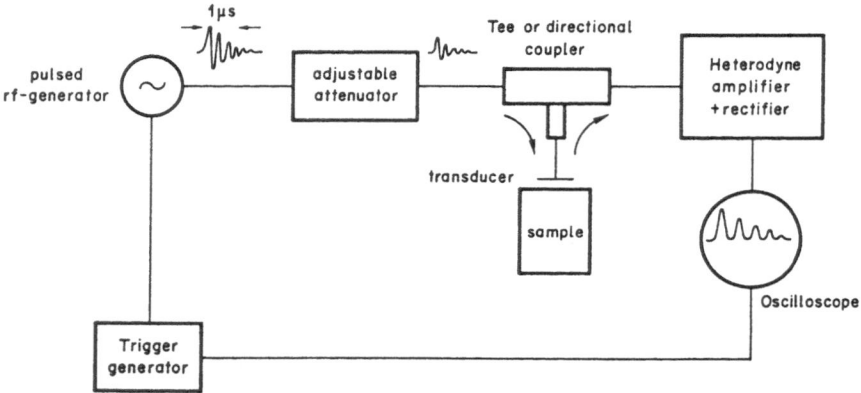

Fig. 7.3. Block diagram of an electronic system used to measure the velocity of sound by the ultrasonic pulse-echo technique [7.2b](*see text*)

Each time the stress wave passes through the sample; hence, each successive echo or pulse signal presented on the oscilloscope is smaller in amplitude than the preceeding one and a pulse echo pattern is obtained, as shown in Fig. 7.2 [7.2a]. The time interval between successive pulses is measured on the oscilloscope. This interval is the round-trip travel time of the ultrasonic pulse. The attenuation of the wave may be found from the decrease in pulse height of successive pulses, with allowance for losses on reflections at the ends. A block diagram of an electronic system used to measure the velocity of sound by the ultrasonic pulse-echo technique is shown in Fig. 7.3 [7.2b]. Triggered with the rate of about 100 Hz the rf generator produces rf pulses of roughly 1 μs pulse duration. The amplitude of those pulses can be adjusted by an attenuator. At the transducer, part of the electrical pulse is transformed into an acoustic pulse of equal duration and frequency. This acoustic pulse travels back and forth between the two parallel faces of the sample. Each time it arrives at the transducer, part of the acoustic energy is coupled out and transformed into an electric signal. These signals are detected and monitored at the oscilloscope as a train of subsequent echos. The time separation between two echos is equal to one round trip in the sample, and the difference in the amplitudes is a measure of the acoustic attenuation of the sample material. More sophisticated measuring systems have been discussed in the book by *Truell* et al. [7.2a].

It is evident that the determination of the velocity of an ultrasonic wave not only requires an accurate determination of the time interval between successive pulses but also an accurate measurement of the length of the crystal. Moreover, in order to determine the elastic constants the density of the crystal must be known, in addition to the velocities of the ultrasonic waves [Ref. 7.1, Problem 3.8.5]. The temperature dependence of the elastic constants of LiCl determined by ultrasonic methods is shown in [Ref. 7.1, Fig. 5.9].

The frequencies Ω of ultrasonic waves are very small compared to the frequencies of thermal vibrations $\omega_{th} \cong k_B T / \hbar$, even at low temperatures ($\Omega \cong 10^6 - 10^9$ Hz, $\nu_{th} \cong 2 \cdot 10^{10}$ Hz at $T = 1$ K). Despite the fact that there is an appreciable frequency difference between ultrasonic and thermal vibrations, there is an interaction between these waves which leads to ultrasonic attenuation (Sect. 7.1.3). Much stronger attenuation is expected for acoustic waves having frequencies comparable to those of thermal vibrations. For this reason it is of great interest to extend the frequency range of acoustic waves towards higher frequencies. Acoustic waves at frequencies above 10^9 Hz are called *hypersonic waves*, distinguishing them from sonic or ultrasonic waves at lower frequencies. References to hypersonic techniques at microwave frequencies ($\sim 10^{10}$ Hz) are given by E.H. *Jacobsen* in the book by *Bak* [7.3]; see also the interesting papers by *Boemmel* and *Dransfeld* [7.4], by *Baranskii* [7.5] and by *Shiren* [7.6].

In an important paper *Grill* and *Weis* [7.7] reported successful piezo-electric surface excitation using laser light which extends the attainable frequencies to $\sim 10^{13}$ Hz. This method is known as *coherent phonon generation;* the main attributes of this method is that the phonons so generated are monochromatic (at the frequency of the laser light), are spatially coherent and propagate along a path given by a cylinder whose axis is predefined by the piezoelectric tensor at the excitation surface [7.8] and has a diameter roughly to that of the laser beam at the surface. Moreover, depending on the orientation of the piezoelectric crystal and on the excitation surface, the phonons generated may be predominantly longitudinally or transversely polarized. It is possible, therefore, to inject by these means a coherent monochromatic phonon distribution with known propagation vector and known polarization. The main disadvantage of the method is that the piezoelectric conversion effeciency is small, of the order of 10^{-7}. Other spectroscopic techniques used for the generation and detection of high-frequency phonons have been discussed in an excellent review article by *Bron* [7.9].

7.1.2 Ultrasonic Attenuation
Due to Phonon-Phonon Interactions

Let us assume that the transducer produces a plane stress wave that is attenuated as it propagates thorough the sample. The stress wave is represented by

$$\sigma(x,t) = \sigma_0 \exp[i(\Omega t - kx)], \tag{7.1}$$

where Ω is the angular frequency and k is the magnitude of the propagation vector. An expression for an attenuated wave is obtained by assuming that k is complex. Taking

$$k = k_1 - i\alpha, \tag{7.2}$$

one obtains the equation of an attenuated wave

$$\sigma(x,t) = \sigma_0 \exp(-\alpha x) \exp[i(\Omega t - k_1 x)]. \tag{7.3}$$

The attenuation α can be determined by measuring the heights of the echos on the oscilloscope, as shown in Fig. 7.2. Since the intensity of the ultrasonic

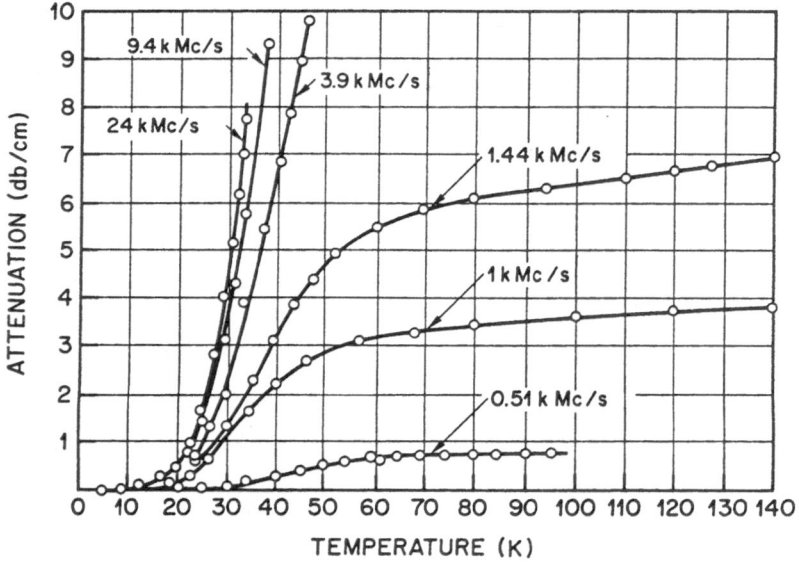

Fig. 7.4. Attenuation of compressional waves as a function of temperature of X-cut crystalline quartz for the frequencies indicated in kMc/s [7.10]. 1 kMc corresponds to 1 Gc in the newer literature

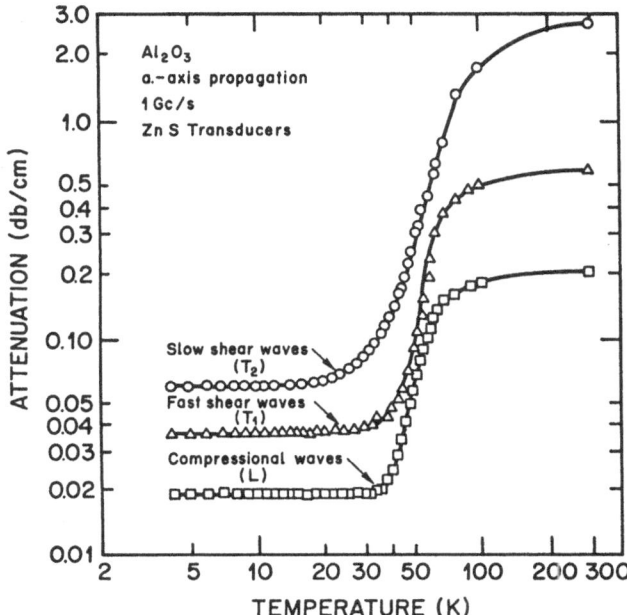

Fig. 7.5. Attenuation versus temperature for Al$_2$O$_3$ [7.11]

beam is proportional to $|\sigma(x,t)|^2$, the decay of the intensity is proportional to $\exp(-2\alpha)$. Figure 7.4 shows the attenuation of compressional waves as a function of temperature and frequency Ω of quartz [7.10]. Figure 7.5 illustrates the temperature dependence of the attenuation of compressional and shear waves of Al$_2$O$_3$ [7.11]. The temperature dependence of the attenuation, $\alpha(T)$, shown in Figs. 7.4 and 5 can be attributed to phonon-phonon interactions. The ultrasonic stress wave interacts in a crystalline solid with the thermal or lattice vibrations. No such interaction should occur if the waves were linearly elastic, that is if the interatomic forces were harmonic. It is the anharmonicity which leads to coupling between different lattice waves or to collisions between phonons [Ref. 7.1, Chap. 5]. The anharmonic behaviour may be approximated by means of cubic terms in the elastic energy density [Ref. 7.1, Sect. 3.6], which means the involvement of third-order elastic constants; higher-order terms may, of course, also be involved.

The actual calculation of the attenuation α in terms of the phonon parameters of the anharmonic system is complicated and involved [7.12–17]. In the following we only outline the main ideas and assumptions which are used for the derivation of the attenuation α and discuss simple approximate expressions for $\alpha(T,\Omega)$.

The ultrasonic wave produces a local strain $\epsilon_{\alpha\beta}(\mathbf{r},t)$ defined by (4.12). Assuming a plane wave we can write

$$\epsilon_{\alpha\beta}(\boldsymbol{r},t) = \epsilon_{\alpha\beta}\exp[i(\Omega t - \boldsymbol{k}\cdot\boldsymbol{r})], \tag{7.4}$$

where Ω is the circular frequency and $\boldsymbol{k}$ the propagation vector. This local strain produces a corresponding change in the frequency $\omega_s[s = (\boldsymbol{q}j)]$ of a thermal phonon, which may be written phenomenologically using the Grüneisen tensor $\gamma_{\alpha\beta}(s)$ [Ref. 7.1, Sects. 5.2.3, 5.3.1, 2]

$$\Delta\omega_s(\boldsymbol{r},t) = \sum_{\alpha\beta}\gamma_{\alpha\beta}(s)\epsilon_{\alpha\beta}(\boldsymbol{r},t)\omega_s. \tag{7.5}$$

This expression corresponds to the linear terms of an expansion of ω_s in terms of the components of the deformation tensor. If a single longitudinal wave is excited in a cubic crystal propagating along (001), only ϵ_{zz} is different from zero in the sum (7.5); similarly, if a single transverse wave is excited propagating along (010) with particle displacement along (001), only ϵ_{yz} is different from zero.

In the absence of strain, the number of thermal phonons in the mode s, with the lattice in thermal equilibrium at a termperature T, is given by the Bose-Einstein distribution

$$\bar{n}_s = [\exp(\hbar\omega_s/k_\mathrm{B}T) - 1]^{-1}, \tag{7.6}$$

If follows from (7.6) that, if the frequency of a phonon is changed as the result of the application of a strain and if the number of phonons in a mode remains fixed, the effective temperature of that mode must change. The frequency shift $\Delta\omega_s(\boldsymbol{r},t)$ of (7.5) corresponds to this temperature change. Since $\gamma_{\alpha\beta}(s)$ varies appreciably from mode to mode and may even change sign [Ref. 7.1, Sect. 5.3.2], the effect of strain is to give different vibrational modes different effective temperatures. The perturbed system tends to equilibrium as these temperature differences are equalized because of collisions among phonons in different modes s having life times τ_s [Ref. 7.1, Sect. 5.5.3]. These life times τ_s are the times between phonon collisions and can be regarded as relaxation times for restoring the equilibrium. The result is an increase in the entropy of the system which is the cause of ultrasonic attenuation.

Based on this idea *Boemmel* and *Dransfeld* [7.4]have derived an approximate expression for the attenuation α. A more elaborate calculation which is based on the Boltzmann equation has been performed by *Woodruff* and *Ehrenreich* [7.12]. Both of these approaches are extensions of the work of *Akhieser* [7.13]. The interaction between sound waves and thermal phonons has also been studied by *Landau* and *Rumer* [7.14], *Simons* [7.15], *Maries* [7.16a], and by *Guyer* [7.16b].

In the following we discuss an approximate expressioin for α, obtained by *Woodruff* and *Ehrenreich* [7.12]. The approximation is based on a single mean Grüneisen parameter γ which represents the coupling between the sound wave and the crystal. Furthermore, a single mean relaxation time τ_{th} for the thermal phonons is introduced, and the approximation is based on the Debye model, for which $\omega_s = vq$, where v is a constant average velocity of the thermal phonons. In the case $\Omega\tau_{th}\ll 1$ (Ω: frequency of the ultrasonic wave), the expression for the attenuation is

$$\alpha = \frac{C_v T \gamma^2 \Omega^2 \tau_{th}}{3\varrho v^3} \ , \tag{7.7}$$

where C_v is the specific heat and ϱ the density. In terms of the thermal conductivity [Ref. 7.17, Chap. 4]

$$\sigma_{th} = \frac{1}{3} C_v v^2 \tau_{th} \ , \tag{7.8}$$

the attenuation α becomes

$$\alpha = \frac{\gamma^2 \Omega^2 T \sigma_{th}}{\varrho v^5} \ . \tag{7.9}$$

Since the mean collision time τ_{th} decreases with increasing temperature, the condition $\Omega\tau_{th}\ll 1$ and hence (7.7, 9) hold at high temperatures. At temperatures greater than the Debye temperature θ_D of the solid, τ_{th} and σ_{th} are proportional to $1/T$ (see the discussion in [Ref. 7.1, Sect. 5.5.3]for the temperature dependence of the damping $\Gamma = \tau^{-1}$), and hence α becomes independent on temperature in agreement with the behaviour shown in Figs. 7.4 und 5.

The limit $\Omega\tau_{th}\gg 1$ corresponds to low temperatures and the classical Boltzmann treatment of *Woodruff* and *Ehrenreich* does not strictly apply because quantum effects become important. Nevertheless the classical expression for α for the case $\Omega\tau_{th}\gg 1$ has some similarity to a quantum mechnaical result of *Landau* and *Rumer* [7.13]in that the dependence of α on T and on Ω is the same in both cases. The result is [7.12]

$$\alpha = \frac{\pi\gamma^2 \Omega C_v T}{4\varrho v^3} . \tag{7.10}$$

From (7.10) it is seen that α tends to zero at very low temperatures. The residual attenuation observed sometimes at low temperatures (Fig. 7.5) is believed to arise from various crystal defects that can scatter ultrasonic waves. According to (7.10) the decrease of α with temperature is proportional to $C_v(T)\cdot T$, i.e. to T^4 for $T<\theta_D$.

7.2 Inelastic Electron Tunneling Spectroscopy

Inelastic electron tunneling spectroscopy (IETS) is a technique that provides a versatile and sensitive method for measuring the vibrational properties of a thin insulating barrier sandwiched between two metals. We present here

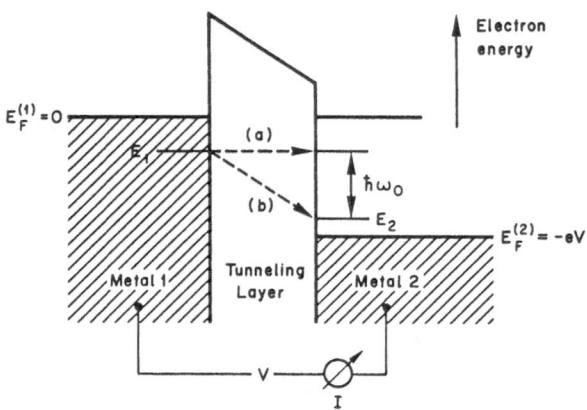

Fig. 7.6. Schematic energy-level diagram for tunneling between normal metals. Elastic processes *(a)* give rise to ordinary tunneling. In the inelastic case *(b)* the electron loses energy $\hbar\omega_0$ to excitation of an oscillator in the barrier region. This can only occur if $eV > eV_0 = \hbar\omega_0$ where V_0 is the threshold voltage. The Fermi energies in the two metals are related by $E_F^{(1)} - E_F^{(2)} = eV$ [7.23]

only a qualitative discussion of IETS; for details the reader is referred to the literature [7.18–21]. The first clear evidence of inelastic tunneling via the excitation of vibrational states of impurities in the barrier was made by *Jaklevic* and *Lambe* [7.22, 23].

The process we call inelastic tunneling can be visualized in terms of a rather simple picture. Figure 7.6 represents a tunneling-junction energy-level diagram where the two Fermi levels of the normal metals are separated by the energy eV with V the applied voltage. According to energy conservation, the electrons in metal 1 are capable of making elastic (horizontal) transitions into empty states in the right-hand metal, provided a voltage V is applied to the junction. To a first approximation the current-voltage characteristic of the elastic tunneling is linear at sufficiently small voltages [7.24]. However, if it is possible for a new transfer mechanism to exist whereby an electron can tunnel from left to right and at the same time give energy to a phonon (due to the electron-phonon interaction) or to a local impurity state of the barrier, a new channel for tunneling will open up (Fig. 7.6). This new contribution can only occur if any empty state on the right is open for the tunneling electron, i.e. if $eV > \hbar\omega_0$, where $\hbar\omega_0$ is a vibrational frequency of the barrier. As V increases still further, current from this process will continue to increase since more and more candidates for inelastic tunneling are brought into play. The onset of a new tunneling process causes an increase of the current, a small change in the conductance dI/dV and a pronounced peak in the second derivative d^2I/dV^2 (Fig. 7.7). Inelastic tunneling of electrons has not only been observed in metal-insulator-metal junctions but also

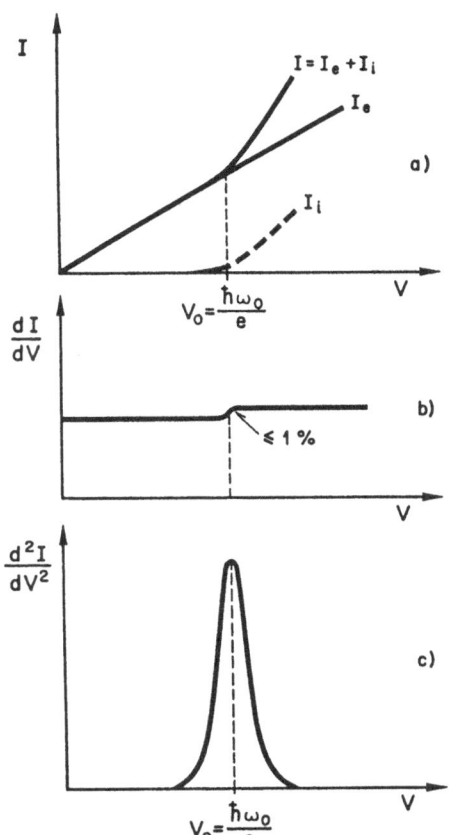

Fig. 7.7. Schematic representation of (a) the I-V characteristic (I_e and I_i are the elastic and inelastic currents, respectively), (b) conductance, dI/dV, versus voltage, and (c) second derivative d^2I/dV^2 versus voltage. ω_0 is the vibrational frequency and V_0 is the threshold voltage

in narrow $p-n$ junctions (Esaki diodes) at low temperatures. Examples will be discussed in Sect. 7.2.3.

7.2.1 Experimental Techniques

The metal-insulator-metal tunnel junctions are prepared as shown in Fig. 7.8 [7.24].

a) A metal film is deposited on a glass slide.
b) A superficial oxide film is formed or, alternatively, a thin film of the barrier material is evaporated on the metal film.
c) Metal films are deposited across the barrier films.
d) Current and voltage connections are made.

Figure 7.8 also shows a schematic circuit for measuring the I-V characteristics. More sophisticated circuits are used in practice. The first and second derivative of the voltage with respect to the current and as a function of

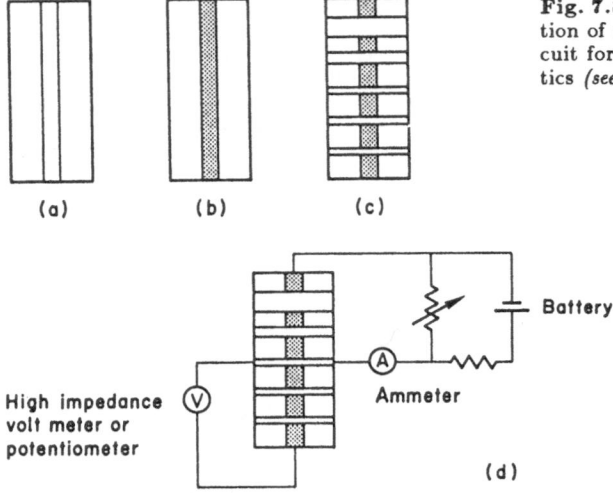

Fig. 7.8a-d. Schematic representation of sample preparation and circuit for measuring I-V characteristics *(see text)* [7.24]

(a)   (b)   (c)

High impedance
volt meter or
potentiometer

Battery

Ammeter

(d)

voltage can be obtained directly by a modulation technique using a lock-in amplifier in a bridge-like circuit [7.25, 26].

As will be discussed in Sect. 7.2.3, the experiments must be performed at low temperatures, usually at 4.2 K (liquid helium temperature). The current flow through the barrier film in a junction is not necessarily due to a tunneling mechanism. Therefore, the first step in analyzing measured current-voltage characteristics is the establishment of the mechanism of the charge transport across the barrier. The most definite verification of the tunneling mechanism is the use of at least one superconducting electrode with energy gap 2Δ and observe that only a small current due to thermally excited quasi-particles flows for $|eV| < 2\Delta$ [7.27].

7.2.2 The Inelastic Electron Current

Independent on the details of the electron-phonon (or electron-adsorbed molecule) interaction some comments on expected line shapes and line widths for IETS experiments can be made. In the simplest theory the inelastic electron current for normal metal-insulator-metal junctions is given by [7.19, 23]

$$I_i = C \int\limits_{-\infty}^{+\infty} f(E)[1 - f(E + eV - \hbar\omega_0)]dE. \tag{7.11}$$

Here we have hidden all the details of the electron-phonon interaction in the coupling parameter C and displayed only the integral over the Fermi functions for the two normal metal electrodes. This integral simply expresses analytically the condition we discussed qualitatively in the introductory subsection: electrons must tunnel from a filled state into an empty state in the other electrode and this transition is assisted by a phonon (or molecular vibration) of energy $\hbar\omega_0$. The bias voltage across the junction is V. A similar expression, but summed over the phonon energies $\hbar\omega_j(\boldsymbol{q})$, is also valid for $p-n$ junctions. The integral (7.11) can be easily evaluated. Using the Fermi distribution function

$$f(E) = (e^{E/k_{\mathrm{B}}T} + 1)^{-1}, \tag{7.12}$$

one obtains

$$I_{\mathrm{i}} = Ck_{\mathrm{B}}Tx\frac{e^x}{e^x - 1}, \quad \text{where} \tag{7.13}$$

$$x = (eV - \hbar\omega_0)/k_{\mathrm{B}}T. \tag{7.14}$$

Taking derivatives we obtain

$$\frac{dI_{\mathrm{i}}}{dV} = eC\frac{e^x(e^x - 1 - x)}{(e^x - 1)^2}, \quad \text{and} \tag{7.15}$$

$$\frac{d^2I_{\mathrm{i}}}{dV^2} = C\frac{e^2}{k_{\mathrm{B}}T}e^x\frac{(x - 2)e^x + x + 2}{(e^x - 1)^3}. \tag{7.16}$$

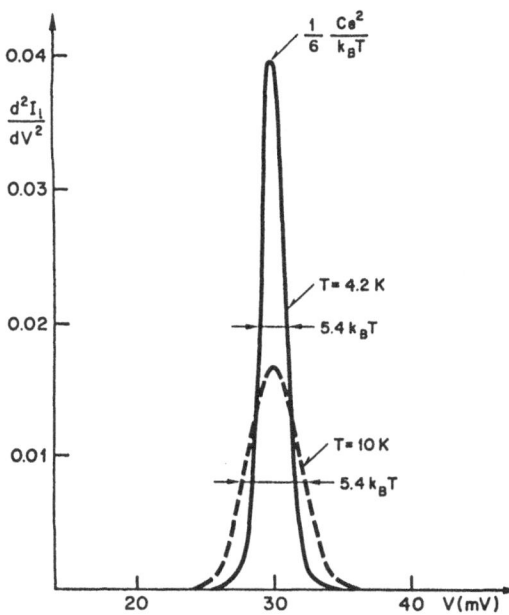

Fig. **7.9.** Second derivative of the tunnel current, d^2I_{i}/dV^2, versus voltage for $T = 10\,\mathrm{K}$ and $\hbar\omega_0 = 30\,\mathrm{meV}$ according to (7.16). The line width due to thermal broadening is $5.4\,k_{\mathrm{B}}T$

For $x = 0$ one obtains $I_i = C k_B T$, $dI_i/dV = eC/2$ and $d^2 I_i/dV^2 = e^2 C/6 k_B T$. From (7.16) and Fig. 7.9 it is obvious that IETS experiments must be performed at very low temperatures; note that $d^2 I_i/dV^2 \sim 1/T$ at $x = 0$ and that the line width due to thermal broadening is given by $5.4 \, k_B T$. This thermal broadening is a rather fundamental property of the inelastic tunneling. Of course, there will be other sources of broadening in the actual vibrational levels. Nevertheless, there is a lower limit on the instrumental line width which places an ultimate limit on the resolving power of a tunneling spectrometer when used with normal metal electrodes. If superconducting electrodes are used the resolving power is improved over that observed with normal metals [7.23].

7.2.3 Some Selected Examples of IETS

As an example Fig. 7.10 shows the IETS for a clean Al/Al$_2$O$_3$/Pb junction, one with negligible molecular impurities [7.23]. The slowly rising background is due to the slowly increasing elastic conductance that is characteristic of aluminum oxide tunneling barriers. The assignment of the structures observed in Fig. 7.10, as given in [7.23], cannot be correct. The present author's assignment is as follows: The prominent peak near 0.12 V (corresponding to $\tilde{\nu} = 960 \, \text{cm}^{-1}$) is predominantly due to LO phonons associated with Al-O stretching modes of the (amorphous) Al$_2$O$_3$. This mode can also be observed by means of infrared reflection absorption spectroscopy (Fig. 7.27) [7.28]. The weak structure near 0.45 V is due to hydroxides (O-H stretching), while the very weak structure near 0.23 V is possibly due to carbon contamination (C=O stretching). The structures originate from the unavoidable impurities present in or at the surface of the aluminum oxide. *Giaever* and *Zeller* have established for the first time that IETS can in-

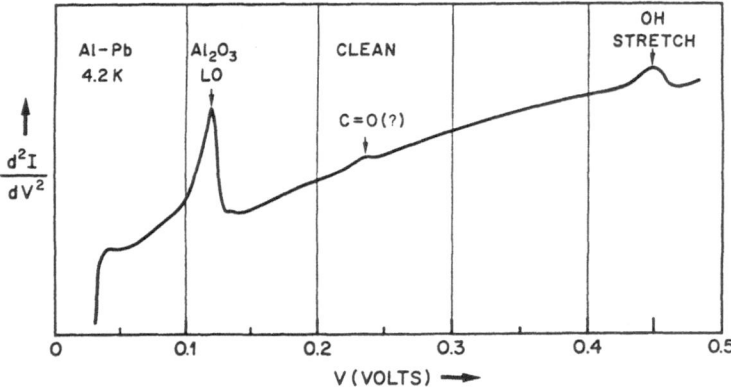

Fig. 7.10. Tunneling spectrum of an undoped Al/Al$_2$O$_3$/Pb junction observed at 4.2 K [7.23] (assignment modified, *see text*)

186

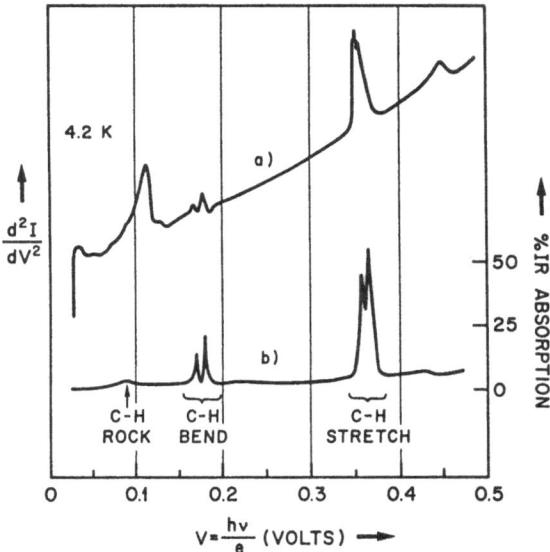

Fig. 7.11. (a) Tunneling spectrum of a large molecular weight hydrocarbon evaporated onto the oxide of an $Al/Al_2O_3/Pb$ junction. (b) Assigned infrared absorption spectrum of the same material [7.23]

deed be used to determine LO phonons of the barrier material [7.26]. They determined LO phonons in some very thin II–VI compound films.

Figure 7.11 shows the tunneling spectrum of a large molecular weight hydrocarbon evaporated onto the oxide of an $Al/Al_2O_3/Pb$ junction [7.23]. Note the additional structures over the clean junction spectrum (Fig. 7.10) near 360 mV (C-H stretching modes), near 175 mV (bending modes) and near 90 mV (rocking mode). The figure also contains the assigned infrared absorption spectrum of the same material for comparison. Figure 7.11 demonstrates that IETS can not only be used to observe phonon frequencies but also for the investigation of the vibrational spectrum of a minute quantity of organic molecules on the insulator of a metal-insulator-metal junction. Sensitivity is the key advantage over the conventional infrared and Raman spectroscopy. One monolayer is the sample thickness that is optimal for most tunneling experiments. An excellent review article about IETS applied to molecular spectroscopy, surface physics, catalysis, radiation physics and biology has been written by *Hansma* [7.19].

As mentioned in the introductory subsection, phonon-assisted tunneling of electrons has also been observed in *p-n* junctions (Esaki diodes) at low temperaturs. The material on one side of the junction is doped to be *p*-type and on the other to be *n*-type, so that well defined Fermi surfaces exist in the valence and conduction bands on opposite sides of the junc-

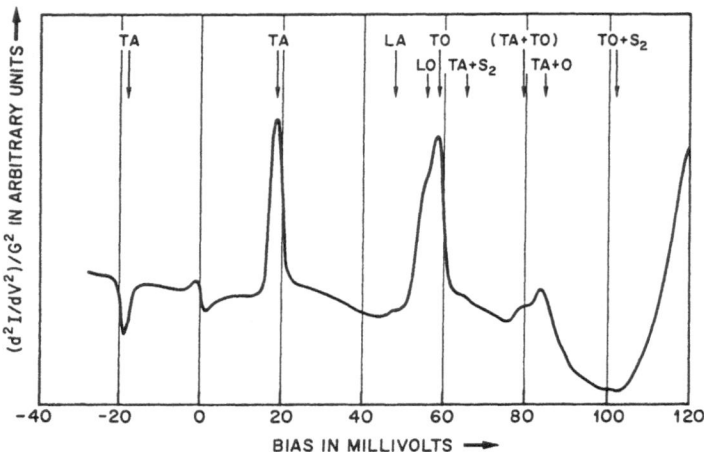

Fig. 7.12. Tunneling spectrum of a silicon *p-n* junction at 4.2 K, showing several of the observed phonon and phonon combination peaks together with their assignments [7.29]

tion. These Fremi surfaces are in well defined regions of reciprocal space, whose wave vector is determined by the particular band structure of the semiconductor. According to a simple theory of the tunneling process the electrons tunnel across the junction from either *p*-type to *n*-type material, or vice versa, assisted by a phonon whose wave vetor is equal to the wave vector difference between the top of the valence band and the minimum of the conduction band. Figure 7.12 shows the IETS experiment of a silicon *Esaki* junction at 4.2 K [7.29]. These experiments give great precision in energy, about 0.1 %. However, the determination of the momentum transfer is more uncertain. The band structure of the semiconductor must already be known, and even then there is some uncertainty about the value of the wave vector *q* for the phonons involved. The doped semiconductors must also be free from impurity levels in order to avoid additional structures arising from such impurities.

7.3 Point Contact Spectroscopy

During the last ten years a new spectroscopic method using point contancts between metals has been developed to study the interaction of electrons with elementary excitations in metals, i.e. phonons, magnons, etc. Up to now most of the experiments peformed in this field deal with the investigation of the electron-phonon interaction in pure metals or dilute alloys. An excellent review has been given by *Jansen* et al. [7.30].

For metallic contacts one usually expects a linear relation between the applied voltage and the current through the contact, according to Ohm's law. However, interesting nonlinear behaviour can be observed in the current-voltage characteristics of metallic constrictions if the linear dimension of the contact becomes comparable with or smaller than the mean-free path of the electrons in the metal. It turns out that the observed phenomena can be used for a fundamental study of the scattering mechanisms of the conduction electrons in a metal, leading to a new experimental tool, nowadays referred to as *point-contact spectroscopy*. As inelastic electron tunneling spectroscopy, point-contact spectroscopy must be peformed at low temperatures, usually liquid helium temperatures.

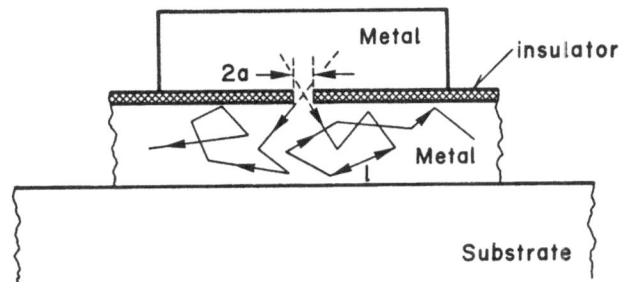

Fig. 7.13. Schematic diagram of cross section of a short-circuited film structure for point-contact spectroscopy. The radius of the point-contact is a and l is the electron mean-free path. In point-contact spectroscopy $l > a$

In an experimental study of metal-insulater-metal tunnel junctions, with a short circuit in the oxide layer between the metal films (Fig. 7.13), *Yanson* [7.31, 32] has found that the current-voltage characteristics of these metallic microcontacts showed a nonlinear behaviour at liquid helium temperatures. To characterize the nonlinearity Yanson has measured the second derivative d^2V/dI^2 of the voltage with respect to the current as a function of the voltage V. As a most remarkable result, he observed structure in the the measured second-derivative signal at applied voltages corresponding to the bulk phonon frequencies of the metal forming the junction. He interpreted his experimental data as a direct measurement of the electron-phonon interaction function $\alpha^2(\omega)g(\omega)$. Roughly speaking, the *Eliashberg function* $\alpha^2(\omega)g(\omega)$ is the product of the phonon density of states $g(\omega)$ [Ref. 7.1, Sect. 3.5] and the squared matrix element α for the electron-phonon interaction, averaged over the Fermi sphere [7.33]. As the radius of the investigated contact is small compared with the mean free path l of the electrons $(a < l)$, the transport of electrons through the contact is ballistic rather than diffusive, as indicated in Figs. 7.13 and 17. Within a mean-free path distance, the electrons are accelerated due to the electric field caused by the applied

189

voltage, and are injected from one metal side to the other by passing the contact. This injection current is linear in the applied voltage, as expected for ballistic transport through a contact. After injection, however, some of the electrons can flow back through the orifice due to inelastic scattering processes (Fig. 7.17); this process gives rise to a negative and voltage-dependent correction to the current. Hence, the nonlinear current-voltage relation contains information about the inelastic collisions of electrons with phonons.

7.3.1 Experimental Techniques

Two types of point-contacts have been investigated, namely shorted-thin film and pressure-type point contacts. The pioneering first point-contact experiments were performed by *Yanson* [7.31, 32]using tunneling-junction geometries, such as discussed in Sect. 7.2.1. Using conventional evaporation techniques, a sandwich of two films of the metal of interest, separated by an oxide layer (Fig. 7.13) is produced. Accidentally or intentionally, a point-contact is produced in the oxide layer by an electric breakdown thus connecting the two metal films by a metallic short-circuit. In later experiments [7.34]a sharply pointed steel needle was pressed onto the surface of

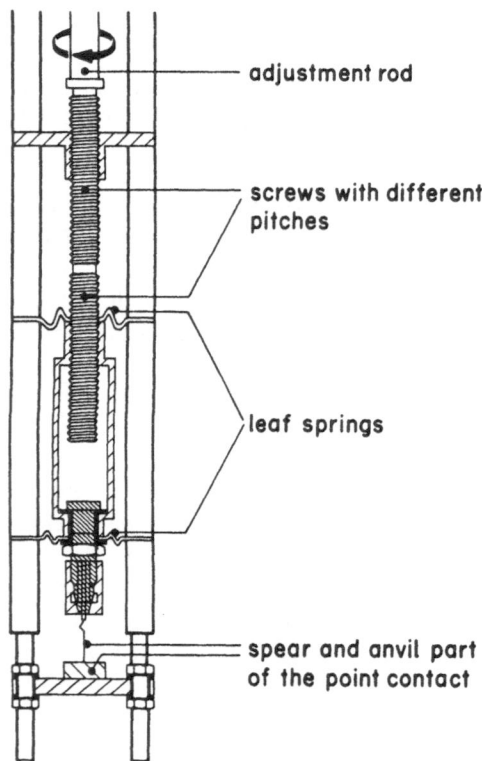

adjustment rod

screws with different pitches

leaf springs

spear and anvil part of the point contact

Fig. 7.14. Schematic view of a pressure-type contact. By means of a differential screw mechanism the spear can be moved towards the anvil in order to adjust a contact while immersed in liquid helium [7.30]

the junction, providing a crack in the dieelectric layer and so forming the contact.

Pressure-type contacts are much simpler to control [7.35]. A sharply etched metal wire ("spear") is carfully pressed against a flat metal surface ("anvil") to form the contact. The spear is fabricated by electrolytically etching a wire (50–100 μm diameter) to a tip with a radius of ~1 μm, as is well known from whisker technology [7.36]. Together with the chemically cleaned anvil the spear is then mounted in a system with a differential screw mechanism which allows a contact to be adjusted while immersed in liquid helium (Fig. 7.14) [7.30]. Pressure-type contacts allow the investigation of a large variety of samples, including studies of anisotropies of single crystals. It is possible to obtain stable spear-anvil contacts with a resistance of up to ~ 50 Ω.

The technique for recording derivatives (dV/dI) and (d^2V/dI^2) in point-contact spectroscopy is the same as in inelastic electron tunneling spectroscopy (Sect. 7.2.1). The quantity which is actually measured is the voltage dependence of $d^2/V/dI^2$, whereas the function of interest, namely the Eliashberg function $\alpha^2(\omega)g(\omega)$, is proportional to dR/dV or to d^2I/dV^2

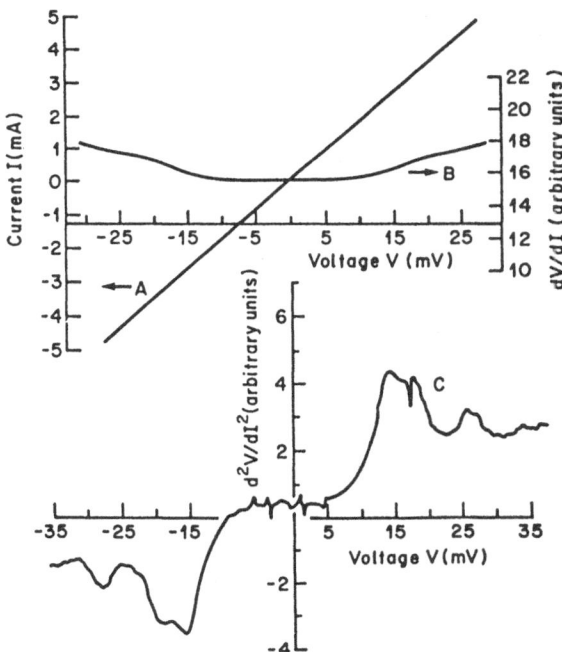

Fig. 7.15. Recorder output of the current-voltage characteristics of a copper contact. The current I (curve A), the first derivative $R = dV/dI$ (curve B) and the second derivative d^2V/dI^2 (curve C) are given as functions of the applied voltage V. Resistance $R_0 = 5.7\,\Omega$, temperature $T = 1.2\,\mathrm{K}$ [7.30]

(Sect. 7.3.2). The different quantities are related by

$$\frac{d^2 I}{dV^2} = -\frac{1}{R^3}\frac{d^2 V}{dI^2} = -\frac{1}{R^2}\frac{dR}{dV},$$ (7.17)

where $R = dV/dI$ is the dynamic resistance. If the voltage dependence of R is small (as for Cu, K and Na where R changes maximally by 10 % for bias voltages equal to the Debye energy), no corrections are needed. In other cases, the change in R is large and has to be taken into account in order to compare the measured $d^2 V/dI^2$ spectra with the theoretical predictions.

Figure 7.15 shows the recorder output of the current-voltage characteristics of a copper point-contact [7.30]. The nonlinearity can hardly be seen in $I(V)$ but is clearly visible in the measured first and second derivatives $R = dV/dI$ and $d^2 V/dI^2$. The structure in the second derivative around 17 and 27 mV occurs at energies corresponding to the typical phonon frequencies of copper.

7.3.2 Qualitative Discussion of Point-Contact Spectra

The behaviour of a point-contact can be characterized by the ratio of the electronic mean free path l to the radius of the contact a. The problem resembles the well known Knudsen problem in the kinetic gas theory: by pumping through a small hole in a gas container, the pressure of the gas will be lowered, however, at the moment where the mean-free path of the molecules becomes comparable with the diameter of the hole, there is no longer diffusive flow and the molecules will pass the orifice ballistically. The problem was first considered by Knudsen and the different regimes are characterized by the Knudsen ratio $K = l/a$.

For electrical contacts with an arbitrary Knudsen number K the resistance R is found to be [7.37]

$$R = R_0 + R_1 = \frac{4\varrho l}{3\pi a^2}\left[1 + \frac{3\pi}{8}\Gamma(K)\frac{a}{l}\right].$$ (7.18)

Here, $\Gamma(K)$ is a slowly varying function of K with $\Gamma(K = 0) = 1$ and $\Gamma(K = \infty) = 0.694$. For large values of K, as is the case for point-contact spectroscopy, the *Sharvin* resistance R_0 [7.38]dominates and the term R_1 is a small correction term. Since the resistivity ϱ is proportional to l^{-1}, ϱl and hence R_0 are independent on l, but R_1 is proportional to l^{-1}. Usually, the mean free path l will depend on the energy of the electrons, $l = l(E)$. From (7.18) one finds

$$\frac{dR}{dV} = \frac{dR_1}{dV} = \Gamma(K)\frac{\varrho l}{2a}\frac{d}{dV}\left(\frac{1}{l(E = eV)}\right).$$ (7.19)

Thus, it is the energy dependence of the mean-free path which gives rise to structure in the point-contact spectra; the Sharvin resistance R_0 leads to no structure, but this term is essential as it allows to be an electric field within the metal. The total scattering length l of the electrons is determined by the elastic scattering with impurities (l_{imp}) and the inelastic scattering with phonons (l_{ep}). Using Matthiessen's rule the total scattering length l can be written as

$$l^{-1} = l_{imp}^{-1} + l_{ep}^{-1}. \tag{7.20}$$

Here, only the electron-phonon length will be energy dependent and give rise to a contribution to the signal via equation (7.19). Using the well known Fermi's rule argument for the transition rate of an electron arising from the electron-phonon interaction, it is possible to calculate the scattering time $\tau_{ep}(E) = l_{ep}(E)/v_F$ of an electron with energy E above the Fermi level. For $T = 0$ the final result is [7.30]

$$\tau^{-1}(E) = v_F/l(E) = q\pi \int_0^{E/\hbar} d\omega\, \alpha^2 g(\omega), \tag{7.21}$$

where, as before, $\alpha^2 g(\omega)$ is the Eliashberg function for the electron-phonon interaction and v_F is the Fermi velocity. Remember that $\alpha(\omega)$ is a measure of the effective electron-phonon coupling and $g(\omega)$ is the phonon density of states. An evaluation of $\alpha^2(\omega)g(\omega)$ based on Born-von Karman force constant models and pseudo-potential theory has been carried out by *Carbotte* and *Dynes* [7.33] for Na, K, and Al. From (7.19 and 21) one obtains the important result that for low temperatures the voltage derivative of the

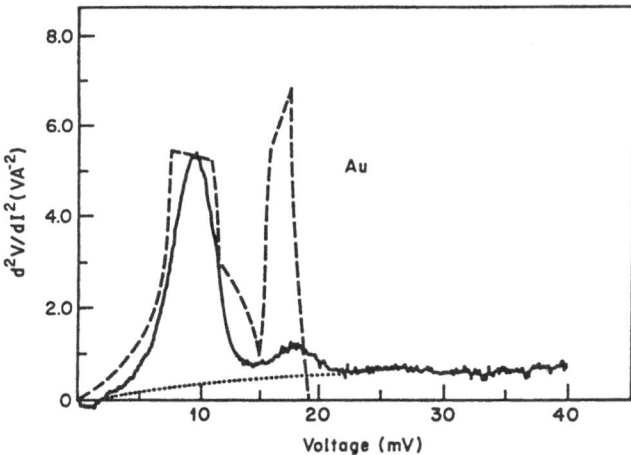

Fig. 7.16. Measured d^2V/dI^2 spectrum of a gold point-contact. The long dashed line gives the phonon density of states $g(\omega)$ obtained from inelastic neutron scattering. The short-dashed line shows the smooth background signals [7.30]

point-contact resistance is given by

$$\frac{dR}{dV}(V) = \frac{\varrho l}{2a} \Gamma(K) \frac{2\pi e}{\hbar v_{\mathrm{F}}} \alpha^2 g(eV).$$ (7.22)

This equation allows a direct determination of $\alpha^2 g$ by measuring the voltage derivative of the resistance of a metallic contact.

Figure 7.16 shows the measured second derivative $d^2V/dI^2 = R\, dR/dV$ as a function of the applied voltage for a gold point-contact [7.30]. To compare the measured signal with the phonon spectrum, the same figure displays also the phonon density of states $g(\omega)$, as obtained from inelastic neutron scattering experiments. As expected, there is a clear agreement in the voltage dependence of the signal with the frequencies of the transverse ($\sim 9\,\mathrm{mV}$) and the longitudinal ($\sim 17\,\mathrm{mV}$) phonons. The comparison between $\alpha^2(\omega)g(\omega)$ and $g(\omega)$ shows that the electron-phonon coupling is smaller for the longitudinal phonons than for the transverse phonons; this was observed for all noble-metal point contacts [7.39, 40]. In other words, the electron-phonon coupling function $\alpha^2(\omega)$ is large for the transverse modes but small for the longitudinal modes in the case of the noble metals.

The short discussion given above is based on a more phenomenological approach based on the simple interpolation formula (7.18) which bridges the two limiting cases for small ($K \ll 1$) and high ($K \gg 1$) Knudsen numbers. It is instructive to discuss some results of a more fundamental approach which is

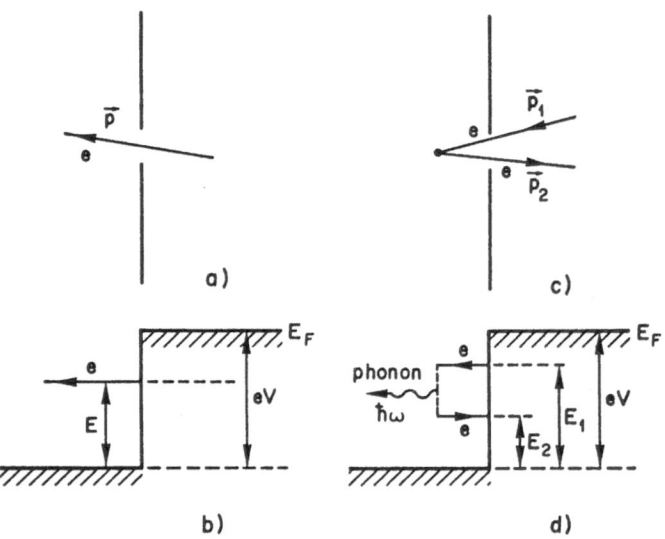

Fig. 7.17. (a) Electron passing ballistically through the orifice without being back scattered; (b) energy level diagram associated with process (a); (c) by collision with phonons, electrons can flow back through the orifice; (d) single-collision back-flow process in energy space, where a phonon with energy $\hbar\omega = E_1 - E_2$ is emitted [7.30]

based on the solution of the full nonlinear Boltzmann equation for the point-contact problem [7.41–46]. If a voltage is applied over the contact, electrons are injected from one metal to the other. Close to the contact, the distribution of the electrons in momentum space will be changed in such a way that there is a difference in kinetic energy (equal to eV) between electrons that have or have not passed through the orifice. Far from the orifice, the electron system has to be in equilibrium again and, through inelastic collisions with phonons, the energy difference will be washed out. Figure 7.17a shows an electron passing ballistically through the orifice without being back scattered, and Fig. 7.17b shows the associated energy level diagram for $T = 0$. This process gives rise to the zero-order current $I_0 = V/R_0$, where R_0 is the Sharvin resistance given by the first term in (7.18) which is independent on the electron mean free path l and hence on V as discussed above. By collisions with phonons, electrons can flow back through the orifice (Fig. 7.17c). Such collisions lead to emission (or absorption) of phonons as shown in Fig. 17.7d for the case of a single-collision back-flow process. Double and higher collision back-flow processes associated with the emission or absorption of two or more phonons are also possible. These back-flow processes are associated with small negative and nonlinear currents which produce the structure observed in the spectra. First-order processes of the type shown in Fig. 7.17d are associated with the correction term R_1 of (7.18) which depends on l and hence on V. The first-order correction to the current associated with single-collision back-flow processes is found for the case of zero temperature to be [7.44]

$$I_1 = -\frac{4\pi a^3}{\hbar e v_F \varrho l} \int\limits_0^{eV} dE_2 \int\limits_{E_2}^{eV} dE_1 \alpha^2(\omega) g_p(E_1 - E_2). \tag{7.23}$$

In this expression E_1 is the energy of a hot injected electron, E_2 that of a cold electron scattered back through the orifice, as shown in Fig. 7.17d. The expression $\alpha^2(\omega) g_p(\omega)$ is a modified electron-phonon interaction function, where the subscript p indicates the difference with respect to the usual Eliashberg function $\alpha^2(\omega) g(\omega)$ occurring in the simpler theory, see (7.22); this difference is due to a geometrical transport efficiency factor which depends on the angle between the moments $\boldsymbol{p}_1$ and $\boldsymbol{p}_2$ of the electron before and after the collision. Differentiating equation (7.23) gives for $I = I_0 + I_1$

$$\frac{d^2 I}{dV^2} = -\frac{4\pi e a^3}{\hbar v_F \varrho l} \alpha^2 g_p(eV). \tag{7.24}$$

Comparison of (7.24) and (7.17) shows that the observed quantity d^2V/dI^2 is approximately proportional to $\alpha^2 g_p(eV)$ if the voltage dependence of R is small. This is the case for copper (curve B in Fig. 7.15) or for potassium and sodium, but not for Li. The energy dependence of d^2V/dI^2, $\alpha^2 g(eV)$

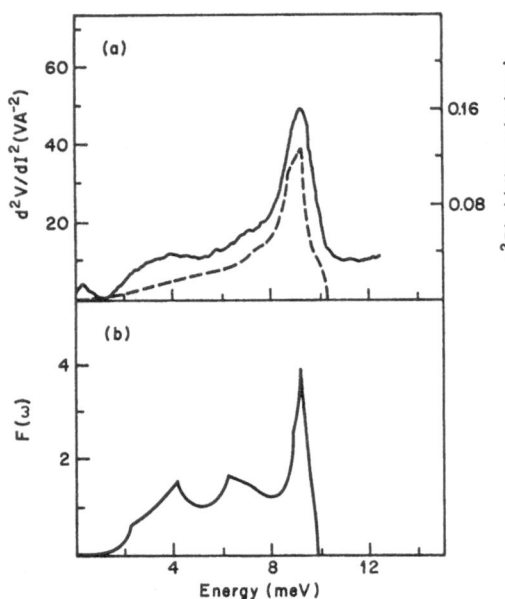

Fig. **7.18a,b**. Point-contact spectroscopy of potassium. The full curve (left-hand scale) in (a) is a measured point-contact spectrum for K (junction resistance at zero voltage is 2.9 Ω; measuring temperature is 1.2 K) [7.47]. The broken curve (right-hand scale) is the function $\alpha^2 F(\omega)$ obtained theoretically [7.33]. The curve in (b) represents the phonon density of states $g(\omega)$ obtained from experiments involving inelastic neutron scattering [7.48]

and $g(eV) = F(\omega)$ of potassium is shown in Fig. 7.18 [7.47]. The function $\alpha^2 g(eV)$ has been obtained by a model calculation based on pseudopotential theory [7.33], while $g(\omega)$ has been evaluated on the basis of lattice dynamical calculations and inelastic neutron scattering experiments [7.48]. Figure 7.18 shows that the theoretical function $\alpha^2 g(eV)$ is qualitatively similar to the observed structure in d^2V/dI^2. Comparison between Fig. 7.18a and b indicates that the electron-phonon interaction for the (high-frequency) longitudinal modes is stronger than for the transverse modes; this is in contrast to the situation found for gold in Fig. 7.16.

Our discussion has been limited to the case $T = 0$ since point-contact experiments are usually done at liquid helium temperatures. With increasing temperature the structures of the derivative spectra are found to broaden considerably, as in inelastic electron tunneling spectra, and are smeared out completely at temperatures of the order of 100 K [7.49].

7.4 Spectroscopy of Surface Phonons, Thin Films and Adsorbates

7.4.1 Acoustical and Optical Surface Phonons

Our discussion of the vibrational properties of crystals has been based on the assumption that the displacements of the atoms obey the cyclic or periodic boundary conditioins of Born and von Karman [Ref. 7.1, Sect. 3.3]. As has been discussed in [Ref. 7.1, Sect. 2.1.1], the choice of cyclic boundary conditions is mathematically convenient and introduces negligible errors in the calculation of bulk properties of large crystals. However, the adoption of cyclic boundary conditions eliminates the possibility of studying the dynamical properties of atoms in the neighbourhood of a free surface of a real crystal.

An understanding of the structural and vibrational properties of surface atoms and adsorbates is of fundamental importance in many areas of the physics and chemistry of surfaces, most notably in the field of heterogeneous catalysis on metals and alloys.

The forces acting on atoms in the surface layers will be different from the forces acting on the atoms in the bulk, since an atom in the surface layers has fewer nearest neighbours, next-nearest neighbours, etc. than an atom in the interior of a crystal. It is therefore not at all necessary that the atomic arrangement in the outermost layers be identical to that in the bulk. The atoms may shift off the lattice positions appropriate to the bulk crystal, with the consequence that the symmetry of the surface layer is lower than the bulk. This phenomenon is referred to as *surface reconstruction*, which can be studied by means of low-energy electron diffraction techniques (LEED) [7.50]. From the different forces acting on the atoms at the surface one would expect that not only the structural but also the dynamical properties of atoms, such as their mean-square displacements are different for atoms in the surface layers of a crystal than for atoms in the bulk.

Theoretical and experimental studies have shown that the presence of a free surface can give rise to localized vibration modes in which the atomic displacements decay exponentially with increasing distance into the crystal from the surface. Because the crystal retains its periodicity in the directions parallel to the crystal surface, the atomic displacement amplitude in these localized surface modes are wavelike in these directions and are characterized by the wave vector $\boldsymbol{q}_{\parallel} = (q_x, q_y)$ parallel to the surface [7.51].

Such localized modes associated with crystal surfaces can have acoustic or optical character. Long-wavelength acoustic surface modes are also called *Rayleigh* waves, after their discoverer [7.51]. The dispersion relation for these waves is

$$\omega(\boldsymbol{q}_{\parallel}) = \varrho v_{\mathrm{T}}(q_x^2 + q_y^2)^{1/2}, \tag{7.25}$$

where v_{T} is the velocity of sound for transverse waves and ϱ is a positive, pure number, smaller than unity, whose value is determined by the values of the Lamé constants λ and μ, which characterize the elastic properties of an isotropic solid [7.52]. There are two classes of acoustic surface modes: The low-frequency Rayleigh surface waves have displacement fields that penetrate deeply into the crystal as $\boldsymbol{q}_{\parallel} \to 0$, while the remaining acoustic surface modes are localized on the outermost atomic layers.

In crystals with more than one atom per unit cell, we encounter surface optical phonons, i.e. surface modes which have finite, nonzero frequency in the limit $q_{\parallel} \to 0$. These modes also come in two distinctly different classes. If we consider a crystal lattice with two or more atoms per unit cell, and only short-range forces enter the lattice dynamics of the crystal, then in general as $q_{\parallel} \to 0$, the atomic displacements associated with the surface optical phonon remains localized in the outermost few atomic layers. Such modes may be encountered in the homopolar semiconductors silicon and germanium, and at the surface of metals of the appropriate crystal structure. On the other hand, in ionic crystals, where the Coulomb interactions between the ions lead to couplings that extend over large distances, we encounter a second class of surface optical phonons which penetrate deeply into the crystal as $q_{\parallel} \to 0$. These optical modes may be described by a macroscopic electromagnetic theory just as the Rayleigh waves can be described by a macroscopic elastic continuum theory. All these modes may also be derived equally well within the lattice-dynamical theory. Optical surface modes of ionic crystals are also called *Fuchs-Kliever modes* [7.53–55], after the authors who described them first with the formalism of lattice dynamics instead of the older description provided by electromagnetic theory. Where the modes are described within the framwork of electromagnetic theory, most particularly when the effect of retardation is included (Appendix B), these modes are called *surface polaritons* [7.56].

For an isotropic dielectric medium with dielectric constant $\varepsilon(\omega)$, the dispersion relation for surface polaritons is given by the relation

$$k^2 = \frac{\omega^2}{c^2} \varepsilon(\omega) / [\varepsilon(\omega) + 1], \qquad (7.26)$$

where k is the wave vector of the electromagnetic field associated with the polar surface phonons [7.56]. This relation is to be compared with the corresponding relation for bulk polaritons $k^2 = (\omega^2/c^2)\varepsilon(\omega)$, see (2.40). From (7.26) it follows that for large values of k $(k \gg \omega/c)$, the frequency of the surface mode is given by $\varepsilon(\omega) + 1 \cong 0$. For a cubic diatomic crystal $\varepsilon(\omega)$ is given by (2.33). Neglecting damping and using the LST relation $\varepsilon_0 = \varepsilon_\infty \omega_{LO}^2 / \omega_{TO}^2$ [Ref. 7.1, Sect. 4.3.2], one obtains for the frequency of the optical surface or Fuchs-Kliever mode of the ionic crystal the following expression:

$$\omega_s = \left(\frac{\varepsilon_0 + 1}{\varepsilon_\infty + 1} \right)^{1/2} \omega_{TO}, \qquad (7.27)$$

where ω_{TO}, ε_0 and ε_∞ are the bulk TO frequency for $q \to 0$, and the static and high-frequency dielectric constant, respectively. Since $\varepsilon_0 > \varepsilon_\infty$ it follows that $\omega_{TO} < \omega_s < \omega_{LO}$.

In the last two decades there has been a remarkable increase in interest in the lattice dynamics of crystal surfaces. This stems in large part from a number of experimental advances such as Electron Energy Loss Spectroscopy (EELS), Inelastic Molecular Beam Spectroscopy (IMBS), Infrared Reflection Absorption Spectroscopy (IRAS) and Attenuated Total Reflection (ATR). In the following we shall give a short discussion of some of these techniques.

7.4.2 Electron Energy Loss Spectroscopy

This subsection is devoted to a short discussion of the vibrational motion of atoms and molecules on and near the surface by the analysis of low-energy electrons backscattered from it. An electron incident on the crystal with energy E_I may excite a quantized vibrational mode with energy $\hbar\omega_s$ befor being backscattered into the vacuum. It thus emerges with energy $E_s = E_I - \hbar\omega_s$, so an analysis of the energy spectrum of the backscattered electrons provides direct information on the vibrational frequencies of the surface, or on those atoms or molecules adsorbed on the surface. The method is sensitive to the surface because, with the incident energy chosen suitably (E_I is of the order of $10\,\mathrm{eV}$), the electron penetrates, at most, two or three atomic layers into the crystal. The backscattered electrons thus carry information on only the near vicinity of the surface [7.57].

In the present context we are interested in *dipole scattering:* As an atom on a surface or in a molecule vibrates, it generally modulates the electric dipole moment of its environment in a time-dependent fashion. Such dipole moments are induced by polar surface modes such as the Fuchs-Kliever modes discussed in Sect. 7.4.1. An electron in the vacuum above the crystal senses a long-rang electric field of dipolar character, and this produces a "lobe" of inelastic scattering intensity which is sharply peaked around the specular direction. At large deflection angles from the specular direction the scattering mechanism is quite different from dipolar scattering and is called impact scattering [Ref. 7.57, p. 102].

Conservation of energy and momentum leads to the following relations [Ref. 7.57, p. 257](Fig. 7.21b):

$$\frac{\hbar^2}{2m_e}(k_I^2 - k_s^2) = \pm\hbar\omega_j(\boldsymbol{q}_{\|}), \tag{7.28}$$

$$\hbar(\boldsymbol{k}_{I,\|} - \boldsymbol{k}_{s,\|}) = \hbar(\boldsymbol{\tau}_{\|}\pm\boldsymbol{q}_{\|}). \tag{7.29}$$

Here, m_e is the mass of electrons with incident and scattered wave vectors $\boldsymbol{k}_I$ and $\boldsymbol{k}_s$, respectively, $\boldsymbol{k}_{I,\|}$ and $\boldsymbol{k}_{s,\|}$ are projections onto the plane parallel to the crystal surface, while $\boldsymbol{q}_{\|}$ and $\boldsymbol{\tau}_{\|}$ are the phonon and reciprocal lattice vectors parallel to the surface, respectively. Due to the lack of translational symmetry perpenducilar to the surface, only the projections of the relevant vectors onto the surface enter the momentum conservation (7.29).

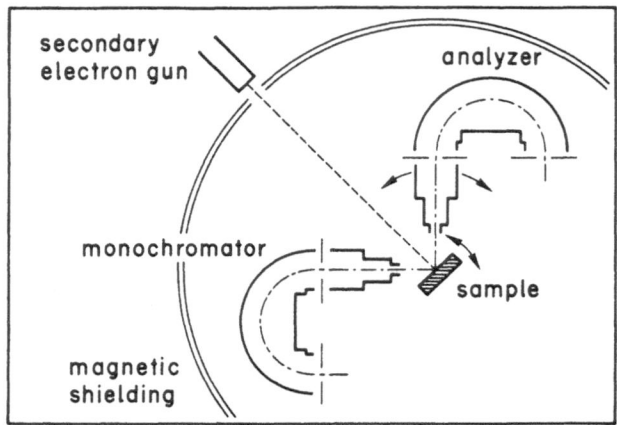

Fig. 7.19. Schematic drawing of a high resoluton energy-loss spectrometer (HREELS) [7.60, 61](see text)

The theory of inelastic electron-phonon scattering is similar to inelastic X-ray or neutron scattering. As we have discussed in Chaps. 5 and 6, the cross sections for inelastic scattering of X-rays and neutrons contain the scalar product of the scattering vector $K = k_s - k_I$ and the polarization vector $e(q, j)$ of the normal mode. For the case of specular reflection, which is relevant for dipole scattering, K is pepernducilar to the surface, which makes the cross section zero for modes polarized parallel to the surface.

A schematic drawing of a High Resolution Energy Loss Spectrometer (HREELS) is shown in Fig. 7.19 [7.58, 59]. To form a monoenergetic beam of primary electrons an electrostatic deflecting system in combination with entrance and exit slits (hemispherical monochromator) is used to select electrons of well-defined energy from those emitted by a cathode. The electrons which pass through the monochromator strike the sample, and a second electrostatic system (the hemispherical analyzer) is used as an analyzer of the energy spectrum of the scattered electrons. The analyzer is mobile in the plane of incidence of the monochromatic electrons in the figure. By this means, the distribution of scattered electrons in angle, as well as energy may be sampled. The resolution limit of the spectrometer is of the order of 3 meV $(24\,\mathrm{cm}^{-1})$. Charging effects on insulating samples can be eliminated by illuminating the crystal with the defocussed beam of an auxiliary electron gun working in the ranges of energy and current of a typical Auger-spectrometer electron gun. This electron gun is mounted outside the magnetic shielding around the HREEL spectrometer.

Fuchs-Kliever phonons were evidenced with EELS on ZnO by the pioneering work of *Ibach* [7.62]. Simultaneously to this measurement a comprehensive theory of surface excitations in solids by electrons was developed

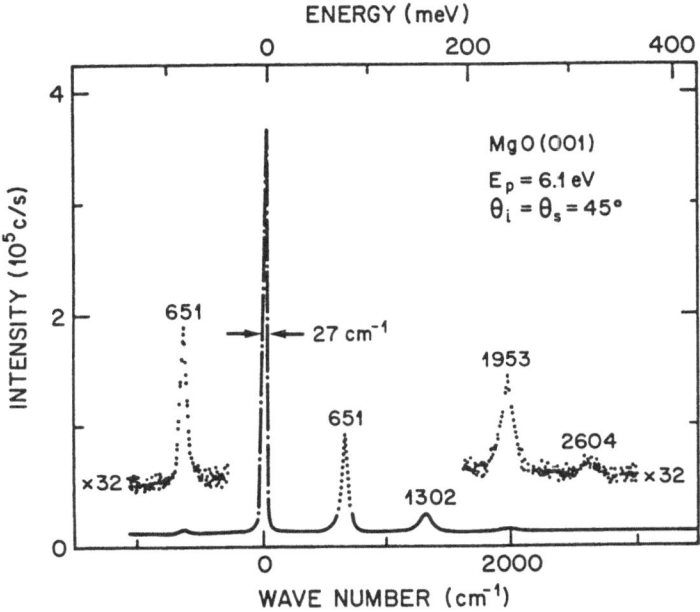

Fig. 7.20. Energy-loss spectrum of 6.1 eV electrons after reflection on a MgO(001) surface [7.69]

by *Lucas* and *Sunjic* [7.63] and has been successfully applied to ZnO. Other measurements have been reported since on compound semiconductors GaAs [7.64], InSb and GaP [7.65], as well as on TiO_2 [7.66], $SrTiO_3$ [7.67] and LiF [7.68]. All these measurements were performed with rather low resolution (around 10 meV or more).

Recently, HREELS has been performed on single crystals of MgO by *Thiry* et al. [7.69] using the type of spectrometer shown in Fig. 7.19, which has a resolution as high as about 3 meV. Figure 7.20 shows a result obtained at room temperature with an angle of incidence and analysis of 45°. The beam energy indicated (6.1 eV) was measured by scanning the energy range up to the cutoff of the spectrum. Up to seven loss peaks (corresponding to absorption of surface phonons) were detected at an angle of incidence of 80°. This allowed the accurate determination of the surface phonon resonance frequency from the distance between the multiple losses; the resulting value is $651\pm2\,\text{cm}^{-1}$. This coincides, within less than 0.2 %, with the Fuchs-Kliever surface phonon frequency ω_s given by the relation (7.27). For MgO, $\omega_{TO} = 393\,\text{cm}^{-1}$, $\varepsilon_0 = 9.8$ and $\varepsilon_\infty = 1.95$.

The differential cross section for EELS one-phonon processes associated with phonon emission is given by [7.57, 70, 71]

201

$$\frac{dS}{d\omega|R|^2} = \frac{8[\bar{n}(\omega)+1]}{\pi^2 a_0 k_I \cos\theta} B(\Psi_c, \theta) P(\omega). \tag{7.30}$$

$P(\omega)$ is the *loss function* discussed below and $|R|^2$ is the intensity of the specularly reflected elastic beam, normalized to the incident intensity. $\bar{n}(\omega)$ is the mean occupation number, a_0 the Bohr radius, and k_I the magnitude of the wave vector $\boldsymbol{k}_I$ of the incident electrons. θ designates the angle of incidence of the electron beam and $B(\Psi_c, \theta)$ is a function that corrects for the finite apertur Ψ_c of the spectrometer. A similar expression holds for phonon absorption (or gain in electron energy) but with $\bar{n}+1$ replaced by $\bar{n}$. The important quantity is the loss function $P(\omega)$ which for an isotropic solid is defined by

$$P(\omega) = \frac{1}{\omega} \operatorname{Im}\left\{\frac{-1}{\varepsilon(\omega)+1}\right\}, \tag{7.31}$$

where $\operatorname{Im}\{\ldots\}$ means the imaginary part and $\varepsilon(\omega)$ the dielectric constant which can be written in the form

$$\varepsilon(\omega) = \varepsilon_\infty + \frac{(\varepsilon_0 - \varepsilon_\infty)\omega_{TO}^2}{\omega_{TO}^2 - \omega^2 - i\gamma\omega} \tag{7.32}$$

for a cubic crystal with a single TO mode for $q \to 0$ having damping γ, see (2.33, 34). The whole spectrum of multiple losses and gains is then obtained by autocorrelating the loss function by use of the Fourier transform; the detailed procedure which also includes normalization and convolution with the instrumental transfer function has been described in [7.63, 70]. The final fit depends on three parameters, ω_s, γ and $Q = I_1/I_0$, the ratio of the single loss intensity to the elastic intensity; this procedure was found to give excellent fits to the experimental data, from which it is then possible to determine the infrared optical constant using (7.32).

It is possible to excite optical surface phonons also on non-infrared active materials such as on a clean cleaved Si (111) surface [7.72]. Although the effective charge vanishes inside the volume of the material, a dipole active surface layer may exist induced by the lower symmetry of the reconstructed surface.

7.4.3 Inelastic Molecular Beam Scattering

Inelastic scattering of low energy neutral particles (e.q. H_2, D_2, He, Ne) is a promising method to study acoustic surface phonons (i.e., Rayleigh modes). Molecular beams with high velocity monochromacy can be produced today with energies and momenta comparable to thermal neutrons (Fig. 1.1, region marked by IMBS). In contrast to neutrons, however, low-energy neutral particles are not penetrating, so they probe exlusively the surface. In the

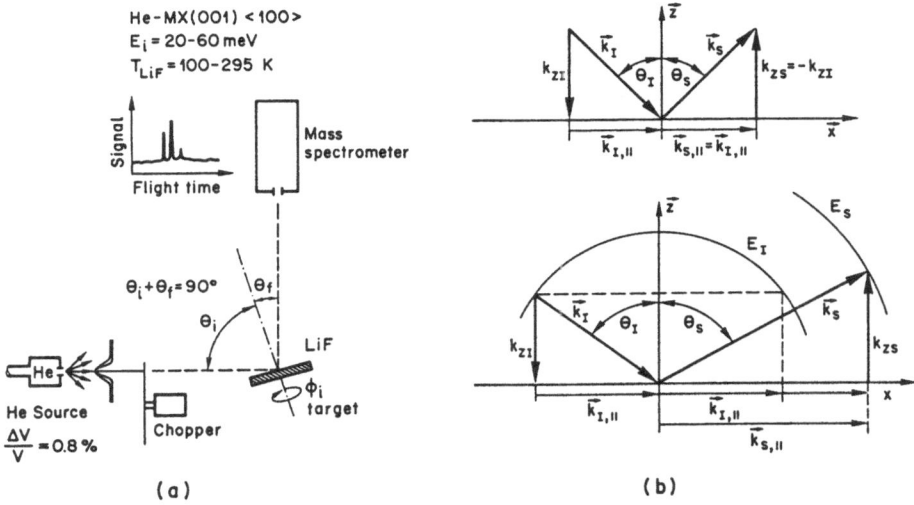

Fig. 7.21. (a) Schematic diagram of a IMBS apparatus [7.75–77]. (b) Geometry and definition of angles used for inplane scattering; *upper part*: elastic or diffractive scattering, *lower part*: inelastic scattering

following we give a short description of the experimental method and of some results. For details the reader is referred to the literature [7.73, 74].

Figure 7.21a shows a schematic diagram of the spectrometer used by *Brusdeylins*, et al. [7.75–77]. A supersonic nozzle system provides a high intensity, nearly monochromatic beam and so combines the functions of a source and a monochromator. Next, a time structure is imposed on the monochromatic beam in order to facilitate time-of-flight detection after scattering; high-frequency choppers serve well for this purpose. Upon scattering on the crystal surface, the beam intensity as function of time is monitored using a mass spectrometer. In the apparatus illustrated in Fig. 7.21a the angle between incident and scattered beams is fixed $(\theta_i + \theta_s = 90°)$, but the crystal can be rotated.

Figure 7.21b shows the usual geometry and definition of angles used for in-plane scattering, i.e. for a scattering geometry in which the three vectors z, k_I and k_s are in the same plane. The upper part of Fig. 7.21b shows the situation for *elastic* or *diffractive scattering*, which yields information about surface structures (surface reconstruction). The lower part of Fig. 7.21b shows the scattering geometry for *inelastic scattering*. Energy and momentum conservation are given by (7.28, 29). For in-plane scattering it is an easy exercise to show that from (7.28, 29) the following equation relating the total momentum transfer $K = \tau_{||} \pm q_{||}$ with the experimental quantities $\hbar\omega$ and θ_i holds

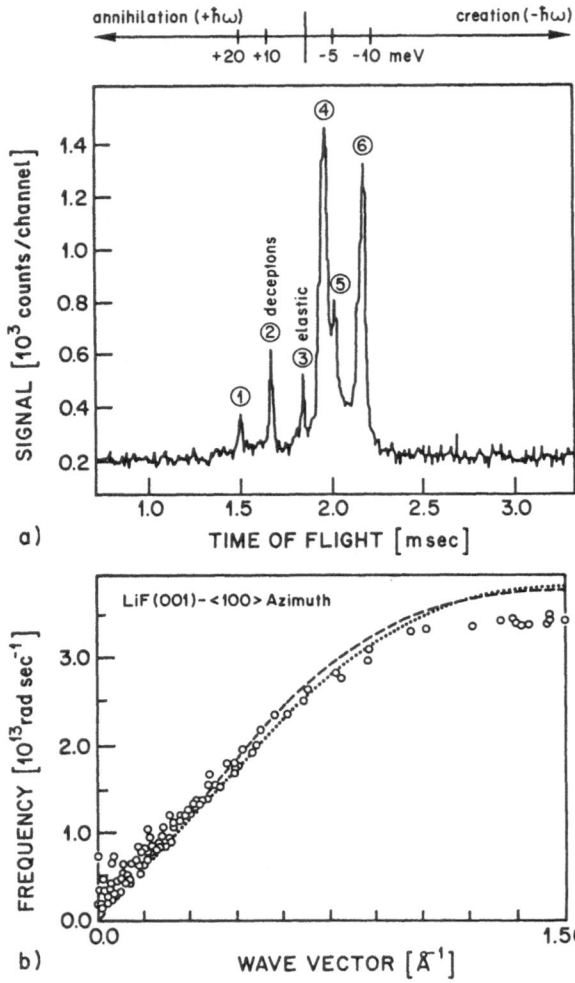

Fig. 7.22. (a) Time-of-flight spectrum for He atoms scattered from LiF⟨100⟩ with $E_I \cong$ 19 meV, $k_I \cong 6.0\,\text{Å}^{-1}$ and $\theta_I = 64.2°$. (b) Dispersion of acoustic surface phonons of LiF⟨100⟩; the dotted and dashed lines are the results of theoretical calculations [7.75]

$$\frac{\hbar\omega}{E_I} = -1 + \left(1 - \frac{K}{k_I}\right)^2 \frac{\sin^2\theta_I}{\sin^2\theta}. \tag{7.33}$$

Imagine a scattering experiment has been performed at a specific angle θ_i (and $\theta_f = 90° - \theta_i$) and at an incident energy E_I. The energy transfer $\Delta E = \hbar\omega$ associated with a particular peak in the time-of-flight spectrum can be calculated from the flight time t and the flight distance d, giving the velocity

v and hence k_s and E_s. With $\hbar\omega$, E_I and θ_I known, the corresponding momentum transfer $\boldsymbol{K} = \boldsymbol{\tau}_{\parallel} \pm \boldsymbol{q}_{\parallel}$ can be calculated from (7.33). As usual $\boldsymbol{q}$ is the vector connecting $\boldsymbol{K}$ with the nearest reciprocal lattice vector $\boldsymbol{\tau}_{\parallel}$ of the two-dimensinal reciprocal lattice. In this way dispersion curves $\omega(\boldsymbol{q}_{\parallel})$ can be constructed.

Figure 7.22a shows a time-of-flight spectrum of the atoms scattered from LiF $\langle 100 \rangle$ with $E_I \cong 19\,\mathrm{meV}$, $\quad k_I \cong 6.0\,\mathrm{\AA}^{-1}$ and $\theta_I = 64.2°$ [7.75]. Peak 1 is due to annihilation of a Rayleigh mode, while peaks 4 and 6 are due to creation of Rayleigh modes. Peak 5 originates from a bulk phonon extending to the surface, while peaks 2 and 3 are spurious peaks discussed in [7.75]. Figure 7.22b shows the dispersion for the acoustic surface phonons of LiF $\langle 100 \rangle$ [7.75]and compares them with the theoretical predictions of *De Wette* et al. [7.78]and of *Benedek* [7.79]. It is gratifying to find good agreement between theory and experiments over most of the range of wave vectors. Starting at about $q_{\parallel} = 1\,\mathrm{\AA}^{-1}$, however, the measured points fall below the theoretical predictions. In this region the phonon frequencies are especially sensitive to small changes in force constants, because of the increase in relative amplitudes of adjacent atoms and the shallow penetration of phonons as the wavelength decreases. This discrepancy has been attributed to relaxation [7.76, 80], but could also be due to small errors in the force constants used in the calculations.

7.4.4 Optical Studies of Surface Modes

Optical surface polaritons of semi-infinite crystals as described by (7.26) are non-radiative, i.e. their fields fall off exponentially outside the slab (evanescent wave). For this reason they do not interact with light in conventional experiments on light absorption and reflection. It is, however, possible to excite surface polaritons optically by using a modification of the well-known attenuated total reflection method (ATR) proposed by *Otto* [7.81]for studying surface plasmons in metals. The OTTO configuration is shown in Fig. 7.23. Reflectivley of an infrared beam in a silicon prism or in a KRS5 hemicylinder is reduced in the range of total reflection by lossy coupling of energy into the sample to be studied through a thin air gap of thickness d. This method is also known as *frustrated total reflection spectroscopy* [7.81]. The

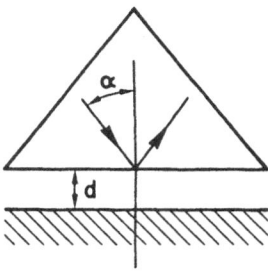

Fig. 7.23. Geometry of the experiment for detection of surface modes by the modified ATR method (OTTO configuration [7.81])

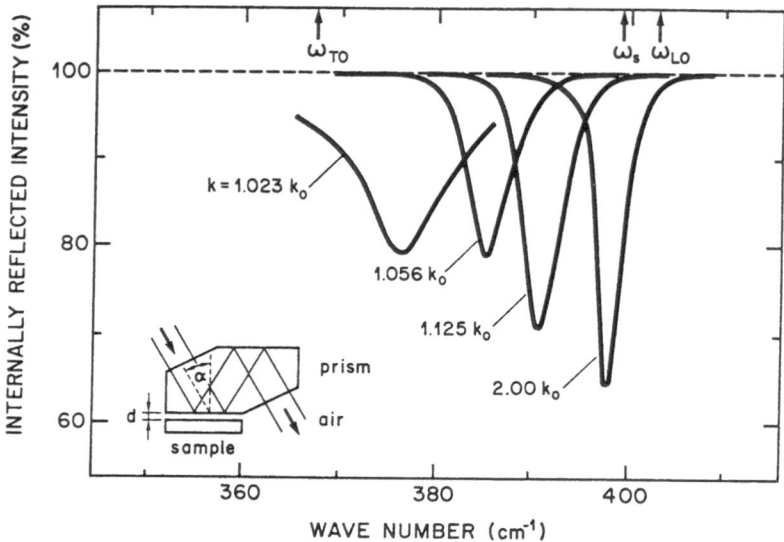

Fig. 7.24. Spectra of frustrated total reflection (parallel polarization) of GaP at room temperature. Different values of k according to (7.34). Spacings: $d = 40$, 25, 12, and 2.5 μm, ordered to increasing values of k. Inset shows experimental arrangement [7.84]

electromagnetic field of the infrared light will penetrate into the air gap as an evanescent wave which can interact with the evanescent wave of the surface optical modes. If the incident angle is α, the evanescent wave in the air gap is able to excite surface modes of the sample at wave vectors.

$$q_{||} = k_{||} = k_0 n \sin \alpha, \tag{7.34}$$

where n is the prism refractive index, and $k_0 = \omega/c$ [7.81, 82]. In the OTTO configuration it is therefore possible to measure the dispersion of surface polaritons simply by changing the incident angle α and observing the frequency of the minimum in the reflected light for each α. Note that in the conventional ATR method the light is multiply reflected within the internal reflection element, which is in direct contact with the sample [7.83], and no surface modes can be observed.

The first dispersion of surface polaritons has been measured by *Marschall* and *Fischer* in GaP [7.84](Fig. 7.24). The minima of the modified ATR spectra correspond to losses of internally reflected energy. These losses are located in the frequency range between ω_{TO} and ω_s, where ω_s is the frequency of the Fuchs-Kliever mode as defined by (7.27). Figure 7.25 shows the dispersion of surface polaritons of GaP obtained on the basis of these measurements [7.84]. The theoretical polariton dispersion curve has been calculated on the basis of (7.26) and using (2.33) for $\varepsilon(\omega)$; the necessary

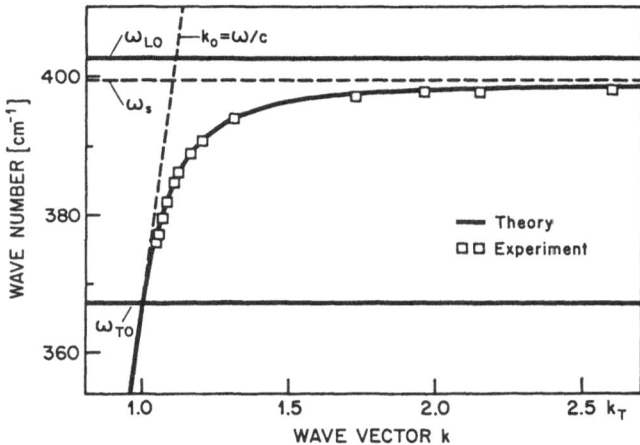

Fig. 7.25. Dispersion of surface polaritons in GaP. Experimental accuracy is given by size of rectangles. Theoretical curve is calculated from (7.26) [7.84]

parameters for GaP are; $\omega_{TO} = 367.3\,\mathrm{cm}^{-1}$, $\omega_{LO} = 403.0\,\mathrm{cm}^{-1}$, $\varepsilon_\infty = 9.091$, $\varepsilon_0 = 10.944$, $\gamma = 0.0035\omega_{TO}$. The agreement between theory and experiment is excellent.

For the study of surface polaritons of other crystals as well as for theoretical considerations the reader is referred to the literature [7.82–88].

7.4.5 Infrared Reflection Absorption Spectroscopy

The vibrational properties of adsorbed molecules on metal surfaces as well as of natural oxide films on metals can be studied by using a special technique, known as infrared reflection absorption spectroscopy (IRAS). In favourable cases the method allows the study of fractions of monolayers.

A schematic representation of the method is illustrated in Fig. 7.26. The infrared beam from the source is multiply reflected between two metal plates which are covered with the dielectric film to be investigated.

As has been discussed in Sect. 2.2.2, only light which is polarized parallel to the plane of incidence (*p*-polarized light) interacts with the thin film and excites LO modes. By definition TO modes are polarized parallel to the film/metal interface and cannot be observed in sufficiently thin films. For

Fig. 7.26. Schematic representation of the IRAS technique *(see text)*

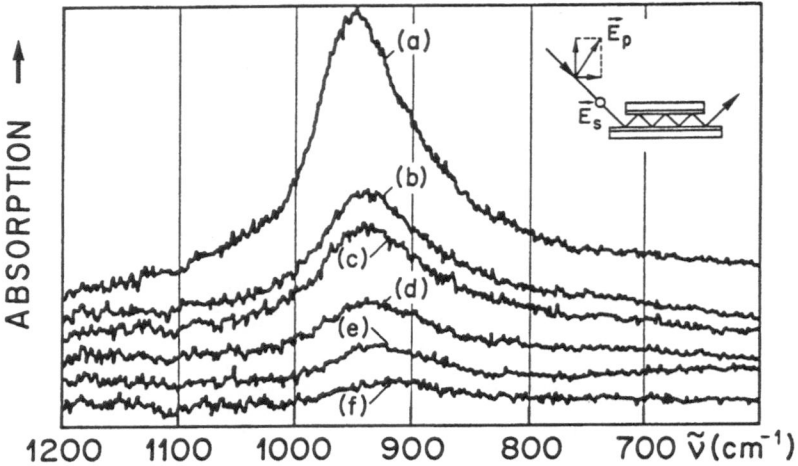

Fig. 7.27. IRAS spectra of Al$_2$O$_3$ films of various thicknesses d sputtered on Au. *(a)* $d = 50$ Å, *(b)* $d = 25$ Å, *(c)* $d = 20$ Å, *(d)* $d = 15$ Å, *(e)* $d = 10$ Å, *(f)* $d = 5$ Å. Note the decrease in the LO frequency with decreasing thickness [7.28]

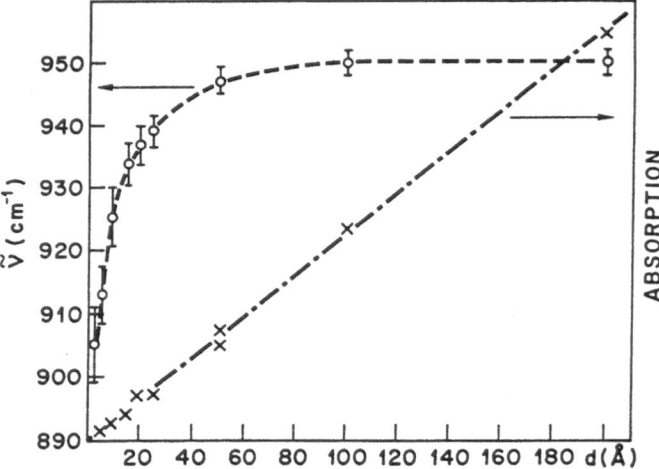

Fig. 7.28. Thickness dependence of IRAS absorption of Al$_2$O$_3$ films on Au (x), and of peak frequency of the LO mode (0); (- - -) model calculation [7.28]

theoretical and experimental considerations of the IRAS method the reader is referred to the literatur [7.89–92].

  Brüesch et al. have studied the IRAS spectra of Al$_2$O$_3$ films sputtered on gold as well as natural Al$_2$O$_3$ oxide films on aluminum [7.28]. Figure 7.27 shows the results for the system Al$_2$O$_3$/Au with film thicknesses ranging from 5 to 50 Å. Larger thicknesses up to 200 Å (not shown in Fig. 7.27) have also been measured. The peak frequency at 950 cm^{-1} observed at films with thicknesses larger than 100 Å corresponds to the bulk LO frequency of the Al-O stretching vibration. The large line widths observed in the spectra of Fig. 7.27 are due to the amorphous nature of the films which gives rise to a distribution in vibrational frequencies and oscillator strengths [7.28].

Figure 7.28 shows the thickness dependence of both the amplitudes of the absorption peaks as well as of the peak frequencies in the range between 3 and 200 Å. The absorption is linear in the whole thickness range. On the other hand, the peak frequency shows a pronounced thickness dependence: The frequency of a 3 Å thick film, corresponding to about one monolayer of Al_2O_3, is 905±6 cm^{-1}; with increasing thickness the frequency increases rapidly and saturates at about 100 Å to the bulk frequency of 950±2 cm^{-1}. A similar thickness dependence of the LO frequency has been observed for amorphous Al_2O_3 films on aluminium [7.28]. This striking thickness dependence reflects the transition from a three-dimensional (bulk) to a two-dimensional (monolayer) system and can be explained by the use of a simple model which is based on the fact that the restoring forces acting on the atoms at the surface are smaller than those acting on the atoms in the bulk [7.28].

Appendices

A. Michelson Interferometer

Referring to Fig. 2.2 we denote by l the optical path between the source 1 and the point P of the beam which is reflected at the fixed mirror 5, and E_1 the electric field of this beam at P. Similarly, $l + x$ and E_2 are the optical path and the electric field of the beam reflected at the moving mirror 6. For a monochromatic source of frequency $\omega = 2\pi c\tilde{\nu}$ we have

$$E_1(\tilde{\nu}) = A_I(\tilde{\nu}) A_S(\tilde{\nu}) e^{2\pi i c\tilde{\nu}t} e^{ikl}. \tag{A.1}$$

$$E_2(\tilde{\nu}) = A_I(\tilde{\nu}) A_S(\tilde{\nu}) e^{2\pi i c\tilde{\nu}t} e^{ik(l+x)}. \tag{A.2}$$

$A_I(\tilde{\nu})$ depends on instrumental conditions of the interferometer (source efficiency, losses by transmission of beam splitter, lenses, filters, windows and by reflection at mirrors, etc.), and $A_S(\tilde{\nu})$ depends on the optical properties of the sample. For a reflectivity experiment $A_S(\tilde{\nu}) = \tilde{r}_s(\tilde{\nu})$ is the complex amplitude of the beam reflected at the sample, for a transmission experiment $A_S(\tilde{\nu}) = \tilde{t}_s(\tilde{\nu})$ is the complex amplitude of the beam transmitted by the sample. Using $k = 2\pi/\lambda = 2\pi\tilde{\nu}$ and the expressions (A. 1, 2) the intensity at the point P is

$$I(x, \tilde{\nu}) = \frac{1}{2}|E_1 + E_2|^2 = S(\tilde{\nu})(1 + \cos 2\pi\tilde{\nu}x), \tag{A.3}$$

where $S(\tilde{\nu}) = |A_I(\tilde{\nu})|^2 \, |A_S(\tilde{\nu})|^2$ is the spectral intensity at wavenumber $\tilde{\nu}$. For a polychromatic source emitting light between $\tilde{\nu} = 0$ and $\tilde{\nu} = \infty$ we obtain for the interferogram

$$I(x) = \int_0^\infty S(\tilde{\nu})(1 + \cos 2\pi\tilde{\nu}x)d\tilde{\nu}$$

$$= \frac{1}{2}I(0) + \int_0^\infty S(\tilde{\nu}) \cos (2\pi\tilde{\nu}x)d\tilde{\nu}, \tag{A.4}$$

where $I(0)$ is the intensity at zero path difference. Applying the classical form of Fourier's integral, we obtain for the desired spectrum

$$S(\tilde{\nu}) = 4 \int_0^\infty [I(x) - \tfrac{1}{2}I(0)] \cos (2\pi\tilde{\nu}x)dx. \tag{A.5}$$

In practice, $S(\tilde{\nu})$ is approximated by the sum (2.3). Let G and $D(\tilde{\nu})$ be the gain of the amplifier and the detector sensitivity, respectively. Then the actual computed transmission spectrum of the sample is

$$P_S(\tilde{\nu}) = G_S D(\tilde{\nu}) |A_I(\tilde{\nu})|^2 T_S(\tilde{\nu}), \tag{A.6}$$

where $T_S(\tilde{\nu}) = |A_S(\tilde{\nu})|^2 = |\tilde{t}_s(\tilde{\nu})|^2$ is the transmission of the sample. The spectrum of the background is

$$P_B(\tilde{\nu}) = G_B D(\tilde{\nu}) |A_I(\tilde{\nu})|^2, \tag{A.7}$$

since $T_B(\tilde{\nu}) = 1$ for a hole at the place of the sample. From (A.6, 7) it follows that

$$T_S(\tilde{\nu}) = G_B P_S(\tilde{\nu}) / G_S P_B(\tilde{\nu}). \tag{A.8}$$

For a reflectivity experiment $|A_S(\tilde{\nu})|^2 = |\tilde{r}_s(\tilde{\nu})|^2 = R_S(\tilde{\nu})$ is the reflectivity of the sample and $|A_B(\tilde{\nu})|^2 = R_B(\tilde{\nu}) = 1$ for in ideal mirror at the place of the sample, and we obtain

$$R_S(\tilde{\nu}) = G_B P_S(\tilde{\nu}) / G_S P_B(\tilde{\nu}). \tag{A.9}$$

B. Classical Optics

B.1 Interaction of Electromagnetic Waves with Lattice Vibrations

Maxwell's equation for a neutral, homogeneous, isotropic and non-magnetic medium are

$$\nabla \times E = -\frac{1}{c}\dot{H}, \tag{B.1}$$

$$\nabla \times H = \frac{1}{c}\dot{D}, \tag{B.2}$$

$$\nabla \cdot D = 0, \tag{B.3}$$

$$\nabla \cdot H = 0. \tag{B.4}$$

The materials equations are

$$D = \varepsilon \cdot E = E + 4\pi P, \tag{B.5}$$

$$P = \frac{1}{4\pi}(\varepsilon - 1)E. \tag{B.6}$$

We consider solutions in terms of plane waves

$$A = A_0 \exp[-i(\omega t - k \cdot r)], \tag{B.7}$$

where A is one of the variables E, H, D, P. We then have

$$\dot{A} = -i\omega A, \tag{B.8}$$

$$\ddot{A} = -\omega^2 A, \tag{B.9}$$

$$\nabla \cdot A = i k \cdot A, \tag{B.10}$$

$$\nabla \times A = i k \times A. \tag{B.11}$$

Maxwell's equation then become

$$k \times E = \frac{\omega}{c} H, \tag{B.12}$$

$$k \times H = -\frac{\omega}{c} D, \tag{B.13}$$

$$k \cdot D = 0, \tag{B.14}$$

$$k \cdot H = 0. \tag{B.15}$$

According to Problem (2.4.1) the polarization for a diatomic cubic crystal is given by

$$P = \frac{1}{v_a} e_T^* w + \frac{1}{4\pi} (\varepsilon_\infty - 1) E, \tag{B.16}$$

where the relative displacement w is given by

$$w = \frac{(e_T^*/\mu) E}{\omega_{TO}^2 - \omega^2 - i\gamma\omega}, \tag{B.17}$$

and the dielectric constant can be written in the form

$$\varepsilon(\omega) = \varepsilon_\infty + \frac{4\pi e_T^{*2}/\mu v_a}{\omega_{TO}^2 - \omega^2 - i\gamma\omega}$$

$$= \varepsilon_\infty \frac{\omega_{LO}^2 - \omega^2 - i\gamma\omega}{\omega_{TO}^2 - \omega^2 - i\gamma\omega}. \tag{B.18}$$

We now consider (B.14), namely

$$k \cdot D = \varepsilon(\omega) k \cdot E = 0.$$

Since $E \neq 0$ this equation has two solutions, one for which $\varepsilon(\omega) = 0$ and the other for which $k \cdot E = 0$.

First Case: $\varepsilon(\omega) = 0$; then $D = 0$ from (B.5) and $k \times H = 0$ from (B.13). This requires that H either vanishes or is parallel to k; on the other hand, (B.15) requires that H either vanishes or is perpendicular to k. From this it follows that $H = 0$. Therefore $k \times E = 0$ according to (B.12). Thus E is parallel to k : $E \| k$. Using (B.6, 17) we see that $w \| E \| P \| k$, which corresponds to a *longitudinal* solution whose frequency is the solution of $\varepsilon(\omega) = 0$. From (B.18) we obtain

$$\omega^2 + i\gamma\omega - \omega_{LO}^2 = 0, \tag{B.19}$$

which yields the complex frequency

$$\omega = \left(\omega_{LO}^2 - \frac{1}{4}\gamma^2\right)^{1/2} - \frac{1}{2}i\gamma. \tag{B.20}$$

This solution is *independent* on k; its real part represents the longitudinal frequency renormalized by the damping. For $\gamma = 0$ we obtain $\omega = \omega_{LO}$, the frequency of the LO mode.

Second Case: $k \cdot E = 0$; this implies that k is perpendicular to E : $k \perp E$. The wave is therefore *transverse*. Let $k = (0, 0, k)$ and $E = (E, 0, 0)$. According to (B.12) it follows that $H = (c/\omega)kEe_y$, where e_y is a unit vector in the direction y. On the other hand (B.13) gives $D = (c/\omega)kHe_x$. Thus we can write

$$H = \frac{c}{\omega}kE, \quad \text{and} \tag{B.21}$$

$$D = \varepsilon E = \frac{c}{\omega}kH. \tag{B.22}$$

Combining these two equations we obtain the result that for transverse modes the dispersion is given by

$$k^2 = \frac{\omega^2}{c^2}\varepsilon(\omega). \tag{B.23}$$

Defining the complex index of refraction by

$$n^2 = \varepsilon, \tag{B.24}$$

we can write (B.23) in the form

$$k = \frac{\omega}{c}n. \tag{B.25}$$

Setting

$$n = n_1 + in_2, \quad \text{and} \tag{B.26}$$

$$\varepsilon = \varepsilon_1 + i\varepsilon_2, \tag{B.27}$$

we obtain from (B.24)

$$\varepsilon_1 = n_1^2 - n_2^2, \tag{B.28a}$$

$$\varepsilon_2 = 2n_1n_2. \tag{B.28b}$$

Substituting (B.18) into (B.23) yields

$$\frac{c^2}{\omega^2}k^2 = \varepsilon_\infty \frac{\omega_{LO}^2 - \omega^2 - i\gamma\omega}{\omega_{TO}^2 - \omega^2 - i\gamma\omega}. \tag{B.29}$$

This relation defines the dispersion $\omega^2(k)$ for the transverse modes. Writing $k = k_1 + ik_2$ for the complex wave vector we obtain from (B.25, 28, 29)

$$k_1 = \frac{\omega}{c}n_1 = \frac{1}{\sqrt{2}}\frac{\omega}{c}[(|\varepsilon| + \varepsilon_1)]^{1/2}, \tag{B.30a}$$

$$k_2 = \frac{\omega}{c}n_2 = \frac{1}{\sqrt{2}}\frac{\omega}{c}[(|\varepsilon| - \varepsilon_1)]^{1/2}, \quad \text{where} \tag{B.30b}$$

$$|\varepsilon| = (\varepsilon_1^2 + \varepsilon_2^2)^{1/2}.$$

Note that the complex wave vector $k = k_1 + ik_2$, if substituted in (B.7), gives rise to damped waves of the form

$$A = A_0\exp[-i(\omega t - k_1 \cdot r)]\exp(-k_2 \cdot r). \tag{B.31}$$

B.2 Reflection at a Surface of an Infinitely Thick Medium

We consider a plane wave propagating along the z-axis and assume that $E = (E_x, 0, 0)$. E_x is now a function of z and t: $E_x = E_x(z, t)$; the general solution is a superposition of two waves, one propagating along $+z$ and one along $-z$:

$$E_x(z, t) = (E_+e^{ikz} + E_-e^{-ikz})e^{i\omega t}. \tag{B.32}$$

From (B.1) we obtain

$$\frac{\partial H_y}{\partial t} = -c\frac{\partial E_x}{\partial z} = -ick(E_+e^{ikz} - E_-e^{-ikz})e^{-i\omega t},$$

and integration yields together with (B.25)

$$H_y(z, t) = n(E_+e^{ikz} - E_-e^{-ikz})e^{-i\omega t}. \tag{B.33}$$

We now suppose that there are two different homogeneous media a and b having the plane $z = 0$ as an interface. Medium a occupies all space with $z < 0$, medium b occupies all space with $z > 0$. A plane wave is generated by a source at $z = -\infty$. This gives an incident wave travelling in the $+z$ direction in medium a. The discontinuity generates a reflected wave in medium a

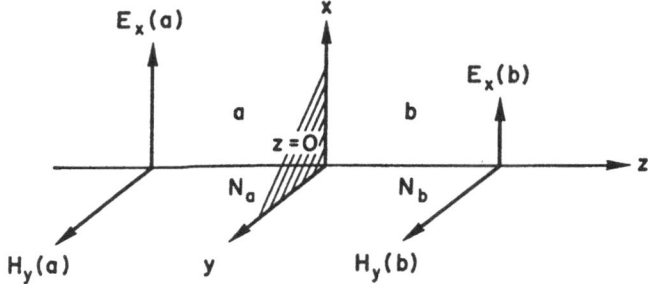

Fig. B.1. Propagation of a plane electromagnetic wave through the two media a and b separated by the interface at $z = 0$

and a transmitted wave in medium b. For simplicity, we consider normal incidence (Fig. B.1). Let $\tilde{r}$ and $\tilde{t}$ be the complex amplitudes of the reflected and transmitted E_x. Then we have $E_- = \tilde{r}E_+$ in (B.32, 33) and therefore

$$E_x(a) = E_+(e^{ik_a z} + \tilde{r}e^{-ik_a z})e^{-i\omega t}, \tag{B.34}$$

$$E_x(b) = \tilde{t}E_+e^{ik_b z}e^{-i\omega t}. \tag{B.35}$$

From (B.33) we obtain similarly

$$H_y(a) = n_a E_+(e^{ik_a z} - \tilde{r}e^{-ik_a z})e^{-i\omega t}, \tag{B.36}$$

$$H_y(b) = n_b \tilde{t} E_+ e^{ik_b z}e^{-i\omega t}. \tag{B.37}$$

Now using the condition that E_x and H_y are continuous at $z = 0$, that is

$$E_x(a) = E_x(b) \text{ and } H_y(a) = H_y(b), \tag{B.38}$$

which can be prooved by applying *Stoke*'s theorem [2.13] we obtain from (B.34, 35) at $z = 0$

$$1 + \tilde{r} = \tilde{t}, \tag{B.39}$$

and from (B.36, 37)

$$n_a(1 - \tilde{r}) = n_b \tilde{t}. \tag{B.40}$$

Combining these two equations yields

$$\tilde{r} = \frac{n_a - n_b}{n_a + n_b} = |\tilde{r}|e^{i\phi}. \tag{B.41}$$

The amplitude of the reflected wave is $\tilde{r}$ times that of the incident wave. ϕ is the phase change of the reflected wave as caused by the reflection. If I_0 is the intensity of the incident light and I_R the intensity of the reflected light, we have

$$I_0 = \tfrac{1}{2}|E_0|^2, \quad I_R = \tfrac{1}{2}|E_R|^2, \tag{B.42}$$

and the reflectivity at the infinitely thick medium is defined as

$$R = I_R/I_0. \tag{B.43}$$

Since according to (B.34) $E_R = \tilde{r}E_0$ we find

$$R = \tilde{r}\tilde{r}^* = \left|\frac{n_a - n_b}{n_a + n_b}\right|^2. \tag{B.44}$$

For the special case where medium a is vacuum with $n_a = 1$ and putting $n_b = n = n_1 + in_2$ we find

$$R = \left|\frac{1-n}{1+n}\right|^2 = \frac{(1-n_1)^2 + n_2^2}{(1+n_1)^2 + n_2^2}, \quad \text{and} \tag{B.45}$$

$$\tan\phi = -\frac{2n_2}{1 - n_1^2 - n_2^2}. \tag{B.46}$$

B.3 Transmission and Reflection for a Plate of Parallel Faces with Thickness d

Because of its practical importance we summarize here the formulas for the transmission and reflection of light by an absorbing plate of parallel faces with thickness d (Fig. 2.4). These formulas are derived in the book of *Stratton* [2.12]. The transmission at normal incidence is given by

$$\tilde{T} = \frac{I_T}{I_0} = \frac{[(1-R)^2 + 4R \sin^2 \phi]T}{(1-RT)^2 + 4RT \sin^2 (\phi + \alpha)}, \tag{B.47}$$

$$\tilde{R} = \frac{I_R}{I_0} = R\frac{(1-T)^2 + 4T \sin^2 \alpha}{(1-RT)^2 + 4RT \sin^2 (\phi + \alpha)}. \tag{B.48}$$

Here R and ϕ are defined by (B.45, 46), respectively, while $T = \exp(-Kd)$. $K = 2n_2\omega/c$ is the absorption coefficient and $\alpha = 2\pi n_1 d/\lambda_0$, where λ_0 is the wavelength in vacuum. The second term in the denominator of (B.47, 48) takes account of interference effects within the sample. Note that if $d \to \infty$ and $K>0$ then $\tilde{R} = R$ and $\tilde{T} = 0$.

B.4 Determination of Optical Constants from Reflectivity Measurements Using Polarized Light at Non-Normal Incidence

We consider a thick anisotropic crystal belonging to the tetragonal, hexagonal, trigonal or orthorhombic system. The crystal symmetry axis are denoted by 1,2,3; (2, 3) is always the plane of incidence (Fig. B.2) [2.35]. $R_\perp$

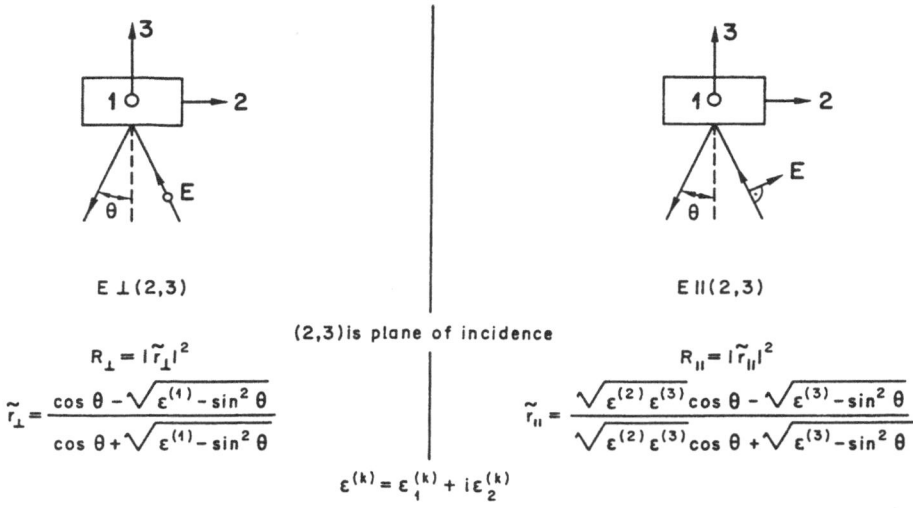

E ⊥ (2,3) | (2,3) is plane of incidence | E ∥ (2,3)

$$R_\perp = |\tilde{r}_\perp|^2$$

$$\tilde{r}_\perp = \frac{\cos\theta - \sqrt{\varepsilon^{(1)} - \sin^2\theta}}{\cos\theta + \sqrt{\varepsilon^{(1)} - \sin^2\theta}}$$

$$R_\| = |\tilde{r}_\||^2$$

$$\tilde{r}_\| = \frac{\sqrt{\varepsilon^{(2)}\varepsilon^{(3)}}\cos\theta - \sqrt{\varepsilon^{(3)} - \sin^2\theta}}{\sqrt{\varepsilon^{(2)}\varepsilon^{(3)}}\cos\theta + \sqrt{\varepsilon^{(3)} - \sin^2\theta}}$$

$$\varepsilon^{(k)} = \varepsilon_1^{(k)} + i\varepsilon_2^{(k)}$$

Fig. B.2. Reflectivity of light at non-normal incidence by an anisotropic crystal

and $R_\|$ is the reflectivity for light polarized perpendicular and parallel, respectively, to the plane of incidence. $\varepsilon^{(k)}$ is the complex dielectric constant with respect to the axis k. If $R_\perp$ and $R_\|$ are measured at various angles of incidence θ, it is possible to determine $\varepsilon^{(k)}$ on the basis of the Fresnel relations for $r_\perp$ and $r_\|$ [2.36] as given in Fig. B.2. The difficulty of obtaining absolute values of the reflectivity, particularly at large angles of incidence, when using a conventional spectrometer, may, in principle, be overcome by using infrared or far infrared lasers [2.27]. In the visible region the method has been applied to graphite [2.34] and to the one-dimensional metal KCP [2.35].

C. Second-Order Dipole Moment

We proove (2.120). Substituting (2.105) into (2.199) we obtain with $B\binom{q}{j} = [\omega_j(q)]^{-1/2}A\binom{q}{j}$

$$M^{(2)} = \frac{\hbar}{2N}\sum_{qj}\sum_{q'j'}B\binom{q}{j}B\binom{q'}{j'}\sum_{l\kappa\kappa'}\frac{e(\kappa|\substack{q\\j})e(\substack{ll'\\\kappa\kappa'})e(\kappa'|\substack{q'\\j'})}{(m_\kappa m_{\kappa'})^{1/2}}$$
$$\times \sum_l \exp\{i(k+q+q')\cdot r\binom{l}{\kappa}\}\exp\{-iq'\cdot[r\binom{l}{\kappa}-r\binom{l'}{\kappa'}]\}. \quad (C.1)$$

Using

$$r\binom{l}{\kappa} - r\binom{l'}{\kappa'} = r(l - l') + r(\kappa) - r(\kappa') = r(L) + r(\kappa) - (\kappa'), \quad (C.2)$$

we have

$$M^{(2)} = \frac{\hbar}{2N} \sum_{qj} \sum_{q'j'} B(_j^q) B(_{j'}^{q'}) \sum_{L\kappa\kappa'} \frac{e(\kappa|_j^q) e(_{\kappa\kappa'}^L) e(\kappa'|_{j'}^{q'})}{(m_\kappa m_{\kappa'})^{1/2}}$$

$$\times \exp\{-iq'\cdot[r(L) + r(\kappa) - r(\kappa')]\} \exp[i(k + q + q')\cdot r(\kappa)]$$

$$\times \sum_l \exp[i(k + q + q')\cdot r(l)]. \tag{C.3}$$

The last sum is equal to $N\Delta(k+q+q')$, where $\Delta(k+q+q') = 1$ if $k+q+q' = \tau$ is a vector of the reciprocal lattice. Since $k = 0$ and restricting ourselves to the first Brillouin zone, that is, setting $\tau = 0$, we have $q' = -q$ and obtain

$$M^{(2)} = \frac{\hbar}{2} \sum_{qjj'} B(_j^q) B(_{j'}^{-q}) \sum_{L\kappa\kappa'} \exp[iq\cdot r(_{\kappa\kappa'}^L)] \frac{q(\kappa|_j^q) e(_{\kappa\kappa'}^L) e(\kappa'|_{j'}^{-q})}{(m_\kappa m_{\kappa'})^{1/2}}. \tag{C.4}$$

This expression can be written in the form

$$M^{(2)} = \frac{\hbar}{2} \sum_{qjj'} \frac{S(_{jj'}^q)}{[\omega_j(q)\omega_{j'}(q)]^{1/2}} A(_j^q) A(_{j'}^{-q}), \tag{C.5}$$

where $S(_{jj'}^q)$ is given by (2.121–123) and $A(_j^q)$ by (2.124).

D. $\varepsilon_2^{(2)}(\omega)$ for Second-Order Dipole Moment

$\varepsilon_2^{(2)}(\omega)$ can be transformed into a form which is more convenient for a discussion of its frequency dependence. The summation over q can be transformed into an integral over q according to the rule [Ref. 2.17, Sect. 3.5.1]

$$\sum_q \ldots = \frac{V}{(2\pi)^3} \int d^3q \ldots ,$$

where V is the volume of the crystal. Introducing the notation $\Omega_{jj'} = \omega_{j'} \pm \omega_j$ and obvious abbreviations we can write (2.127) in the form

$$\varepsilon_2^{(2)}(\omega) = F(\omega) + G(\omega)$$

$$= \frac{\pi^2\hbar}{(2\pi)^3} \sum_{jj'} \int d^3q [F_{jj'}(q)\delta(\omega - \Omega_{jj'}) + G_{jj'}(q)\delta(\omega - \Omega_{jj'})]. \tag{D.1}$$

Applying the results of [Ref. 2.17, Sect. 3.5.1] we have

$$d^3q = \frac{1}{|\nabla\Omega_{jj'}(q)|} dS_{\Omega_{jj'}} d\Omega_{jj'}, \tag{D.2}$$

where $dS_{\Omega_{jj'}}$ is the surface element in q space. Substituting (D.2) into (D.1), the first term of (D.1) can be written in the form

$$F(\omega) = \frac{\pi^2 \hbar}{(2\pi)^3} \sum_{jj'} \int \left(\int dS_{\Omega_{jj'}} \frac{F_{jj'}(q)}{|\nabla \Omega_{jj'}(q)|} \right) \delta(\Omega_{jj'} - \omega) d\Omega_{jj'}. \qquad (D.3)$$

Using the notation

$$f_{jj'}(\Omega_{jj'}) = \int dS_{\Omega_{jj'}} \frac{F_{jj'}(q)}{|\nabla \Omega_{jj'}(q)|}, \qquad (D.4)$$

and the relation

$$\int f_{jj'}(\Omega_{jj'}) \delta(\Omega_{jj'} - \omega) d\Omega_{jj'} = f_{jj'}(\omega)$$

$$= \int\limits_{S_\omega} dS_\omega \frac{F_{jj'}(q)}{|\nabla \Omega_{jj'}(q)|_{\Omega_{jj'}=\omega}}, \qquad (D.5)$$

where the integration extends over the surface in q space for which $\Omega_{jj'} = \omega$, we obtain

$$F(\omega) = \frac{\pi^2 \hbar}{(2\pi)^3} \sum_{jj'} dS_\omega \frac{F_{jj'}(q)}{|\nabla \Omega_{jj'}(q)|_{\Omega_{jj'}=\omega}}. \qquad (D.6)$$

A similar representation is possible for $G(\omega)$. Using (2.127), $\varepsilon_2^{(2)}(\omega)$ can therefore be written in the form

$$\varepsilon_2^{(2)}(\omega) = \frac{\hbar}{8\pi} \sum_{jj'} \int\limits_{S_\omega} |C(\substack{q\\jj'})|^2 \frac{(\bar{n}_j + \frac{1}{2}) \pm (\bar{n}_{j'} + \frac{1}{2})}{|\nabla \Omega_{jj'}(q)|_{\Omega_{jj'}=\omega}} dS_\omega. \qquad (D.7)$$

E. Power Spectrum

Using (3.4, 13) one obtains

$$P(\omega) = 2A \lim_{\tau \to \infty} \frac{1}{\tau} |J(\tau)|^2, \quad \text{where} \qquad (E.1)$$

$$J(\tau) = J_0(\tau) + J_1(\tau) + J_2(\tau), \qquad (E.2)$$

$$J_0(\tau) = k_0 \left(\frac{\sin\left[\frac{1}{2}(\omega_L - \omega)\tau\right]}{\omega_L - \omega} + \frac{\sin\left[\frac{1}{2}(\omega_L + \omega)\tau\right]}{\omega_L + \omega} \right) \qquad (E.3)$$

$$J_1(\tau) = k_1 \left(\frac{\sin\left[\frac{1}{2}(\omega_L - \omega_s - \omega)\tau\right]}{\omega_L - \omega_s - \omega} + \frac{\sin\left[\frac{1}{2}(\omega_L - \omega_s + \omega)\tau\right]}{\omega_L - \omega_s + \omega} \right) \qquad (E.4)$$

$$J_2(\tau) = k_2 \left(\frac{\sin\left[\frac{1}{2}(\omega_L + \omega_s - \omega)\tau\right]}{\omega_L + \omega_s - \omega} + \frac{\sin\left[\frac{1}{2}(\omega_L + \omega_s + \omega)\tau\right]}{\omega_L + \omega_s + \omega} \right). \quad \text{(E.5)}$$

Defining

$$P_n(\omega) = 2A \lim_{\tau \to \infty} \frac{1}{\tau} \{J_n^2(\tau)\} \quad \text{(E.6)}$$

one finds

$$P_0(\omega) = 2Ak_0^2 \lim_{\tau \to \infty} \left(\frac{\sin^2\left[\frac{1}{2}(\omega_L - \omega)\tau\right]}{\tau(\omega_L - \omega)^2} + \frac{\sin^2\left[\frac{1}{2}(\omega_L + \omega)\tau\right]}{\tau(\omega_L + \omega)} \right.$$

$$\left. + 2\frac{\sin\left[\frac{1}{2}(\omega_L - \omega)\tau\right] \sin\left[\frac{1}{2}(\omega_L + \omega)\tau\right]}{\tau(\omega_L^2 - \omega^2)} \right). \quad \text{(E.7)}$$

The third term of (E.7) is zero for $\tau \to \infty$. Using

$$\sin^2 \alpha = \frac{1}{2}(1 - \cos 2\alpha)$$

and [3.28],

$$\delta(x \pm x_0) = \frac{1}{\pi} \lim_{s \to \infty} \frac{1 - \cos s(x \pm x_0)}{s(x \pm x_0)^2},$$

one obtains

$$P_0(\omega) = \pi Ak_0^2 [\delta(\omega - \omega_L) + \delta(\omega + \omega_L].$$

Since ω and ω_L are positive, the second term vanishes with the result

$$P_0(\omega) = \pi Ak_0^2 (\omega - \omega_L). \quad \text{(E.8)}$$

The analogous derivation gives

$$P_{1,2}(\omega) = \pi Ak_{1,2}^2 \delta(\omega - \omega_L \pm \omega_s). \quad \text{(E.9)}$$

The cross terms of the form

$$P_{nm}(\omega) = 2A \lim_{\tau \to \infty} \frac{1}{\tau}[J_n(\tau)J_m(\tau)]$$

all vanish, and the final result is

$$P(\omega) = \pi A[k_0^2 \delta(\omega - \omega_L) + k_1^2 \delta(\omega - \omega_L + \omega_s)$$

$$+ k_2^2 \delta(\omega - \omega_L - \omega_s)]. \quad \text{(E.10)}$$

220

F. Placzek's Approximation

F.1 Perturbed Wavefunction

Substitution of (3.51) into (3.50), neglecting second-order terms of the form $(E \cdot M)\Psi_k^{(1)}$ and using (3.44) gives

$$\left(H_0 - i\hbar \frac{\partial}{\partial t}\right)\Psi_k^{(1)} = (E \cdot M)\Psi_k^{(0)}. \tag{F.1}$$

For the solution of (F.1) we make the Ansatz

$$\Psi_k^{(1)} = \Xi_k^+ \exp[-i(\omega_k + \omega_L)t] + \Xi_k^- \exp[-i(\omega_k - \omega_L)t]. \tag{F.2}$$

Substituting (F.2) into (F.1) and using (3.47, 48) we obtain by comparing the coefficients of equal time dependence

$$H_0 \Xi_k^+ - \hbar(\omega_k + \omega_L)\Xi_k^+ = (A \cdot M)\Xi_k, \tag{F.3a}$$

$$H_0 \Xi_k^- - \hbar(\omega_k - \omega_L)\Xi_k^- = (A^* \cdot M)\Xi_k. \tag{F.3b}$$

We develop $(A \cdot M)\Xi_k$ and $(A^* \cdot M)\Xi_k$ in terms of the unperturbed eigenfunctions Ξ_r :

$$(A \cdot M)\Xi_k = \sum_r (A \cdot M_{kr})\Xi_r, \tag{F.4a}$$

$$(A^* \cdot M)\Xi_k = \sum_r (A^* \cdot M_{kr})\Xi_r, \quad \text{where} \tag{F.4b}$$

$$M_{kr} = M_{rk}^* = \int \Xi_r^* M \Xi_k d\tau. \tag{F.5}$$

Substitution of (F.4) into (F.3) gives

$$\Xi_k^+ = \frac{1}{\hbar} \sum_r \frac{(A \cdot M_{kr})}{\omega_{rk} - \omega_L} \Xi_r, \tag{F.6a}$$

$$\Xi_k^- = \frac{1}{\hbar} \sum_r \frac{(A^* \cdot M_{kr})}{\omega_{rk} + \omega_L} \Xi_r, \quad \text{where} \tag{F.6b}$$

$$\omega_{rk} = \omega_r - \omega_k. \tag{F.7}$$

That $\Xi_k^{\pm}$ given by (F.6) are indeed solutions of (F.3) can be verified by direct substitution and using (3.46). This proves (3.52, 53).

F.2 Placzek's Approximation

The unperturbed system is described by (3.46), namely

$$H_0 \Xi_r(q) = E_r \Xi_r(q). \tag{F.8a}$$

Here $r = (e, n)$ denotes collectively the electronic quantum numbers e and the nuclear (vibrational) quantum numbers n, while $q = (r, R)$ specifies the electronic coordinates r and the nuclear coordinates R. Written out explicitly (F.8a) reads

$$H_0 \Xi_{en}(r, R) = E_{en} \Xi_{en}(r, R). \tag{F.8b}$$

In the adiabatic approximation [Ref. 3.21, Sect. 1.3] we imagine the nuclei to be fixed in some configuration R for which we construct the Schroedinger equation for the electrons

$$[T_{el} + \Phi_{el,el}(r) + \Phi_{i,el}(r, R)] \chi_e(r, R) = E_e(R) \chi_e(r, R). \tag{F.9}$$

Substituting the Ansatz

$$\Xi_{en}(r, R) = \chi_e(r, R) \psi_{en}(R), \tag{F.10}$$

into (F.8b) and making use of (F.9) one obtains in the adiabatic approximation

$$[T_i + \Phi_{ii}(R) + E_e(R)] \psi_{en}(R) = E_{en}(R) \psi_{en}(R). \tag{F.11}$$

In (F.10) $\chi_e(r, R)$ is the electronic wavefunction of state e for fixed nuclear configuration R, and $\psi_{en}(R)$ is the vibrational wavefunction of the electronic state e and vibrational state n. For E_{en} we write

$$E_{en} = E_e(0) + W_{en}, \tag{F.12a}$$

where $E_e(0) = E_e(R = R_0)$ is the eigenvalue of (F.9) if all nuclei are fixed in the equilibrium configuration $R = R_0$. For a system with fixed nuclear configuration R we can easily write down the eigenfunction of the electronic ground state $e = 0$, perturbed by the incident light; since in this case $\Xi_r = \chi_e(r, R)$ we obtain from (3.47, 51–53)

$$\begin{aligned}
\tilde{\Psi}_0 = \Psi_0^{(0)} + \Psi_0^{(1)} &= \chi_0(r, R) \exp[-i\omega_0(R)t] \\
&\quad + \tilde{\Xi}_0^+ \exp\{-i[\omega_0(R) + \omega_L]t\} \\
&\quad + \tilde{\Xi}_0^- \exp\{-i[\omega_0(R) - \omega_L]t\},
\end{aligned} \tag{F.12b}$$

where

$$\tilde{\Xi}_0^+(r, R) = \frac{1}{\hbar} \sum_e {}' \frac{(A \cdot M_{oe})}{\omega_{eo} - \omega_L} \chi_e(r, R), \tag{F.13a}$$

$$\tilde{\Xi}_0^-(r, R) = \frac{1}{\hbar} \sum_e {}' \frac{(A^* \cdot M_{oe})}{\omega_{eo} + \omega_L} \chi_e(r, R). \tag{F.13b}$$

According to (3.54), the elements M_{oe} in (F.13) are given by

$$M_{oe}(R) = M_{eo}^*(R) = \int \chi_e^*(r, R) M \chi_0(r, R) dr. \qquad (F.14)$$

In the sums of (F.13) the terms with $e = 0$ are excluded. For a system with vibrating nuclei we obtain from (3.47, 51–53) with $k = (0, n)$

$$\Psi_{on} = \Psi_{on}^{(0)} + \Psi_{on}^{(1)} = \Xi_{on}(r, R)\exp(-i\omega_{on}t)$$

$$+ \Xi_{on}^+\exp[-i(\omega_{on} + \omega_L)t] + \Xi_{on}^-\exp[-i(\omega_{on} - \omega_L)t], \quad (F.15)$$

where

$$\Xi_{on}^+ = \frac{1}{\hbar} \sum_e \sum_{n'}' \frac{(\boldsymbol{A}\cdot\boldsymbol{M}_{on,en'})}{\omega_{en',on} - \omega_L} \chi_e(r, R)\psi_{en'}(R), \qquad (F.16a)$$

$$\Xi_{on}^- = \frac{1}{\hbar} \sum_e \sum_{n'}' \frac{(\boldsymbol{A}^*\cdot\boldsymbol{M}_{on,en'})}{\omega_{en',on} + \omega_L} \chi_e(r, R)\psi_{en'}(R). \qquad (F.16b)$$

Now $\omega_{en',on} = \omega_{eo} + \omega_{n'n}$; we now assume that $\omega_{eo} - \omega_L \gg \omega_{n'n}$. We can therefore replace $\omega_{en',on}$ by ω_{eo} in the denominators of (F.16). In addition we assume that $\omega_L \gg \omega_{n'n}$; we can then neglect $\omega_{on',on} = \omega_{n'n}$ in the terms with $e = 0$. Using (F.12a) in the form $\omega_{on} = \omega_0(0) + \hat{\omega}_{on}$ where $\hbar\hat{\omega}_{on} = W_{on}$ we obtain from (F.15, 16)

$$\Psi_{on}(r, R, t) \cong \exp[-i\omega_0(0)t]\{\chi_0(r, R)\psi_{on}(R)$$

$$+\Xi_{on}^+\exp[-i(\hat{\omega}_{on} + \omega_L)t] + \Xi_{on}^-\exp[-i(\hat{\omega}_{on} - \omega_L)t]\}, \qquad (F.17)$$

where

$$\Xi_{on}^+ = \frac{1}{\hbar} \sum_e' \sum_{n'} \frac{(\boldsymbol{A}\cdot\boldsymbol{M}_{on,en'})}{\omega_{eo} - \omega_L} \chi_e(r, R)\psi_{en'}(R)$$

$$= \frac{1}{\hbar} \sum_e' \left(\frac{\sum_{n'}(\boldsymbol{A}\cdot\boldsymbol{M}_{on,en'})\psi_{en'}}{\omega_{eo} - \omega_L} \right) \chi_e(r, R) \qquad (F.18a)$$

$$\Xi_{on}^- = \frac{1}{\hbar} \sum_e' \left(\frac{\sum_{n'}(\boldsymbol{A}^*\cdot\boldsymbol{M}_{on,en'})\psi_{en'}}{\omega_{eo} + \omega_L} \right) \chi_e(r, R). \qquad (F.18b)$$

By definition we obtain from (F.10)

$$\boldsymbol{A}\cdot\boldsymbol{M}_{on,en'} = \int \psi_{en'}^*(R)[\boldsymbol{A}\cdot\boldsymbol{M}_{oe}(R)]\psi_{on}(R)dR, \qquad (F.19)$$

where $M_{oe}(R)$ is given by (F.14).

Developing $[A \cdot M_{eo}]\psi_{on}(R)$ in terms of $\psi_{en'}(R)$:

$$[A \cdot M_{eo}(R)]\psi_{on}(R) = \sum_{n'} B_{on,en'}\psi_{en'}(R),$$

multiplying with $\psi^*_{en'}(R)$, integrating over R and using the orthogonallity of the functions $\psi_{en'}$ gives

$$B_{on,en'} = (A \cdot M_{on,en'}),$$

and we obtain

$$\sum_{n'}(A \cdot M_{on,en'})\psi_{en'}(R) = [A \cdot M_{oe}(R)]\psi_{on}(R). \tag{F.20}$$

Substituting (F.20) into (F.18) gives

$$\Xi^+_{on}(r, R) = \tilde{\Xi}^+_0(r, R)\psi_{on}(R), \tag{F.21a}$$

$$\Xi^-_{on}(r, R) = \tilde{\Xi}^-_0(r, R)\psi_{on}(R), \tag{F.21b}$$

where $\tilde{\Xi}^\pm_0$ are given by (F.13).

Substituting (F.21) into (F.17) we obtain

$$\begin{aligned}
\Psi_{on} = \Big(&\chi_0(r, R)\exp[-i\omega_0(0)t] + \tilde{\Xi}^+_0(r, R)\exp\{-i[\omega_0(0) + \omega_L]t\} \\
&+ \tilde{\Xi}^-_0(r, R)\exp\{-i[\omega_0(0) - \omega_L]t\}\Big)\psi_{on}(R)\exp(-i\hat{\omega}_{on}t). \quad \text{(F.22)}
\end{aligned}$$

Comparing (F.22) with (F.10) it is seen that only the electronic part of the eigenfunction is modified by the incident light wave. Moreover, the electronic part of Ψ_{on} (large bracket in (F.22)) is, apart from a constant factor of the form $\exp[-i\Delta\omega_0(R)]$, the same as $\tilde{\Psi}_0(r, R)$, the perturbed wavefunction for fixed nuclei as given by (F.12b). Thus we can write

$$\Psi_{on}(r, R) = \text{const } \tilde{\Psi}_0(r, R)\psi_{on}(R)\exp(-i\hat{\omega}_{on}t). \tag{F.23}$$

This form shows explicitly that the perturbation of the vibrating system by the incident light is the same for each nuclear configuration R as for the system with fixed nuclei in the configuration R. This proves the assumption made in Sect. 3.3.3.

In analogy to (3.56) we now form the matrix element

$$M^{(1)}_{on,on'} = \int\int \Psi^*_{on'}(r, R)M\Psi_{on}(r, R)dr\,dR, \tag{F.24}$$

associated with the transition $(on) \rightarrow (on')$. Using (F.15, 16) and neglecting terms proportional to E^2 we obtain

$$M^{(1)}_{on,on'}$$

$$= \int \psi^*_{on'}\psi_{on} \int \{\chi^*_0\chi_0 \exp(-i\omega_{nn'}t) + (\chi^*_0\tilde{\Xi}^+_0 + \tilde{\Xi}^{-*}_0\chi_0)$$

$$\times \exp[-i(\omega_{nn'}+\omega_L)t] + (\chi^*_0\tilde{\Xi}^-_0 + \tilde{\Xi}^{+*}_0\chi_0)$$

$$\times \exp[-i(\omega_{nn'}-\omega_L)t]\} M\,dr\,dR. \tag{F.25}$$

Using (F.13) we can write for the second term in (F.25)

$$\int (\chi^*_0\tilde{\Xi}^+_0 + \tilde{\Xi}^{-*}_0\chi_0)M\,dr$$

$$= \frac{1}{\hbar}\sum_e{}' \left(\frac{(\boldsymbol{A}\cdot\boldsymbol{M}_{oe}(R))\boldsymbol{M}_{eo}(R)}{\omega_{eo}-\omega_L} + \frac{\boldsymbol{M}_{eo}(R)(\boldsymbol{A}\cdot\boldsymbol{M}_{eo}(R))}{\omega_{eo}+\omega_L} \right) = C_{00}(R),$$

$$\tag{F.26}$$

where we have used (F.14) and (3.58). In a similar way one obtains for the third term in (F.25) and by using (3.59)

$$\int (\chi^*_0\tilde{\Xi}^-_0 + \tilde{\Xi}^{+*}_0\chi_0)M\,dr = D_{00}(R). \tag{F.27}$$

Defining the permanent dipole moment of the electronic ground state by

$$M_{00}(R) = \int \chi^*_0(r,R)M\chi_0(r,R)dr, \tag{F.28}$$

Eq. (F.25) can be written in the form

$$M^{(1)}_{on,on'} = \exp(-i\omega_{nn'}t)\int \psi^*_{on'}(R)M_{00}(R)\psi_{on}(R)dr$$

$$+ \exp[-i(\omega_{nn'}+\omega_L)t]\int \psi^*_{on'}(R)C_{00}(R)\psi_{on}(R)dR$$

$$+ \exp[-i(\omega_{nn'}-\omega_L)t]\int \psi^*_{on'}(R)D_{00}(R)\psi_{on}(R)dR. \tag{F.29}$$

As in the general formulas (3.57, 66) the first term in (F.29) represents the spontaneous emission; in the present approximation it is determined by the permanent dipole moment $M_{00}(R)$ of the electronic ground state. This term radiates light with frequency $\omega_{nn'} = \omega_n - \omega_{n'} > 0$. Since $\omega_{nn'}$ is a vibrational transition frequency, this term corresponds to spontaneous emission in the infrared region. The second term in (F.29) represents Raman scattering; it is determined by the scattering amplitude $C_{00}(R)$, which represents the electronic dipole moment of the electronic ground state for a system with fixed nuclear configuration R, induced by the incident light. There will be radiation only if $\omega_{nn'} + \omega_L > 0$ or $\hbar\omega_{n'} < \hbar\omega_n + \hbar\omega_L$, corresponding to Stokes and anti-Stokes lines (Figs. 3.20, 26). The last term of (F.29) represents induced double quantum emission, but only if $\omega_{nn'} - \omega_L > 0$ or $\omega_L < \omega_{nn'}$. Since in the above derivation we have assumed that $\omega_L \gg \omega_{nn'}$, this term has no significance.

Finally, we can replace the scattering amplitude by the scattering tensor. According to (3.68) we write

$$(C_\varrho(R))_{00} = \sum_\sigma (c_{\varrho\sigma}(R))_{00} A_\sigma, \tag{F.30}$$

and obtain for the ϱ-component of the second integral in (F.29)

$$\int \Psi_{on'}^*(R)[C_\varrho(R)]_{00}\Psi_{on}(R)dR$$
$$= \sum_\sigma \int \Psi_{on'}^*(R)[c_{\varrho\sigma}(R)]_{00}\Psi_{on}(R)dR = \sum_\sigma [c_{\varrho\sigma}(R)]_{on,on'}. \tag{F.31}$$

We have seen in Sect. 3.3.2, (3.72), that $(c_{\varrho\sigma})_{kk}$ is real and symmetric and is identical with the electronic polarizability $\alpha_{\varrho\sigma}^{(k)}$ of the state k. Thus, $[c_{\varrho\sigma}(R)]_{00}$ is identical with the electronic polarizability of the electronic ground state:

$$[c_{\varrho\sigma}(R)]_{00} = \alpha_{\varrho\sigma}^{(0)}(R) = \alpha_{\varrho\sigma}(R) = \alpha_{\sigma\varrho}(R), \tag{F.32}$$

and we obtain from (F.31)

$$[c_{\varrho\sigma}(R)]_{on,on'} = [\alpha_{\varrho\sigma}(R)]_{on,on'}, \tag{F.33}$$

which proves (3.77).

G. The Laue Interference Function

Consider a Bravais array of atoms of a lattice of dimensions $N_1 a_1$, $N_2 a_2$, $N_3 a_3$. The volume of the unit cell is v_a and $N = N_1 N_2 N_3$. From $r(l) = \sum l_i a_i$ the sum (5.27) can be written in the form

$$S(\boldsymbol{K}) = \prod_{i=1}^3 \sum_{l_i} \exp(i\boldsymbol{K}\cdot l_i \boldsymbol{a}_i) = \prod_{i=1}^3 S_i(\boldsymbol{K}), \tag{G.1}$$

where l_i runs from 0 to $N_i - 1$, or alternatively from $-(N_i-1)/2$ to $(N_i-1)/2$ if we place the origin in the centre of the crystal. Setting $2M_i = N_i - 1$ we have

$$S_i = \sum_{l_i=-M_i}^{M_i} \exp(il_i\alpha_i) = \exp(-iM_i\alpha_i) \sum_{l_i=0}^{2M_i} \exp(il_i\alpha_i)$$

$$= \exp(-iM_i\alpha_i)\frac{1 - \exp[i(2M_i + 1)\alpha_i]}{1 - \exp(i\alpha_i)}$$

$$= \frac{\exp(-i\beta_i) - \exp(i\beta_i)}{\exp(-i\alpha_i/2) - \exp(i\alpha_i/2)} = \frac{\sin\,[(M_i + 1/2)\alpha_i]}{\sin\,(\alpha_i/2)},$$

where $\beta_i = (M_i + 1/2)\alpha_i$ and $\alpha_i = \mathbf{K}\cdot\mathbf{a}_i$. From $M_i + \frac{1}{2} = N_i/2$ we obtain

$$S(\mathbf{K}) = \prod_{i=1}^{3} \frac{\sin\,[N_i(\mathbf{a}_i\cdot\mathbf{K})/2]}{\sin\,[(\mathbf{a}_i\cdot\mathbf{K})/2]}. \tag{G.2}$$

If $\mathbf{K} = \boldsymbol{\tau} = \sum m_j \mathbf{b}_j$ is a vector of the reciprocal lattice we have $\mathbf{a}_i\cdot\mathbf{b}_j = 2\pi\delta_{ij}$ and $\mathbf{a}_i\cdot\mathbf{K} = 2\pi m_i$. If $\mathbf{K}$ is very close to $\boldsymbol{\tau}$ we write $\mathbf{a}_i\cdot\mathbf{K} = 2\pi m_i + 2\varepsilon$. Substitution in (G.2) and forming the limit $\varepsilon \to 0$ one obtains

$$S(\boldsymbol{\tau}) = \prod_{i=1}^{3} N_i = N.$$

If N_i is large the quantity on the right-hand side of (G.2) has the characteristic of a δ-function: it is essentially zero unless $\mathbf{K} = \boldsymbol{\tau}$ and we therefore write

$$S(\mathbf{K}) = N\,\Delta(\mathbf{K} - \boldsymbol{\tau}), \quad \text{where} \tag{G.3a}$$

$$\Delta(\mathbf{K} - \boldsymbol{\tau}) = \begin{cases} 1 & \text{for} \quad \mathbf{K} = \boldsymbol{\tau} \\ 0 & \text{for} \quad \mathbf{K} \neq \boldsymbol{\tau} \end{cases}. \tag{G.3b}$$

The Laue interference function is defined by

$$|S(\mathbf{K})|^2 = \prod_{i=1}^{3} \frac{\sin^2\,[N_i(\mathbf{a}_i\cdot\mathbf{K})/2]}{\sin^2\,[(\mathbf{a}_i\cdot\mathbf{K})/2]}. \tag{G.4}$$

For $\mathbf{K} = \boldsymbol{\tau}$ we have $|S(\boldsymbol{\tau})|^2 = N^2$ and $|S(\mathbf{K})|^2$ falls rapidly to very small values in a distance in reciprocal space about equal to the reciprocal of a linear dimension of the crystal. This can be verified by calculating the position of the first zero near a reciprocal lattice point. Although the square of $\Delta(\mathbf{K} - \boldsymbol{\tau})$ is not the same function as $\Delta(\mathbf{K} - \boldsymbol{\tau})$ it shows again the characteristics of a δ-function if N is large and for practical purposes we write symbolically

$$|S(\mathbf{K})|^2 = N^2\Delta(\mathbf{K} - \boldsymbol{\tau}). \tag{G.5}$$

H. Thermal Diffuse Scattering of X-Rays

According to (5.35) the intensity of the scattered radiation is given by

$$I(K)_{\mathrm{dyn}} = |f(K)|^2 \sum_{ll'} \exp(iK \cdot r_{ll'}) \langle \exp(ip_{ll'}) \rangle, \tag{H.1}$$

where we have used the notation $r(l) = r_l$, $r(l) - r(l') = r_{ll'}$, $u(l) = u_l$, $u(l) - u(l') = u_{ll'}$, and

$$p_{ll'} = K \cdot u_{ll'}. \tag{H.2}$$

It can be shown that

$$\langle \exp(ip_{ll'}) \rangle = \exp\left(-\tfrac{1}{2} \langle p_{ll'}^2 \rangle \right). \tag{H.3}$$

This result can be roughly justified by expanding $\exp(ip_{ll'})$; if we write, for the moment, p for any of the quantities $p_{ll'}$,

$$\langle e^{ip} \rangle = 1 + i\langle p \rangle - \frac{1}{2!} \langle p^2 \rangle - \frac{i}{3!} \langle p^3 \rangle + \frac{1}{4!} \langle p^4 \rangle - \dots$$

$$= 1 - \tfrac{1}{2} \langle p^2 \rangle + \frac{1}{4!} \langle p^4 \rangle - \dots ; \tag{H.4}$$

the mean value of the terms involving odd powers of p will be zero, since positive and negative values of the difference of the displacements parallel to any given direction will be equally likely. If the displacements and hence p is small (H.4) may, to a close approximation, be written in the form (H.3), a result given by Debye and Waller, but first rigorously derived by *Ott* [5.35]. More recent derivations are found in the literature [5.36, 37]. Substituting (H.3) into (H.1) we obtain

$$I(K)_{\mathrm{dyn}} = |f(K)|^2 \sum_{ll'} \exp(iK \cdot r_{ll'}) \exp(-\tfrac{1}{2} \langle p_{ll'}^2 \rangle), \tag{H.5}$$

where

$$\langle p_{ll'}^2 \rangle = \langle (K \cdot u_{ll'})^2 \rangle = \langle [K \cdot (u_l - u_{l'})]^2 \rangle. \tag{H.6}$$

Let $K \cdot u_l = u_{lK}$, $K \cdot u_{l'} = u_{l'K}$. Then

$$\langle p_{ll'}^2 \rangle = \langle u_{lK}^2 \rangle + \langle u_{l'K}^2 \rangle - 2\langle u_{lK} u_{l'K} \rangle. \tag{H.7}$$

The assumptions made by Debye in his first papers about the temperature effect on X-ray scattering were equivalent to assume that the oscillations of the different atoms were independent, and that all possessed the same energy. If this were so we could put

$$\langle u_{lK}^2 \rangle = \langle u_{l'K}^2 \rangle, \quad \langle u_{lK} u_{l'K} \rangle = 0.$$

Now, as Debye himself pointed out in a later paper, the assumption of the independence of the vibrating atoms is not justifiable; the atoms of the lattice are coupled together by the interatomic forces, and the direction of vibration of one atom must influence those of its neighbours. We cannot therefore put $\langle u_{lK} u_{l'K} \rangle = 0$. We shall calculate in Appendix H.4 general correlation functions of the form $\langle u(\begin{smallmatrix} l \\ \kappa \end{smallmatrix}) u(\begin{smallmatrix} l' \\ \kappa' \end{smallmatrix}) \rangle$ for a pair of atoms $(\begin{smallmatrix} l \\ \kappa \end{smallmatrix})$ and $(\begin{smallmatrix} l' \\ \kappa' \end{smallmatrix})$.

H.1 The Evaluation of $\langle p_{ll'}^2 \rangle$

The evaluation of $\langle p_{ll'}^2 \rangle$ is simplified by expressing u_l and $u_{l'}$ in terms of creation and annihilation operators. From (5.36) we have

$$u_l = \left(\frac{\hbar}{2Nm}\right)^{1/2} \sum_{qj} \frac{e(\begin{smallmatrix} q \\ j \end{smallmatrix})}{[\omega_j(q)]^{1/2}} \exp(i\boldsymbol{q}\cdot\boldsymbol{r}_l)[a^+(\begin{smallmatrix} -q \\ j \end{smallmatrix}) + a(\begin{smallmatrix} q \\ j \end{smallmatrix})]. \tag{H.8}$$

Substituting (H.8) into (H.6) gives

$$\langle p_{ll'}^2 \rangle = \frac{\hbar}{2Nm} \left\langle \left\{ \sum_{qj} \frac{K\cdot e(\begin{smallmatrix} q \\ j \end{smallmatrix})}{[\omega_j(q)]^{1/2}} [\exp(i\boldsymbol{q}\cdot\boldsymbol{r}_l) - \exp(i\boldsymbol{q}\cdot\boldsymbol{r}_{l'})][a^+(\begin{smallmatrix} -q \\ j \end{smallmatrix}) + a(\begin{smallmatrix} q \\ j \end{smallmatrix})] \right\}^2 \right\rangle$$

$$= \frac{\hbar}{2Nm} \sum_{qj} \sum_{q'j'} \frac{[K\cdot e(\begin{smallmatrix} q \\ j \end{smallmatrix})][K\cdot e(\begin{smallmatrix} q \\ j' \end{smallmatrix})]}{[\omega_j(q)\omega_{j'}(q')]^{1/2}} [\exp(i\boldsymbol{q}\cdot\boldsymbol{r}_l) - \exp(i\boldsymbol{q}\cdot\boldsymbol{r}_{l'})]$$

$$\times [\exp(i\boldsymbol{q}'\cdot\boldsymbol{r}_l) - \exp(i\boldsymbol{q}'\cdot\boldsymbol{r}_{l'})] \times \langle [a^+(\begin{smallmatrix} -q \\ j \end{smallmatrix}) + a(\begin{smallmatrix} q \\ j \end{smallmatrix})][a^+(\begin{smallmatrix} -q' \\ j' \end{smallmatrix}) + a(\begin{smallmatrix} q' \\ j' \end{smallmatrix})] \rangle.$$
$$\tag{H.9}$$

The average $\langle \cdots \rangle$ in (H.9) gives terms of the form $\langle a_{-s}^+ a_{-s'}^+ \rangle, \langle a_s a_{s'} \rangle, \langle a_s a_{-s'}^+ \rangle$, $\langle a_{-s}^+ a_{s'} \rangle$, where $s = (\begin{smallmatrix} q \\ j \end{smallmatrix}), -s = (\begin{smallmatrix} -q \\ j \end{smallmatrix})$, etc. Using the definition of the thermal average we have, for example,

$$\langle a_s a_{s'}^+ \rangle = \sum_n p_n \langle n | a_s a_{s'}^+ | n \rangle, \tag{H.10}$$

where $\langle n | a_s a_{s'}^+ | n \rangle$ is the eigenvalue of the operator $a_s a_{s'}^+$ in the state $|n\rangle = |n_1 \ldots n_s \ldots \rangle$, and p_n is the probability to find the system in the state $|n\rangle$. Using the relations in [Ref. 5.23, Eqs. (2.107, 120, 121)] one obtains

$$\langle a_s a_{s'}^+ \rangle = \langle a_s^+ a_{s'} \rangle = 0 \quad \text{for} \quad s \neq s',$$

$$\langle a_s a_{s'} \rangle = \langle a_s^+ a_{s'}^+ \rangle = 0,$$

$$\langle a_s^+ a_s \rangle = \sum_n p_n n_s = \bar{n}_s = [\exp(\hbar\omega_s/k_B T) - 1]^{-1},$$

$$\langle a_s a_s^+ \rangle = \sum_n p_n (n_s + 1) = 1 + \bar{n}_s.$$

For the average in (H.9) we therefore obtain

$$\langle \cdots \rangle = \langle a^+ \binom{-q}{j} a \binom{q'}{j'} \rangle + \langle a \binom{q}{j} a^+ \binom{-q'}{j'} \rangle$$

$$= \Delta(-q - q')\delta_{jj'}\bar{n}_j(q) + \Delta(q + q')\delta_{jj'}[1 + \bar{n}_j(q)]$$

$$= \Delta(q + q')\delta_{jj'}[1 + 2\bar{n}_j(q)].$$

On the other hand, using (5.39), namely

$$\bar{E}\binom{q}{j} = \hbar\omega_j(q)[\bar{n}_j(q) + \tfrac{1}{2}],$$

we obtain for the average in (H.9)

$$\langle \cdots \rangle = 2\Delta(q + q')\delta_{jj'} \frac{\bar{E}\binom{q}{j}}{\hbar\omega_j(q)}. \tag{H.11}$$

Substituting (H.11) into (H.9) and using [Ref. 5.23, Eq. (3.31)]

$$e\binom{-q}{j} = e^*\binom{q}{j} \ , \quad \text{we find}$$

$$\langle p_{ll'}^2 \rangle = \frac{2}{Nm} \sum_{qj} |\boldsymbol{K} \cdot e\binom{q}{j}|^2 \frac{\bar{E}\binom{q}{j}}{\omega_j^2(q)} [1 - \cos(\boldsymbol{q} \cdot \boldsymbol{r}_{ll'}]. \tag{H.12}$$

We can therefore write

$$\tfrac{1}{2}\langle p_{ll'}^2 \rangle = 2W(\boldsymbol{K}) - \sum_{qj} G_j(q) \cos(\boldsymbol{q} \cdot \boldsymbol{r}_{ll'}), \qquad \text{where} \tag{H.13}$$

$$2W(\boldsymbol{K}) = \sum_{qj} G_j(q), \tag{H.14}$$

is the Debye-Waller factor, and

$$G_j(q) = \tfrac{1}{2} |\boldsymbol{K} \cdot e\binom{q}{j}|^2 \frac{\bar{E}\binom{q}{j}}{\omega_j^2(q)}. \tag{H.15}$$

H.2 The Evaluation of the Scattered Intensity

Substituting (H.13) into (H.5) gives

$$I(\boldsymbol{K})_{\mathrm{dyn}} = |f(\boldsymbol{K})|^2 \exp[-2W(\boldsymbol{K})]$$
$$\times \sum_{ll'} \exp(i\boldsymbol{K}\cdot\boldsymbol{r}_{ll'}) \exp\left[\sum_{qj} G_j(\boldsymbol{q}) \cos(\boldsymbol{q}\cdot\boldsymbol{r}_{ll'})\right]. \tag{H.16}$$

Setting $x = \sum G_j(\boldsymbol{q}) \cos(\boldsymbol{q}\cdot\boldsymbol{r}_{ll'})$ and using $e^x = 1 + x + x^2/2 + \ldots$ we obtain terms describing zero-, first-, second-order scattering etc., which we denote by $I_0, I_1, I_2, \ldots$. We restrict ourselves to I_0 and I_1, and obtain

$$I(\boldsymbol{K})_{\mathrm{dyn}} = |f(\boldsymbol{K})|^2 \exp[-2W(\boldsymbol{K})] \sum_{ll'} \exp(i\boldsymbol{K}\cdot\boldsymbol{r}_{ll'})$$
$$\times \left[1 + \sum_{qj} G_j(\boldsymbol{q}) \cos(\boldsymbol{q}\cdot\boldsymbol{r}_{ll'})\right]. \tag{H.17}$$

Using the result (G.5) we find

$$\sum_{ll'} \exp(i\boldsymbol{K}\cdot\boldsymbol{r}_{ll'}) = \left|\sum_{l} \exp(i\boldsymbol{K}\cdot\boldsymbol{r}_l)\right|^2 = N^2 \Delta(\boldsymbol{K} - \boldsymbol{\tau}).$$

We therefore obtain

$$I(\boldsymbol{K})_{\mathrm{dyn}} \cong I_0 + I_1, \quad \text{where} \tag{H.18}$$

$$I_0 = N^2 |f(\boldsymbol{K})|^2 \Delta(\boldsymbol{K} - \boldsymbol{\tau}) e^{-2W(\boldsymbol{K})}, \quad \text{and} \tag{H.19}$$

$$I_1(\boldsymbol{K}) = |f(\boldsymbol{K})|^2 e^{-2W(\boldsymbol{K})} \sum_{qj} G_j(\boldsymbol{q}) e^{ix_{ll'}} \cos y_{ll'}, \tag{H.20}$$

where $x_{ll'} = \boldsymbol{K}\cdot\boldsymbol{r}_{ll'}$ and $y_{ll'} = \boldsymbol{q}\cdot\boldsymbol{r}_{ll'}$. Using $\cos y_{ll'} = (1/2)\times[\exp(iy_{ll'}) + \exp(-iy_{ll'})]$, (H.29) can be written in the form

$$I_1(\boldsymbol{K}) = \tfrac{1}{2}|f(\boldsymbol{K})|^2 e^{-2W(\boldsymbol{K})} \sum_{qj} G_j(\boldsymbol{q})$$
$$\times \sum_{ll'} \{\exp[i(x_{ll'} + y_{ll'})] + \exp[i(x_{ll'} - y_{ll'})]\} \quad .$$

From

$$\sum_{ll'} \exp[i(x_{ll'} \pm y_{ll'})] = \left|\sum_{l} \exp[i(\boldsymbol{K}\pm\boldsymbol{q})\cdot\boldsymbol{r}_l]\right|^2 = N^2 \Delta(\boldsymbol{K}\pm\boldsymbol{q} - \boldsymbol{\tau}),$$

we obtain

$$I_1(\boldsymbol{K}) = \tfrac{1}{2}N^2 |f(\boldsymbol{K})|^2 e^{-2W(\boldsymbol{K})} \sum_{qj} G_j(\boldsymbol{q})[\Delta(\boldsymbol{K} + \dot{\boldsymbol{q}} - \boldsymbol{\tau}) + \Delta(\boldsymbol{K} - \boldsymbol{q} - \boldsymbol{\tau})].$$

The sum over q extends over the positive and negative values in the Brillouin zone, and since $G_j(-q) = G_j(q)$, one finally obtains

$$I_1(K) = N^2 |f(K)|^2 e^{-2W(K)} \sum_{qj} G_j(q)\Delta(K + q - \tau),\qquad (\text{H.21})$$

which proves (5.40).

H.3 The Debye-Waller Factor

We prove (5.59), namely

$$2W(K) = \langle |K\,u_l|^2 \rangle. \qquad (\text{H.22})$$

Substituting (H.8) into (H.22) gives

$$2W(K) = \frac{\hbar}{2Nm} \sum_{qj} \sum_{q'j'} \frac{[(K \cdot e\binom{q}{j})][K \cdot e\binom{q'}{j'})]}{[\omega_j(q)\omega_{j'}(q')]^{1/2}} \exp[i(q + q')\cdot r_l]$$

$$\times \langle [a^+(\genfrac{}{}{0pt}{}{-q}{j}) + a\binom{q}{j})][a^+(\genfrac{}{}{0pt}{}{-q'}{j'}) + a\binom{q'}{j'})]\rangle.$$

Using (H.11) for the thermal average in the above expression, as well as $e(\genfrac{}{}{0pt}{}{-q}{j}) = e^*\binom{q}{j}$, we obtain

$$2W(K) = \frac{1}{Nm} \sum_{qj} |K \cdot e\binom{q}{j}|^2 \frac{\overline{E}\binom{q}{j}}{\omega_j^2(q)}. \qquad (\text{H.23})$$

This expression is identical with (H.14, 15).

H.4 The Correlation Function

We calculate the correlation function $\langle u\binom{l}{\kappa} \cdot u\binom{l'}{\kappa'} \rangle$. Expressing $u\binom{l}{\kappa}$ and $u\binom{l'}{\kappa'}$ in terms of creation and annihilation operators according to [Ref. 5.23, Eq. (3.74)], namely

$$u\binom{l}{\kappa} = \left(\frac{\hbar}{2Nm_\kappa}\right)^{1/2} \sum_{qj} \frac{e(\kappa|\genfrac{}{}{0pt}{}{q}{j})}{[\omega_j(q)]^{1/2}} \exp[iq \cdot r\binom{l}{\kappa})][a^+(\genfrac{}{}{0pt}{}{-q}{j}) + a\binom{q}{j})], \quad (\text{H.24})$$

we obtain

$$\langle u\binom{l}{\kappa}) u\binom{l'}{\kappa'}) \rangle = \frac{\hbar}{2N(m_\kappa m_{\kappa'})^{1/2}} \sum_{qj} \sum_{q'j'} \frac{e(\kappa|\genfrac{}{}{0pt}{}{q}{j}) \cdot e(\kappa'|\genfrac{}{}{0pt}{}{q'}{j'})}{[\omega_j(q)\omega_{j'}(q')]^{1/2}}$$

$$\times \exp[iq \cdot r\binom{l}{\kappa})]\exp[iq' \cdot r\binom{l'}{\kappa'})]$$

$$\times \langle [a^+(\genfrac{}{}{0pt}{}{-q}{j}) + a\binom{q}{j})][a^+(\genfrac{}{}{0pt}{}{-q'}{j'}) + a\binom{q'}{j'})]\rangle.$$

Using (H.11) gives

$$\langle u(^l_\kappa) \cdot u(^{l'}_{\kappa'}) \rangle = \frac{1}{N(m_\kappa m_{\kappa'})^{1/2}} \sum_{qj} e(\kappa|^q_j) \cdot e^*(\kappa'|^q_j)$$

$$\times \exp\{i q \cdot [r(^l_\kappa) - r(^{l'}_{\kappa'})]\} \frac{\overline{E}(^q_j)}{\omega_j^2(q)}. \tag{H.25}$$

For a Bravais lattice we can omit the index κ and using $|e(^q_j)|^2 = 1$ according to [Ref. 5.23, Eq. (3.29)], we obtain

$$\langle u(l) \cdot u(l') \rangle = \frac{1}{Nm} \sum_{qj} \exp\{i q \cdot [r(l) - r(l')]\} \frac{\overline{E}(^q_j)}{\omega_j^2(q)}. \tag{H.26}$$

This expression diverges for the linear chain.

I. Constants and Units

General Physical Constants

Gas constant	$R = 8.31 \cdot 10^7 \text{ erg mole}^{-1} \text{K}^{-1}$
Loschmidt's number	$N_L = 6.0225 \cdot 10^{23} \text{ mole}^{-1}$
Speed of light in vacuum	$c = 2.998 \cdot 10^{10} \text{ cm s}^{-1}$

Atomic Constants

Boltzmann's constant	$k_B = 1.381 \cdot 10^{-16} \text{ erg K}^{-1}$
Planck's constant	$h = 6.626 \cdot 10^{-27} \text{ erg s}$
Planck's constant $/2\pi$	$\hbar = 1.054 \cdot 10^{-27} \text{ erg s}$
Electron rest mass	$m = 0.911 \cdot 10^{-27} \text{ g}$
Proton rest mass	$M_p = 1.6725 \cdot 10^{-24} \text{ g}$
Neutron rest mass	$M_n = 1.6747 \cdot 10^{-24} \text{ g}$
Proton mass / electron mass	$M_p/m = 1836$
Atomic mass unit ($\equiv \frac{1}{16}$ mass of O^{16})	$1 \text{ amu} = 1.657 \cdot 10^{-24} \text{ g}$
Charge on electron	$e = 4.803 \cdot 10^{-10} \text{ esu}$ $= 1.602 \cdot 10^{-19} \text{ C}$
Bohr radius of the ground state of hydrogen	$a_0 = \hbar^2/me^2 = 0.529 \text{ Å}$
Rydberg's constant	$1.09737 \cdot 10^5 \text{ cm}^{-1}$

Length

Angström	$1 \text{ Å} = 10^{-8} \text{ cm}$
Micrometer	$1 \, \mu\text{m} = 10^{-4} \text{ cm}$

Energy conversion table

E	erg	J	cal	eV
1 erg =	1	10^{-7}	$2.3892 \cdot 10^{-8}$	$6.242 \cdot 10^{11}$
1 Joule =	10^7	1	$2.3892 \cdot 10^{-1}$	$6.242 \cdot 10^{18}$
1 cal =	$4.1855 \cdot 10^7$	4.1855	1	$2.613 \cdot 10^{19}$
1 eV =	$1.602 \cdot 10^{12}$	$1.602 \cdot 10^{-19}$	$3.827 \cdot 10^{-20}$	1

From $E = hc\tilde{\nu} = k_B T$: $\quad$ 1 eV $\hat{=}$ 8066 cm^{-1} $\hat{=}$ 11605 K
$\qquad$ 1 cm^{-1} $\hat{=}$ 1.438 K; $\quad$ 1 K $\hat{=}$ 0.695 cm^{-1}
$\qquad$ ($\hat{=}$ means: corresponds to).

Force, Force Constants and Elastic Constants

Newton	$1\,N = 1\,kg\,ms^{-2} = 10^5 dyn$
dyn	$1\,dyn = 1\,g\,cm\,s^{-2}$
mdyn	$1\,mdyn = 10^{-3}\,dyn$
Force constants	in dyn/cm or mdyn/Å
Elastic constants	in dyn/cm^2

Energy E

erg	$1\,erg = 1\,dyn\,cm$
Joule	$1\,J = N\,m = 1\,W\,s$

Frequency ν

Hertz	$1\,Hz = 1\,s^{-1}$
Megahertz	$1\,MHz = 10^6\,Hz$
Gigahertz	$1\,GHz = 10^9\,Hz$
Terahertz	$1\,THz = 10^{12}\,Hz$

Wave Number $\tilde{\nu}$

= number of wavelenghts λ per cm
$\tilde{\nu} = 1/\lambda = \nu/c$, in units of cm^{-1}
($1\,THz \hat{=} 33.3\,cm^{-1}$)

References

Chapter 1

1.1   P. Brüesch: *Phonons: Theory and Experiments I*, Springer Ser. Solid-State Sci., Vol. 34 (Springer, Berlin, Heidelberg 1982)
1.2   P. Brüesch: *Phonons: Theory and Experiments III*, Springer Ser. Solid-State Sci., Vol. 66 (Springer, Berlin, Heidelberg 1987)

Chapter 2

2.1   H. Bilz, D.Strauch, R.K. Wehner: *Vibrational Infrared and Raman Spectra of Non-Metals*, Encyclopedia of Physics, Vol. XXV/2d, Licht und Materie, ed. by L. Genzel (Springer, Berlin, Heidelberg 1984)
2.2   A. Hadni: *Essentials of Modern Physics Applied to the Study of the Infrared* (Pergamon, Oxford 1967)
2.3   R.A. Smith, F.E. Jones, R.P. Chasmar: *The Detection and Measurements of Infrared Radiation* (Clarendon, Oxford 1968)
2.4   P.L. Richards: *Spectroscopic Techniques for the Far Infrared, Submillimetre and Millimetre Waves*, ed. by D.H. Martin (North-Holland, Amsterdam 1967) p. 33
2.5   K.D. Möller, W.G. Rothschild: *Far Infrared Spectroscopy* (Wiley, New York 1971)
2.6   R. Geick: *Topics in Current Chemistry* **58**, 73 (Springer, Berlin, Heidelberg 1975)
2.7   D.H. Martin, E.Puplett: Infrared Phys. **10**, 105 (1969)
2.8   F.E. Bates, J.E. Eldridge, M.R. Bryce: Canadian J. Phys. **59**, 339 (1981)
2.9   R.R. Joyce, P.L. Richards: Phys. Rev. Lett. **24**, 1007 (1970)
2.10   D.G. Mead, G.R. Wilkinson: Proc. Roy. Soc. (London) **354**, 349 (1977);
       D.G. Mead: Infrared Phys. **17**, 257 (1977)
2.11   H.O. McMahon: J. Opt. Soc. Am. **40**, 376 (1950)
2.12   J.A. Stratton: *Electromagnetic Theory* (McGraw-Hill, New York 1941)
2.13   F.S.Crawford, Jr.: *Waves*, Berkeley Physics Course, Vol. 3 (McGraw-Hill, New York 1968) p. 575
2.14   A. Mitshuishi, Y. Yamada, H. Hoshinaga: J. Opt. Soc. Am. **52**, 14 (1962)
2.15   R.Geick: Z. Physik **166**, 122 (1962)
2.16   J.R. Jasperse, A. Kahan, J.N. Plendl, S.S. Mitra: Phys. Rev. **146**, 526 (1966);
       D.W. Berreman: Phys. Rev. **130**, 2193 (1963)
2.17   P. Brüesch: *Phonons: Theory and Experiments I*, Springer Ser. Solid-State Sci., Vol. 34 (Springer, Berlin, Heidelberg 1982)
2.18   J.C. Decius, R. Frech, P. Brüesch: J. Chem. Phys. **58**, 4056 (1973)
2.19   P. Brüesch: Solid State Commun. **13**, 13 (1973)
2.20   L.N. Bulaevski, Yu.A. Kukharenko: Sov. Phys. Solid State **14**, 2076 (1973)
2.21   L.S.Agroskin, R.M. Vlasova, A.I. Gutman, R.N. Lyubovskaya, G.V. Papayan, L.P. Rautian, L.D. Rozenshtain: Sov.Phys. Solid State **15**, 1189 (1973)
2.22   A.J. Alister, E.A. Stern: Bull. Am. Phys. Soc. **8**, 392 (1963)
2.23   H.G. Tompkins: In *Methods of Surface Analysis*, Vol. 1, ed. by A.W. Czanderna (Elsevier, New York 1975) p. 447
2.24   F. Stern: In *Solid State Physics* **15**, 299 (Academic, New York 1963)
2.25   F. Wooton: *Optical Properties of Solids* (Academic, New York 1972)
2.26   J.L. Verble, R.F. Wallis: Phys. Rev. **182**, 783 (1960)
2.27   I.F. Chang, S.S. Mitra, J.N. Plendl, L.C. Mansur: Phys. Stat. Sol. **28**, 663 (1968)

2.28	H.D. Lutz, G. Kliche, H. Haeussler: Z. Naturforsch. **36A**, 184 (1981)
2.29	P. Brüesch, W. Bührer, H.J. Smeets: Phys. Rev. B**22**, 970 (1980)
2.30	S.G. Tomlin: Brit. J. Appl. Phys. (J. Phys. D) **1**, 1667 (1968)
2.31	R.D. Bringans: J. Phys. D**10**, 1855 (1977)
2.32	R.E. Denton, R.D. Campbell, S.G. Tomlin: J. Phys. D**5**, 852 (1972)
2.33	G. Hellmann: Z. Physik **152**, 368 (1958)
2.34	D.L. Greenaway, G. Harbecke: Phys. Rev. **178**, 1340 (1969)
2.35	J. Bernasconi, P. Brüesch, D. Kuse, H.R. Zeller: J. Phys. Chem. Solids **35**, 145 (1974)
2.36	P. Drude: Ann. Phys. **32**, 584 (1887)
2.37	D.T. Hodges: Infrared Phys. **18**, 375 (1978)
2.38	E.E. Bell: Infrared Phys. **6**, 57 (1966)
2.39	E.E. Russell, E.E. Bell: Infrared Phys. **6**, 75 (1966)
2.40	K.W. Johnson, E.E. Bell: Phys. Rev. **187**, 1044 (1969)
2.41	J.I. Berg, E.E. Bell: Phys. Rev. B**4**, 3572 (1971)
2.42	T.J. Parker, W.G. Chambers, J.F. Angress: Infrared Phys. **14**, 207 (1974)
2.43	J.R. Birch, G.D. Price, J. Chamberlin: Infrared Phys. **16**, 311 (1976)
2.44	D.G. Mead, L. Genzel: Infrared Phys. **18**, 555 (1978)
2.45	J.E. Eldridge, P.R. Staal: Phys. Rev. B**16**, 4608 (1977)
2.46	P.R. Staal, J.E. Eldridge: Can. J. Phys. **57**, 1784 (1979)
2.47	A.S. Barker: Phys. Rev. B**12**, 4071 (1975)
2.48	E. Burstein: In *Phonons and Phonon Interactions*, ed. by T.A. Bak (Benjamin, New York 1964) pp. 296–297; The values for AgF are from G.L. Bottger, A.L. Geddes: J. Chem. Phys. **56**, 3735 (1972); The data for MnO are from J.N. Plendl, L.C. Mansur, S.S. Mitra, I.F. Chang: Solid State Commun. **7**, 109 (1969)
2.49	D.H. Martin: Adv. Phys. **14**, 39 (1965)
2.50	[Ref. 2.5, p. 439] and references given there; The data for CuCl, CuBr and CuI are from P. Lawaetz: Phys. Status Solidi (b) **63**, 485 (1974)
2.51	M. Alterelli, D.L. Dexter, H.M. Nussenzveig, D.Y. Smith: Phys. Rev. B**6**, 4502 (1972)
2.52	M. Alterelli, D.Y. Smith: Phys. Rev. B**9**, 1290 (1974)
2.53	K. Huang: Proc. Roy. Soc. (London) A**208**, 352 (1951); M. Born, K. Huang: *Dynamical Theory of Crystal Lattices* (Oxford U. Press, New York 1954)
2.54	T.H.K. Barron: Phys. Rev. **123**, 1995 (1961)
2.55	J.R. Hardy: Phil. Mag. **7**, 315 (1962)
2.56	T.N. Casselman, S.S. Mitra, H.N. Spector: J. Phys. Chem. Solids **26** , 529 (1965)
2.57	J. Callaway: *Quantum Theory of the Solid State* (Academic, New York 1974) Pt. A, p. 70
2.58	B. Szigeti: Proc. Roy. Soc. A**258**, 377 (1960)
2.59	R.A. Cowley: Adv. Phys. **12**, 421 (1963)
2.60	R.A. Cowley: In *Phonons in Perfect Lattices and in Lattices with Point Interactions*, ed. by R.W.H. Stevenson (Oliver and Boyd, Edinburgh and London 1966) p. 170
2.61	H. Bilz: In *Phonon in Perfect Lattices and Lattices with Point Imperfections*, ed. by R.W.H. Stevenson (Oliver and Boyd, Edinburgh and London 1966) p. 208
2.62	V.S. Vinogradov: Sov. Phys. Sol. **4**, 519 (1962)
2.63	R.Wehner: Phys. Status Solidi **15**, 725 (1966)
2.64	M. Born, M. Blackman: Z. Physik **82**, 551 (1933)
2.65	M. Blackman: Z. Physik **86**, 421 (1933)
2.66	L.I. Schiff: *Quantum Mechanics* (McGraw-Hill, New York 1968) p. 66
2.67	F. Seitz: *The Modern Theory of Solids* (McGraw-Hill, New York 1940) p. 210
2.68	H. Eyring, J. Walter, G.E. Kimball: *Quantum Chemistry* (Wiley, New York 1963) p. 107
2.69	L.D. Landau, E.M.Lifshitz: *Quantum Mechanics* (Pergamon, Oxford 1965) p. 31

2.70   M. Lax, E. Burstein: Phys. Rev. **97**, 39 (1955)
2.71   P.N. Keating, G. Rupprecht: Phys. Rev. **138**, A866 (1965)
2.72   J.C. Decius, R.M. Hexter: *Molecular Vibrations in Crystals* (McGraw-Hill, New York 1977)
2.73   E. Burstein: [Ref. 2.48a, p. 310].
2.74   M. Balkanski: In *Optical Properties of Solids*, ed. by F. Abelès (North-Holland, Amsterdam 1972) p. 608
2.75   J.L. Birman: Phys. Rev. **131** , 1489 (1963)
2.76   H. Bilz, R. Geick, K.F. Renk: In *Proceedings Int. Conf. on Lattice Dynamics*, ed. by R.F. Wallis (Pergamon, Oxford 1965) p. 355
2.77   R.J. Collins, H.J. Fan: Phys. Rev. **93**, 674 (1954)
2.78   F.A. Johnson: Proc. Phys. Soc. **73**, 265 (1959)
2.79   G. Dolling: In *Inealstic Scattering of Neutrons in Solids and Liquids*, Vol. II (Intern. Atomic Energy Agency, Vienna 1963) p. 37
2.80   A. Ghose: Phys. Rev. **113**, 49 (1959)
2.81   M. Ikezawa, M. Ishigame: J. Phys. Soc. Japan **50**, 3734 (1981)
2.82   R. Wehner, H. Borik, W. Kress, A.R. Goodwin, S.O. Smith: Solid State Commun. **5**, 307 (1967)
2.83   S. Go, H. Bilz, M. Cardona: Phys. Rev. Lett. **34**, 580 (1975)
2.84   R. Loudon: Adv. Phys. **13**, 423 (1964);
Phys. Rev. **137**, A1784 (1965);
Proc. Phys. Soc. London **85**, 379 (1964)
2.85   L. Genzel, H. Happ, R. Weber: Z. Physik **154**, 13 (1959)
2.86   E.R. Cowley: J. Phys. C**5**, 1345 (1972)

Chapter 3

3.1   S.P.S. Porto: In *Light Scattering Spectra of Solids*, ed. by G.B. Wright (Springer, Berlin, Heidelberg 1969) p. 1
3.2   M. Balkanski (ed.): *Light Scattering in Solids* (Flammarion, Paris 1971)
3.3   M. Balkanski, R.C.C. Leite, S.P.S. Porto (eds.): *Light Scattering in Solids* (Flammarion, Paris, Wiley and Sons, New York 1976)
3.4   J.L. Birman, H.Z. Cummins, K.K. Rebane (eds): *Light Scattering in Solids* (Plenum, New York 1979)
3.5   J.A. Koningstein: *Introduction to the Theory of the Raman Effect*, (Reidel, Dordrecht, Holland 1972)
3.6   W. Hayes, R. Loudon: *Scattering of Light by Crystals* (Wiley, New York 1978)
3.7   R. Loudon: Adv. Phys. **13**, 423 (1964)
3.8   M. Cardona (ed.): *Light Scattering in Solids I*, 2nd ed., Topics Appl. Phys., Vol. 8 (Springer, Berlin, Heidelberg 1983)
3.9   M. Cardona, G. Güntherodt (eds.): *Light Scattering in Solids II*, Topics Appl. Phys., Vol. 50 (Springer, Berlin, Heidelberg 1982)
3.10   R. Claus, L. Merten, J. Brandmüller: *Light Scattering by Phonon-Polaritons*, Springer Tracts Mod. Phys., Vol. 75 (Springer, Berlin, Heidelberg 1975)
3.11   H.A. Szymanski (ed.): *Raman Spectroscopy* (Plenum, New York 1967)
3.12   T.C. Damen, S.P.S. Porto, B. Tell: Phys. Rev. **142**, 570 (1966)
3.13   J.C. Decius, R.M. Hexter: *Molecular Vibrations in Crystals* (McGraw-Hill, New York 1977)
3.14   S.P.S. Porto, J.A. Giordaine, T.C. Damen: Phys. Rev. **147**, 608 (1966)
3.15   F.S. Crawford, Jr.: *Waves*, Berkeley Physics Course, Vol. 3 (McGraw-Hill, New York 1968) p. 366
3.16   R.W. Terhune, P.D. Maker, C.M. Savage: Phys. Rev. Lett. **14**, 681 (1965)
3.17   S.J. Cyvin, J.E. Rauch, J.C. Decius: J. Chem. Phys. **43**, 4083 (1965)
3.18   W. Jones, N.H. March: *Theoretical Solid State Physics*, Vol. 2 (Wiley, London 1973) p. 754
3.19   J.F. Nye: *Physical Properties of Crystals* (Clarendon Press, Oxford 1957)

3.20   G. Herzberg: *Molecular Spectra and Molecular Structure* II. Infrared and Raman Spectra of Polyatomic Molecules (D. Van Nostrand, Princeton, NJ 1966)

3.21   P. Brüesch: *Phonons: Theory and Experiments I*, Springer Ser. Solid-State Sci., Vol. 34 (Springer, Berlin, Heidelberg 1982)

3.22   O. Brafman, S.S. Mitra: Phys. Rev. **171**, 931 (1968); S. Bhagavantam, T. Venkatarayudu: Proc. Ind. Acad. Sci. **10A**, 224 (1939)

3.23   S.L. Cunningham, J.R. Hardy, M. Hass: In [Ref. 3.2, p. 257]

3.24   O. Klein: Z. Physik **41**, 407 (1927)

3.25   L.D. Landau, E.M. Lifshitz: *Quantum Mechanics* (Pergamon, Oxford 1965) p. 136

3.26   G. Placzek: In *Handbuch der Radiologie*, Vol. 6, Pt. 2, ed. by E. Marx (Akademische Verlagsgesellschaft, Leipzig 1934) p. 205

3.27   C.H. Henry, J.J. Hopfield: Phys. Rev. Lett. **15**, 964 (1965)

3.28   A. Messiah: *Quantum Mechanics*, Vol. I (North-Holland, Amsterdam 1965) p. 468

3.29   G. Placzek: Z. Physik **70**, 84 (1931)

Chapter 4

4.1   P. Brüesch: *Phonons: Theory and Experiments I*, Springer Ser. Solid-State Sci., Vol. 34 (Springer, Berlin, Heidelberg 1982)

4.2   W. Hayes, R. Loudon: *Scattering of Light by Crystals* (Wiley, New York 1978) p. 80

4.3   J.F. James, R.S. Sternberg: *The Design of Optical Spectrometers* (Chapman and Hall, London 1969) p. 95

4.4   J.R. Sandercock: In *Light Scattering in Solids*, ed. by M. Balkanski (Flammarion, Paris 1971) p. 9

4.5   J.R. Sandercock: Opt. Commun. **2**, 76 (1970)

4.6   J.F. Nye: *Physical Properties of Crystals* (Clarendon, Oxford 1957) p. 243

4.7   M. Born, K. Huang: *Dynamical Properties of Crystal Lattices* (Clarendon, Oxford 1954) p. 373

4.8   I.L. Fabelinski: *Molecular Scattering of Light* (Plenum, New York 1968) p. 139

4.9   G.B. Benedek, K. Fritsch: Phys. Rev. **149**, 647 (1966)

4.10   L. Benedek, G. Bäckström: Phys. Rev. B**8**, 5888 (1973)

4.11   J.F. Nye: *Physical Properties of Crystals* (Clarendon, Oxford 1957) p. 98

4.12   J. Shanker, T.S. Varma: Phys. Status Solidi (b) **99**, 359 (1980)

4.13   V.V.S. Nirwal, R.K. Singh: Phys. Rev. B**20**, 5379 (1979)

4.14   L.V. Chebotarev: Sov. Phys. Solid State **18**, 1956 (1976)

4.15   K.G. Aggarwal, B. Szigety: J. Phys. C**3**, 1097 (1970)

4.16   H. Müller: Phys. Rev. **47**, 947 (1935)

4.17   O. Theimer: Proc. Phys. Soc. **64A**, 1012 (1952)

4.18   M. Leontovich, S. Mandelstam, Jr.: Phys. Z. Sowjet. **1**, 317 (1931)

4.19   H. Braul, C.A. Plint: Solid State Commun. **38**, 227 (1981)

4.20   H.P. Sharma, J. Shanker, M.P. Varma: J. Phys. Chem. Solids **38**, 255 (1977)

Chapter 5

5.1   Z.W. Wilchinsky: J. Appl. Phys. **15**, 806 (1944)

5.2   H. Lambot: Rev. Metall. **47**, 709 (1950)

5.3   A. Guinier: Bull. Soc. Fr. Minéral. Cristallogr. **77**, 680 (1954)

5.4   R. Mort, M. Huber, R. Comès: Phys. Status Solidi A**38** , 695 (1976)

5.5   R. Comès, N. Lambert, H. Launois, H.R. Zeller: Phys. Rev. B**8**, 571 (1973)

5.6   F. Denoyer, R. Comès, A.F. Garito, A.J. Heeger: Phys. Rev. Lett **35**, 445 (1975)

5.7   Y. Le Cars, R. Comès, L. Dechamps, J. Théry: Acta Cryst. A**30**, 305 (1974)

5.8 D.B. McWhan, S.J. Allen, J.P. Remeika, P.D. Dernier: Phys. Rev. Lett. **35**, 953 (1975)
5.9 T. Hibma, H.U. Beyeler, H.R. Zeller: J. Phys. C**9**, 169 (1976)
5.10 H.U. Beyeler: Phys. Rev. Lett. **37**, 1557 (1976)
5.11 H.U. Beyeler, S. Strässler: Phys. Rev. B**20**, 1980 (1979)
5.12 H.U. Beyeler, L. Pietronero, S. Strässler, H.J. Wiesmann: Phys. Rev. Lett. **38**, 1532 (1977)
5.13 D.B. McWhan, P.D. Dernier, C.Vettier, A.S. Cooper, J.P. Remeika: Phys. Rev. B**17**, 4043 (1978)
5.14 H.U. Beyeler, P. Brüesch, T. Hibma, W. Bührer: Phys. Rev. B**18**, 4570 (1978)
5.15 T. Hibma: Phys. Rev. B**28**, 568 (1983)
5.16 R.W. James: *The Optical Principles of the Diffraction of X-Rays*, The Crystalline State, Vol. II, ed. by L.W. Bragg (G. Bell, London 1948)
5.17 A. Guinier: *X-Ray Diffraction in Crystals, Imperfect Crystals, and Amorphous Bodies* (Freeman, San Francisco 1963)
5.18 G. Shirane: Physics **120B**, 108 (1983)
5.19 M. Born: Repts. Progr. Phys. **9**, 294 (1942–43)
5.20 A.A. Maradudin, E.W. Montroll, G.H. Weiss, I.P. Ipatova: *Theory of Lattice Dynamics in the Harmonic Approximation* (Academic, New York 1971) p. 311
5.21 W. Cochran: Repts. Progr. Phys. **26**, 1 (1963); W. Cochran: In *Phonons and Phonon Interactions*, ed. by T.A. Bak (Benjamin, New York 1964) p. 102
5.22 C. Kittel: *Introduction to Solid State Physics* (Wiley, New York 1967) p. 56
5.23 P. Brüesch: *Phonons: Theory and Experiments I*, Springer Ser. Solid-State Sci., Vol. 34 (Springer, Berlin, Heidelberg 1982)
5.24 W. Bührer, R.M. Nicklov, P. Brüesch: Phys. Rev. **17**, 3362 (1978)
5.25 I.S. Grant, W.R. Phillips: *Electromagnetism* (Wiley, New York 1975) p. 423
5.26 A. Sommerfeld: *Elektrodynamik* (Akadem. Verlagsgesell., Geest u. Portig K.G., Leipzig 1961) p. 133
5.27 K. Lonsdale, C.H. MacGilavry, G.D. Rieck (eds.): *International Tables for X-Ray Crystallography*,Vol. III (Physical and Chemical Tables) (Kynoch Press, Birmingham 1968) p. 213
5.28 [Ref. 5.27, p. 247]
5.29 W.J.L. Buyers, J.D. Pirie, T. Smith: Phys. Rev. **165**, 999 (1968)
5.30 C.B. Walker: Phys. Rev. **103**, 547 (1956)
5.31 W. Cochran, R.A. Cowley: *Phonons in Perfect Crystals*, Encyclopedia of Physics, Vol. XXV/2a, ed. by L. Genzel (Springer, Berlin, Heidelberg 1967) p. 74
5.32 J.D. Pirie, T. Smith: J. Phys. C (Proc. Phys. Soc.) **1**, 648 (1968)
5.33 R.M. Nicklow, R.A. Young: Phys. Rev. **152**, 591 (1966)
5.34 D.P. Jackson, B.M. Powell, G. Dolling: Phys. Lett. **51A**, 87 (1975)
5.35 H. Ott: Ann. Phys. (Leipzig) **23**, 169 (1935)
5.36 M. Born, K. Sarginson: Proc. Roy. Soc. A**179**, 69 (1941)
5.37 [Ref. 5.20, p. 304]

Chapter 6

6.1 M. Sakamoto, J. Chihara, H. Kadotai, T. Sekiya, Y. Gotoh, K. Inone, K. Katiyama, O. Yoda, Jaeri-Report-M 8417 (1979)
6.2 H. Bilz, W. Kress: *Phonon Dispersion Relation in Insulators*, Springer Ser. Solid-State Sci., Vol. 10 (Springer, Berlin, Heidelberg 1979)
6.3 G.L. Squires:*Thermal Neutron Scattering* (Cambridge U. Press, Cambridge 1978)
6.4 M. Marshall, S. Lovesey: *Theory of Thermal Neutron Scattering* (Clarendon, Oxford 1971)
6.5 G. Dolling: "Neutron Spectroscopy" in *Dynamical Properties of Solids I*, ed. by G.K. Horton, A.A. Maradudin (North Holland, Amsterdam 1974)
6.6 D.J. Hughes, R.B. Schwartz: *Neutron Cross Sections*, BNL-325, Brookhaven N.Y. (1958)

6.7   L. Koester, H. Rauch, M. Herker, K. Schroeder: KFA-Juelich Report 1755 (1981)
6.8   L.I. Schiff: *Quantum Mechanics*, 3rd ed. (McGraw-Hill, New York 1968)
6.9   P. Brüesch: *Phonons: Theory and Experiments I*, Springer Ser. Solid-State Sci.,
        Vol. 34 (Springer, Berlin, Heidelberg 1982)
6.10   R.J. Elliott, M.F. Thorpe: Proc. Phys. Soc. **91** , 903 (1967)
6.11   A.A. Maradudin, S.H. Vosko: Rev. Mod. Phys. **40**, 1 (1968);
        J.L. Warren: Rev. Mod. Phys. **40**, 38 (1968)
6.12   T.G. Worlton, J.L. Warren: Comp. Phys. Commun. **8**, 71 (1974)
6.13   B.N. Brockhouse: *Inelastic Scattering of Neutrons in Solids and Liquids*, IAEA
        Vienna (1961)
6.14   M.J. Cooper, R. Nathans: Acta Cryst. **23**, 357 (1967)
6.15   P. Brüesch: *Phonons: Theroy and Experiments III*, Springer Ser. Solid-State Sci.,
        Vol. 66 (Springer, Berlin, Heidelberg 1987)
6.16   E.H. Jacobsen: Phys. Rev. **97**, 654 (1955);
        E. Svensson, B. Brockhouse, J. Rowe: Phys. Rev. **155**, 619 (1967)
6.17   W. Buehrer, R.M. Nicklow, P. Brüesch: Phys. Rev. B**17**, 3362 (1978)
6.18   R.A. Cowley: Phys. Rev. **134**, A981 (1964)
6.19   J. Harada, J.D. Axe, G.Shirane: Acta Cryst. A**26** , 608 (1970)
6.20   J.C.Slater: Phys. Rev. **78**, 748 (1950)
6.21   J.T. Last: Phys. Rev. **105**, 1740 (1957)
6.22   W. Glaeser, F. Carvalho, G.Ehret: *Inelastic Scattering of Neutrons in Solids and
        Liquids*, IAEA Vienna (1974)
6.23   H. Boutin, S. Yip: *Molecular Spectroscopy with Neutrons* (MIT Press , Cam-
        bridge, MA 1968)
6.24   L. Reinhard, W. Haelg, W. Buehrer, A. Eschenmoser, R.Schwyzer: LNS ETH
        Zürich, Annual Rept. AF-SSP-127 (1984)

Chapter 7

7.1   P. Brüesch: *Phonons: Theory and Experiments I*, Springer Ser. Solid-State Sci.,
        Vol. 34 (Springer, Berlin, Heidelberg 1982)
7.2   R. Truell, Ch. Elbaum, B.B. Chick:*Ultrasonic Methods in Solid State Physics*
        (Academic, New York 1969) p. 55;
        G.A. Alers, J.R. Neighbours: J. Phys. Chem. Solids **7**, 58 (1958)
7.3   E.H. Jacobson: In *Phonons and Phonon Interactions*, ed. by T.A. Bak (Benjamin,
        New York 1964) p. 505
7.4   H.E. Bömmel, K. Dransfeld: Phys. Rev. **117**, 1245 (1960)
7.5   K.N. Baranskii: Soviet Phys. Doklady **2**, 237 (1957)
7.6   N.S. Shiren: Phys. Rev. Lett. **11**, 3 (1963)
7.7   W. Grill, O. Weis: Phys. Rev. Lett. **35**, 588 (1975)
7.8   O. Weis: Z. Physik B**21**, 1 (1975)
7.9   W.E. Bron: Rep. Prog. Phys. **43**, 301 (1980)
7.10   E.H. Jacobsen: In *Quantum Electron. Symp.*, High View, N.Y. 1959 (Columbia
        U. Press, New York 1960) p. 468
7.11   J. de Klerk: Phys. Rev. **139**, A1635 (1965)
7.12   T.O. Woodruff, H. Ehrenreich: Phys. Rev. **123**, 1553 (1961)
7.13   A. Akhieser: J. Phys. (USSR) **1**, 277 (1939)
7.14   L. Landau, G. Rumer: Physik. Z. Sowjetunion **11**, 18 (1937)
7.15   S. Simons: Proc. Phys. Soc. (London) **83**, 749 (1964)
7.16   H.J. Maris: Phil. Mag. **9**, 901 (1964):
        R.A. Guyer: Phys. Rev. **148**, 789 (1966)
7.17   P. Brüesch: *Phonons: Theory and Experiments III*, Springer Ser. Solid-State Sci.,
        Vol. 66 (Springer, Berlin, Heidelberg 1987)
7.18   T. Wolfram (ed.) : *Inelastic Electron Tunneling Spectroscopy*, Springer Ser. Solid-
        State Sci., Vol. 4 (Springer, Berlin, Heidelberg 1978)
7.19   K. Hansma: Phys. Repts. **30**, 195 (1977)

7.20   C.B. Duke: *Tunneling in Solids*, Solid State Physics, Supplement **10** (Academic, New York 1969)

7.21   R.A. Logan: In *Tunneling Phenomena in Solids*, ed. by E. Burstein and S. Lundqvist (Plenum, New York 1969) p. 149

7.22   R.C. Jaklevic, J. Lambe: Phys. Rev. Lett. **17**, 1139 (1966)

7.23   J. Lambe, R.C. Jaklevic: Phys. Rev. **165**, 821 (1968)

7.24   J.C. Fischer, I. Giaever: J. Appl. Phys. **32**, 172 (1961)

7.25   J.G. Adler, J.E. Jackson: Rev. Sci. Instr. **37**, 1049 (1966)

7.26   I. Giaever, H.R. Zeller: Phys. Rev. Lett. **21**, 1385 (1968)

7.27   I. Giaever: Phys. Rev. Lett. **5**, 147 (1960); **5**, 464 (1960)

7.28   P. Brüesch, R. Kötz, H. Neff, L. Pietronero: Phys. Rev. B**29**, 4691 (1984)

7.29   A.G. Chynoweth, R.A. Logan, D.E. Thomas: Phys. Rev. **125**, 877 (1962)

7.30   A.G.M. Jansen, A.P. van Gelderm, P. Wyder: J. Phys. C**13**, 6073 (1980)

7.31   I.K. Yanson: Sov. Phys. JETP **39**, 506 (1974)

7.32   I.K. Yanson: Sov. Phys. Solid State **16**, 2337 (1975)

7.33   J.P. Carbotte, R.C. Dynes: Phys. Rev. **172**, 476 (1968)

7.34   I.K. Yanson, Yu.N. Shalov: Sov. Phys. JETP **44**, 148 (1977)

7.35   A.G.M. Jansen, F.M. Mueller, P. Wyder: *Proc. 2nd Rochester Conf. on Superconductivity and d- and f-Band Metals*, ed. by D.H. Douglas (Plenum, New York 1976)

7.36   J.W. Dozier, J.D. Rogers: IEEE Trans. MTT **12**, 360 (1964)

7.37   G. Wexler: Proc. Phys. Soc. **89**, 927 (1966)

7.38   Yu.V. Sharvin: Sov. Phys. JETP **21**, 655 (1965)

7.39   A.G.M. Jansen, F.M. Mueller, P. Wyder: Phys. Rev. B**16**, 1325 (1977)

7.40   A.G.M. Jansen, F.M. Mueller, P. Wyder: Science **199**, 1037 (1978)

7.41   I.O. Kulik, A.N. Omel'yanchuk, R.I. Shekhter: Sov. J. Low Temp. Phys. **3**, 740 (1977)

7.42   I.O. Kulik, R.I. Shekhter, A.N. Omel'yanchuk: Solid State Commun. **23**, 301 (1977)

7.43   I.O. Kulik, I.K. Yanson: Sov. J. Low Temp. Phys. **4**, 596 (1978)

7.44   A.P. van Gelder: Solid State Commun. **25**, 1097 (1978)

7.45   A.P. van Gelder: Solid State Commun. **35**, 19 (1980)

7.46   A.P. van Gelder, A.G. Jansen, P. Wyder: Phys. Rev. B**22**, 1515 (1980)

7.47   A.G.M. Jansen, J.H. van den Bosch, H. van Kempen, J.H.J.M. Ribot, P.H.H. Smeets, P. Wyder: J. Phys. F**10**, 265 (1980)

7.48   R.A. Cowley, A.D.B.Woods, G. Dolling: Phys. Rev. **150**, 487 (1966)

7.49   A.P. van Gelder, A.G.M. Jansen, P. Wyder: Phys. Rev. B**22**, 1515 (1980)

7.50   E. Bauer: In *Interactions on Metal Surfaces*, ed. by R. Gomer, Topics Appl. Phys., Vol. 4 (Springer, Berlin, Heidelberg 1975) p. 227

7.51   A.A. Maradudin, E.W. Montroll, G.H. Weiss, I.P. Ipatova: *Theory of Lattice Dynamics in the Harmonic Approximation* (Academic, New York 1971) p. 520; E.A. Ash, E.G.S. Paige (eds.): *Rayleigh-Wave Theory and Application*, Springer Ser. Wave Phenomena, Vol. 2 (Springer, Berlin, Heidelberg 1985)

7.52   For an isotropic solid there are only two independent elastic constants λ and μ. The shear constant μ replaces C_{44}, λ replaces C_{12}, and $\lambda + 2\mu$ replaces C_{11}, where C_{11}, C_{12}, C_{44} are the elastic constants of a cubic crystal

7.53   R. Fuchs, K.L. Kliever: Phys. Rev. **140**, A2076 (1965)

7.54   K.L. Kliever, R. Fuchs: Phys. Rev. **144**, 495 (1966)

7.55   K.L. Kliever, R. Fuchs: Phys. Rev. **150**, 573 (1966)

7.56   R. Claus, L. Merten, J. Brandmüller: *Light Scattering by Phonon-Polaritons*, Springer Tracts Mod. Phys., Vol. 75 (Springer, Berlin, Heidelberg 1975) p. 182

7.57   H. Ibach, D.L. Mills. *Electron Energy Loss Spectroscopy and Surface Vibrations* (Academic, New York 1982)

7.58   [Ref. 7.57, p. 102]

7.59   [Ref. 7.57, p. 257]

7.60   P.A. Thiry, J.J. Pireaux, R. Caudano: Physicalia Mag. **4** (1), 35 (1981)

7.61   M. Liehr, P.A. Thiry, J.J. Pireaux, R. Caudano: J. Vac. Sci. Technol. A**2**, 1079 (1984)

7.62 H. Ibach: Phys. Rev. Lett **24**, 1416 (1970)
7.63 A.A. Lucas, M. Sunjić: Prog. Surf. Sci. **2**, Pt2 (1972)
7.64 R. Matz, H. Lüth: Phys. Rev. Lett. **46**, 500 (1981)
7.65 L.H. Dubois, G.P. Schwartz: Phys. Rev. B**26**, 794 (1982)
7.66 L.L. Kesmodel, J.A. Gates. Y.W. Chung: Phys. Rev. B**23**, 489 (1981)
7.67 A.D. Baden, P.A. Cox, R.G. Egdell, A.F. Orchard, R.J.D. Willmer: J. Phys. C**14**, L1081 (1981)
7.68 H. Onuki, H. Iwamoto, R. Onaka: Solid State Commun. **34**, 941 (1980)
7.69 P.A. Thiry, M. Liehr, J.J. Pireaux, R. Caudano: Phys. Rev. B**29**, 4824 (1984)
7.70 E. Evans, D.L. Mills: Phys. Rev. B**5**, 4126 (1972)
7.71 A.A.Lucas, J.P. Vigneron: Solid State Commun. **9**, 397 (1984)
7.72 H. Froitzheim: In *Electron Spectroscopy for Surface Analysis*, ed. by H. Ibach, Topics Current Phys., Vol. 4 (Springer, Berlin, Heidelberg 1977) p. 205
7.73 G. Benedek, U. Valbusa (eds.): *Dynamics of Gas-Surface Interactions*, Springer Ser. Chem. Phys., Vol. 21 (Springer, Berlin, Heidelberg 1982)
7.74 B. Feuerbacher: In *Vibrational Spectroscopy of Adsorbates*, ed. by R.F. Willis, Springer Ser. Chem. Phys., Vol. 15 (Springer, Berlin, Heidelberg 1980) p. 91
7.75 J.P. Toennis: In [Ref. 7.73, p. 208]
7.76 G. Bursdeylins, R.B.Doak, J.P. Toennis: Phys. Rev. Lett **46**, 437 (1981)
7.77 G. Brusdeylins, R.B. Doak, J.P. Toennis: J. Chem. Phys. **75**, 1784 (1981)
7.78 T.S. Chen, F.W. de Wette, G.P. Aldredge: Phys. Rev. B**15**, 1167 (1977)
7.79 G. Benedek: Surf. Sci. **61**, 603 (1976)
7.80 G. Benedek, N. Garcia: Surf. Sci. **103**, 1143 (1981)
7.81 A. Otto: Z. Physik **216**, 398 (1968)
7.82 V.V. Bryksin, Yu.M. Gerbstein, D.N. Mirlin: Phys. Status Solidi (b) **51**, 901 (1972)
7.83 N.J. Harrick: *Internal Reflection Spectroscopy* (Interscience, New York 1967)
7.84 N. Marshall, B. Fischer: Phys. Rev. Lett. **28**, 811 (1972)
7.85 R. Ruppin: Solid State Commun. **8**, 1129 (1970)
7.86 H.J. Falge, A. Otto: Phys. Status Solidi (b) **56**, 523 (1973)
7.87 H.J. Falge, A. Otto, W. Sohler: Phys. Status Solidi (b) **63**, 259 (1974)
7.88 V.M. Agranovich, D.L. Mills (eds.): *Surface Polaritons* (North-Holland, Amsterdam 1982)
7.89 R.G. Greenler: J. Chem. Phys. **44**, 310 (1966)
7.90 R.G. Greenler: J. Chem. Phys. **50**, 1963 (1969)
7.91 H.G. Tompkins: In *Method of Surface Analysis*, Vol. 1, ed. by A.W. Czanderna (Elsevier, Amsterdam 1975) p. 447
7.92 P. Hollins, J. Pritchard: In [Ref. 7.74, p. 125]

Subject Index

244

248

250

266

271

272

274

Errata of Phonons: Theory and Experiment I

(Springer Series in Solid-State Sciences, Vol. 34)

p. 8   10th line from top; replace Equation (1.8)... by Equation (1.7)... .

p. 19   In (2.13); replace $D_{22}(1)$ by $D_{22}(q)$.

p. 25   7th line from bottom; replace Fig.2.5 by Fig.2.4;
6th line from bottom; replace TO by LA.

p. 27   2nd line from bottom; replace (2.47) by (2.31,47).

p. 30   in (2.63); replace $q > 0$ by $q \geqq 0$ in the first sum of the first equation and in the sum of the second equation.

p. 31   9th line from top; replace $\dot{n}(q)$ by $\dot{n}\binom{q}{j}$.

p. 36   Equation (2.94) should read: $N_n = 2^{\frac{1}{2}}(n+1)^{\frac{1}{2}}N_{n+1} = \dots$.
In (2.96); replace $-i$ by $+i$.

p. 44   In Fig.2.10; replace $\theta_s = \hbar\omega_s/k_B T$ by $\theta_s = \hbar\omega_s/k_B$.

p. 48   In Fig.2.11; replace axis label g by q, g_D by q_D and $\tilde{g}_D$ by $\tilde{q}_D$.

p. 51   The sentence below (2.161) should read: ..., the specific heat of the linear chain is linear in T at low temperatures.

p. 56   In (3.1); replace $r(\ell)$ by $\vec{r}(\ell)$.

p. 61   The sentence below (3.23) should read: This form shows explicitly that the elements of the dynamical matrix $D(\vec{q})$ are independent of L.

p. 65 5th line from bottom; replace Using (3.36)... by Using (3.35)... .

p. 77 15th line from top; the sentence should read: ... the surface of constant frequency is a sphere of radius vq, the Debye sphere.

p. 85 The last sentence should read: ... and the second because of (F.12a) with g = [I|0] and introducing... .

p. 96 In figure caption of Fig.3.17: ... b) $\vec{q} = (\pi/a) (1,1,0)$;... .

p. 98 Problem 3.8.4, Result: $C_D(T) \sim T^2$ at low temperatures.

p. 108 Equation (4.23b) should read: $\Phi_{\alpha\beta}^{(R)}(ii) = -\sum_k' \Phi_{\alpha\beta}^{(R)}(ik)$.

p. 115 In (4.62); replace $R_{\alpha\alpha}(\genfrac{}{}{0pt}{}{\vec{q}\rightarrow 0}{\kappa\kappa'})$ by $|R_{\alpha\alpha}(\genfrac{}{}{0pt}{}{\vec{q}\rightarrow 0}{\kappa\kappa'})|$.

p. 117 Figure 4.5 for LO: small arrows in intermediate layer should have opposite directions.

p. 122 Equation (4.94) should read: $k_L = (1 + \frac{8\pi\alpha^*}{3v_a})^{-1}$.

p. 126 In (4.116); replace ε by ε_∞ in the denominator.

p. 141 Equation (4.151) is valid only in the approximation h = 0; there are errors in the dispersion shown in Fig.4.22. Based on (4.151,152) and the parameters given in the figure caption of Fig.4.22, the correct limiting frequencies are: $\tilde{\nu}_{LO}(q=0) = 350$ cm^{-1}, $\tilde{\nu}_{LO}(q=\pi/a) = 298$ cm^{-1}; $\tilde{\nu}_{LA}(q=0) = 0$, $\tilde{\nu}_{LA}(q=\pi/a) = 186$ cm^{-1}; $\tilde{\nu}_{LIB}(q=0) = 269$ cm^{-1}, $\tilde{\nu}_{LIB}(q=\pi/a) = 410$ cm^{-1}.

p. 212 Below matrix σ_{xz} it should read: $\Phi_{yz} = \Phi_{zy} = 0$.

278